Stability and Chaos in Celestial Mechanics

Alessandra Celletti

Stability and Chaos in Celestial Mechanics

Springer

Published in association with
Praxis Publishing
Chichester, UK

PRAXIS

Professor Alessandra Celletti
Department of Mathematics
University of Rome "Tor Vergata"
Rome
Italy

SPRINGER–PRAXIS BOOKS IN ASTRONOMY AND PLANETARY SCIENCES
SUBJECT *ADVISORY EDITORS*: Philippe Blondel, C.Geol., F.G.S., Ph.D., M.Sc., Senior Scientist, Department of Physics, University of Bath, UK; John Mason, M.B.E., B.Sc., M.Sc., Ph.D.

ISBN 978-3-642-26156-5 Springer-Verlag Berlin Heidelberg New York

Springer is part of Springer-Science + Business Media (springer.com)

Softcover reprint of the hardcover 1st edition 2010

Cover design: Jim Wilkie
Project copy editor: Mike Shardlow
Author-generated LaTex, processed by EDV-Beratung Herweg, Germany

Printed on acid-free paper

To the memory of my mother Mirella,
my best teacher and friend,
and in fond remembrance
of my grandmother Emilia.
Without them I would never have had
the privilege of studying the sky.

Table of Contents

Preface

The last decades have marked the beginning of a new era in Celestial Mechanics. The challenges came from several different directions. The stability theory of nearly–integrable systems (a class of problems which includes many models of Celestial Mechanics) profited from the breakthrough represented by the Kolmogorov–Arnold–Moser theory, which also provides tools for determining explicitly the parameter values allowing for stability. A confinement of the actions for exponential times was guaranteed by Nekhoroshev's theorem, which gives much information about the geography of the resonances. Performing ever-faster computer simulations allowed us to have deeper insights into many questions of Dynamical Systems, most notably chaos theory. In this context several techniques have been developed to distinguish between ordered and chaotic behaviors. Modern tools for computing spacecraft trajectories made possible the realization of many space missions, especially the interplanetary tours, which gave a new shape to the solar system with a lot of new satellites and small bodies. Finally, the improvement of observational techniques allowed us to make two revolutions in the sky: the solar system does not end with Pluto, but it extends to the Kuiper belt, and the solar system is not unique, but the universe has plenty of extrasolar planetary systems.

Cooking all these ingredients together with the classical theories developed from the 17th to the 19th centuries, one obtains the *modern Celestial Mechanics*. In this context the main goal of the CELMEC meetings, conferences which are held every four years on Celestial Mechanics (which I have had the opportunity to co–organize since 1993), is to bring together the analytical aspects (from perturbation theory to stability analysis), the dynamics of the solar system and of stellar interactions, flight dynamics for near–Earth and interplanetary missions.

In the light of these innovations, the aim of this book is to present some topics of Celestial Mechanics, often adopting the point of view of Dynamical Systems theory. The interplay between these two disciplines is definitely rich in exciting discoveries, which lead to a better understanding of the dynamics of the solar system bodies, like the stability of the N–body problem, the role of the resonances, the study of rotational dynamics, etc. In this framework the **first chapter** is devoted to the introduction of basic notions, like continuous and discrete systems, their linear stability, the definition of attractors and the discussion of some paradigmatic models, both conservative and dissipative. The **second chapter** presents some numerical tools, which are used to distinguish among the different kinds of dynamics. After defining the Poincaré mapping we introduce the Lyapunov exponents and some methods for computing the dimension of the attractors. We also review some results concerning discrete time series, derived for example from experimental data. Frequency analysis and the computation of the Fast Lyapunov Indicators have been revealed to be extremely powerful tools for distinguishing between regular and chaotic motions.

The **third chapter** is devoted to Kepler's problem. We start by discussing elliptic motion and we provide the solutions of the hyperbolic and parabolic dynamics. After introducing the action–angle Delaunay variables, we discuss the two–body problem with variable mass. In particular we are interested in Gylden's problem and its Hamiltonian formulation. The three–body problem is the content of the **fourth chapter**. We introduce the planar, circular, restricted three–body problem; in terms of the Delaunay variables this model is described by a nearly–integrable Hamiltonian function, whose perturbing parameter represents the primaries' mass ratio. We also provide explicit formulae for the expansion of the perturbing function. The Hamiltonians of the elliptic and inclined cases are also presented. A staging post is the derivation of the Lagrangian solutions and the discussion of their stability. After dealing in full detail with the planar, circular, restricted case, we pass on to examine the Lagrangian solutions in the elliptic, restricted model as well as in the unrestricted case.

At this point we switch to rotational dynamics, which is the content of the **fifth chapter.** Action–angle Andoyer–Deprit variables are introduced to describe the free and perturbed rigid body motions, where the perturbation is due to the gravitational attraction of a central mass. A simplified model is provided by the spin–orbit problem in which it is assumed that the spin–axis, the body principal axis and the orbit normal coincide. Moreover the orbit is assumed to be a Keplerian ellipse. This problem is described by a one–dimensional, time–dependent Hamiltonian system, which will be taken as the paradigmatic model in many sections of the subsequent chapters. We also introduce a dissipative spin–orbit model, with a dissipation varying linearly with the angular velocity. After a short discussion of the motion of a satellite around an oblate planet and of the interaction within two bodies of finite dimensions, we conclude with the description of the tether and dumbbell satellite systems.

Perturbation theory is introduced in the **sixth chapter**; classical, resonant and degenerate perturbation theories are discussed together with some examples, like the computation of the precession of the perihelion or of the precession of the equinoxes. Then we present the Birkhoff normal form around equilibrium positions and closed trajectories. We conclude with an introduction to the averaging theorem. The existence and fate of invariant surfaces is the content of the **seventh chapter**. The Kolmogorov–Arnold–Moser (KAM) theorem is proved in detail for the specific case of the spin–orbit model of Chapter 5. Moreover we discuss the choice of the frequency as well as a computer–assisted implementation. We also provide a review of the applications of the theory to some models of Celestial Mechanics, both in rotational dynamics and in the context of the three–body problem. We present a numerical technique due to J. Greene for the computation of the breakdown threshold of invariant tori and we shortly describe the existence of lower–dimensional tori. Always taking the spin–orbit problem as a sample model, we provide a brief introduction to a dissipative KAM theorem and to non–existence criteria of invariant tori, valid both in the conservative and in the dissipative settings. Finally we dedicate the last section to the discussion of cantori, which are the remnants of KAM surfaces soon after their breakdown.

The **eighth chapter** is devoted to an introduction to the long–term stability of nearly–integrable Hamiltonian systems. The long proof of Nekhoroshev's theorem is summarized in some main steps, which should give a global idea of the interplay between resonant and non–resonant motions. We also provide a review of some applications of such theory to the three–body problem and to the stability of the Lagrangian points. The determination of periodic orbits is the content of the **ninth chapter**. The implicit function theorem is implemented to construct periodic orbits for both the conservative and dissipative spin–orbit problems. We also review some other methods, like the Lindstedt–Poincaré and KBM techniques, as well as Lyapunov's theorem to prove the existence of families of periodic orbits.

Finally, the **tenth chapter** deals with regularization theory. Precisely, the Levi–Civita transformation is implemented to regularize collisions with one body in the framework of the planar, restricted, circular, three–body problem. The regularization of the spatial case is obtained through the Kustaanheimo–Steifel (KS) transformation. Both Levi–Civita and KS regularizations are local techniques, since they allow us to regularize collisions with one of the primaries. A global regularization is attained through the implementation of the Birkhoff regularization procedure, which concludes the last chapter.

A set of Appendices is included; they are not intended to be exhaustive of the topics dealt with, but they are just a quick reference which definitely needs specific literature for a complete coverage of each argument. Some astronomical data on planets, dwarf planets and satellites are added, as Appendix G, for reference.

A remark on the notation. The following notation has been adopted throughout the text:

- $\mathbf{R}$ and $\mathbf{R}_+$ denote, respectively, the set of real and positive real numbers;
- $\mathbf{Z}$ and $\mathbf{Z}_+$ denote, respectively, the set of integer and positive integer numbers;
- $\mathbf{N}$ denotes the set of natural numbers;
- $\mathbf{T}$ denotes the standard one–dimensional torus;
- e denotes the orbital eccentricity;
- $\mathcal{G}$ denotes the gravitational constant (equal to $6.673 \cdot 10^{-11}\ \mathrm{m}^3\,\mathrm{kg}^{-1}\,\mathrm{s}^{-2}$);
- the derivative with respect to time is, as usual, denoted with a dot, i.e. $\dot{r} \equiv \frac{dr}{dt}$.

Acknowledgments

Since this work represents most of what I have learned and discovered during my career up to now, I am taking the opportunity to thank many people. So please, be patient.

First of all, during the accomplishment of this book I was kindly assisted by several collaborators, who gave me invaluable suggestions, and corrected typos and mistakes. They helped me a lot and I express my deep gratitude to all of them: Luca Biasco, Laura Di Gregorio, Sara Di Ruzza, Corrado Falcolini, Thomas Kotoulas, Christoph Lhotka, Vladislav Sidorenko, and Letizia Stefanelli. Double thanks to Christoph Lhotka for having created with vivid imagination the cover of the book.

I want to thank all my research collaborators within the fields of Dynamical Systems and Celestial Mechanics; a very special thank to Luigi Chierchia (with whom I have been collaborating since we met in Arizona in 1985), Claude Froeschlé (who always transmitted to me his enthusiasm for research) and Antonio Giorgilli (who first taught me how to combine Mechanics and computers in a very elegant way). I learned a lot working with all of them.

Writing popular books with Ettore Perozzi has always been an exciting adventure, which allows me to have a wider panorama of the main problems of Celestial Mechanics. With Ettore I have also shared the activity of organizing the CELMEC meetings since the first one in 1993.

Needless to say the referees of my Master and PhD theses gave me unique opportunities to study with them: their guide has been inestimable; a grateful thankyou to Giovanni Gallavotti and Jörg Waldvogel, and to the late Jürgen Moser.

I have also profited from many discussions with and suggestions of my colleagues (and friends), and especially Giancarlo Benettin, Alain Chenciner, Rudolf Dvorak, Sylvio Ferraz–Mello, Luigi Galgani, Massimiliano Guzzo, Jacques Laskar, Elena Lega, Anne Lemaitre, Rafael de la Llave, Ugo Locatelli, Robert McKay, Andrea Milani, Piero Negrini, Carles Simó, Bonnie Steves, and Giovanni Valsecchi. A special thankyou to Maria Cristina Luiso and to Rossella Petreschi for their help at the very beginning of my studies, and to Velleda Baldoni and to Elisabetta Strickland, with whom I enthusiastically share many projects.

I heartily thank all the students of the past years, both from the University of L'Aquila and the University of Roma Tor Vergata. I owe very much also to the CELMEC participants: thanks to them I have always been kept informed on the latest novelties in Celestial Mechanics.

I want to express my gratitude to the publisher, Clive Horwood, and the editor, John Mason, for having enthusiastically supported the idea of this book since the beginning. I also thank Frank Herweg and Mike Shardlow for their precious help with the preparation of the final draft of the book.

My husband Enrico Romita played a very special role in the accomplishment of this project: he drew most of the preliminary versions of the figures, he helped me in checking some parts of the text and, most of all, he was always very tolerant of my pace of work. I have been extremely lucky to meet such a special person: to him, all my love. I want also to thank my wonderful sister Paola for her continuous and invaluable enthusiasm for my studies, and my nephew Raffaele for making my life brighter.

I have dedicated this book to my mother and to my grandmother. No book of any length could include all the acknowledgments I owe them for all the love and support they gave me during our wonderful life together. My mother played a unique role in my education. She was the most curious person, in all fields of knowledge, that I have ever met. She was very enthusiastic about mathematics and astronomy; though not being an expert, she tried to follow the recent advances of Celestial Mechanics, often coming with me to popular and technical conferences. I am sure, absolutely sure, that if she were alive, she would have taken this book, and slowly turning the pages she would have tried to find, within the labyrinth of formulae, an explanation to her many questions about the mysteries of the sky.

Roma, 21 September 2009 Alessandra Celletti

1 Order and chaos

The description of dynamical systems requires the preliminary definition of some basic notions as described in this chapter, which introduces the language, the notation and some enlightening examples. As a first step it is necessary to distinguish between continuous and discrete systems (Section 1.1), as well as to discriminate between stable and unstable motions (Section 1.2). Next we need to discern between conservative and dissipative dynamical systems (Section 1.3). In the latter case we introduce the notion of attractors as well as of their basins of attraction (Section 1.4).

These concepts are made clear by means of paradigmatic examples, like the logistic map (Section 1.5), the standard mapping both in its conservative (Section 1.6) and dissipative (Section 1.7) version, and Hénon's mapping (Section 1.8).

1.1 Continuous and discrete systems

The description of a physical problem is usually given in terms of the variation with time of a set of coordinates, which can be provided as continuous functions of time or they can be given at discrete intervals of time. The corresponding dynamical system is called *continuous* or *discrete.* To give a concrete example, let us think to a flow of water from a tap. If the stream is constant and without interruptions, we say that the flow is continuous as time goes on. On the other hand, if the water drips we can assume that the system is discrete, since each drop leaks at some interval of time. Having in mind this *experimental* description, a mathematical definition of continuous or discrete systems is given as follows.

A continuous system is described by a differential equation whose solution is a function of a quantity, the time, which varies continuously in the set of the real numbers. The number of degrees of freedom of the system is the minimal number of independent variables apt to describe the system. In general, a continuous system with ℓ degrees of freedom is described by a set of differential equations of the type

$$\dot{\underline{x}} = \underline{f}(\underline{x}) \ , \tag{1.1}$$

where $\underline{x} \in \mathbf{R}^\ell$ and $\underline{f} = (f_1, \dots, f_\ell)$ is a vector function from $\mathbf{R}^\ell$ into itself. The system is called *autonomous* whenever the vector function $\underline{f}$ does not depend explicitly on time, otherwise it is called *non–autonomous* and it is described by differential equations of the form

$$\dot{\underline{x}} = \underline{g}(\underline{x}, t) \ , \tag{1.2}$$

for some vector function $\underline{g} = (g_1, \ldots, g_\ell)$ from $\mathbf{R}^{\ell+1}$ to $\mathbf{R}^\ell$. The solution of (1.1) or (1.2) at time t with initial datum $\underline{x}_0$ is denoted by

$$\underline{x} = \underline{x}(t; \underline{x}_0) , \qquad t \in \mathbf{R} .$$

An *equilibrium point* $\underline{x} = \underline{x}_0$ for the continuous system (1.1) is a solution of the set of equations

$$\underline{f}(\underline{x}_0) = \underline{0} .$$

A *discrete system* is described by a mapping whose evolution beats time according to an iteration index running over the set of integer numbers. An ℓ–dimensional discrete system is represented by a set of equations of the form

$$\underline{x}_{n+1} = \underline{f}(\underline{x}_n) , \qquad n \in \mathbf{N} , \tag{1.3}$$

where $\underline{x}_n \in \mathbf{R}^\ell$ and $\underline{f} = (f_1, \ldots, f_\ell)$. Equation (1.3) provides the description of the dynamics at the "discrete time" $n + 1$ as a function of the solution at step n; in other words, the solution of the discrete system at step n with initial datum $\underline{x}_0$ is given by the sequence $\underline{x}_n$ computed iterating the mapping (1.3) starting from $\underline{x}_0$.

Using an alternative notation which avoids indexing, the map (1.3) can be equivalently written as

$$\underline{x}' = \underline{f}(\underline{x}) \tag{1.4}$$

(notice that the prime is used to denote the iterated point; elsewhere the prime will be used for the derivative as it will be properly specified). In the following we will equally use the notation (1.3) or (1.4) as describing the same dynamical system. A *fixed point* $\underline{x} = \underline{x}_0$ for the discrete system (1.4) is a solution of the equation

$$\underline{f}(\underline{x}_0) = \underline{x}_0 .$$

Let us provide concrete examples of continuous and discrete dynamical systems. We start by introducing the *harmonic oscillator* which is described by the continuous system of two first–order differential equations

$$\begin{aligned} \dot{x}_1 &= x_2 \\ \dot{x}_2 &= -\omega^2 x_1 , \end{aligned} \tag{1.5}$$

where ω is a real parameter. In compact form we can write (1.5) as $\underline{\dot{x}} = \underline{f}(\underline{x})$, where $\underline{x} \equiv (x_1, x_2)$ and $\underline{f}(\underline{x}) \equiv (x_2, -\omega^2 x_1)$. The origin is an equilibrium point being a solution of $\underline{f}(\underline{x}) = \underline{0}$. The system (1.5) can be equivalently written as the second–order differential equation

$$\ddot{x}_1 + \omega^2 x_1 = 0 ,$$

whose solution is given by

$$x_1(t) = A \cos(\omega t + B)$$

for some real constants A and B depending on the initial conditions at $t = 0$. Indeed, if $(x_1(0), x_2(0))$ denotes the initial position, then one obtains that

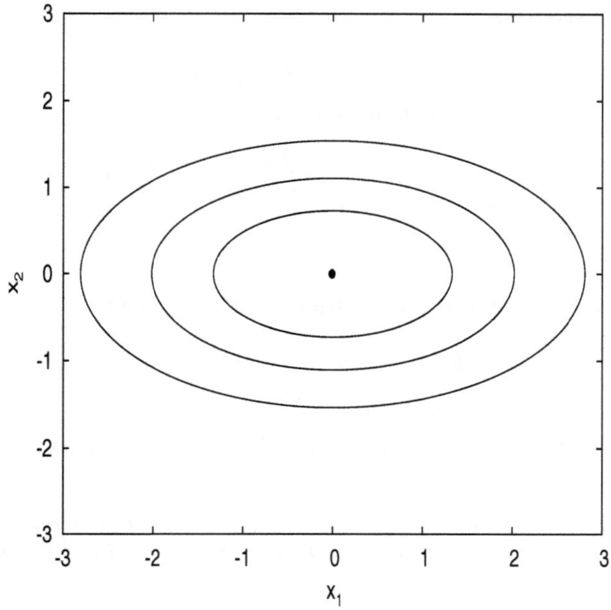

Fig. 1.1. Solutions of the harmonic oscillator with frequency $\omega = \sqrt{0.3}$ and for the initial conditions $(x_1(0), x_2(0)) = (1.33, 0)$, $(2.02, 0)$, $(2.81, 0)$.

$A = \sqrt{x_1(0)^2 + \frac{x_2(0)^2}{\omega^2}}$ and $B = \arctan\left(-\frac{x_2(0)}{\omega x_1(0)}\right)$. The solution in the (x_1, x_2)–plane as the initial condition varies is provided by the ellipses shown in Figure 1.1 centered around the origin; the ellipses reduce to circles for $\omega = 1$.

An example of a discrete system is provided by the one–dimensional model described by the linear map

$$x_{n+1} = \alpha x_n + \beta \ ,$$

where α and β are real parameters. When $\alpha = 0$ any point is a fixed point. Let us suppose that $\alpha \neq 0$ and let $f(x_n) \equiv \alpha x_n + \beta$; the fixed point of the mapping is the solution of the equation $f(x_n) = x_n$ which yields $x_n^* = \frac{\beta}{1-\alpha}$. For $|\alpha| < 1$ the fixed point is attracting, since any solution with initial condition different from x_n^* will tend to x_n^*; for $|\alpha| > 1$ the fixed point is repelling, since any other solution with initial condition different from x_n^* will move away from x_n^*. There are two cases still to consider, namely $\alpha = 1$ and $\alpha = -1$. When $\alpha = 1$ the fixed point condition $f(x_n) = x_n$ yields $\beta = 0$, which means that the map reduces to the identity, namely $x_{n+1} = x_n$. For $\alpha = -1$ whatever the initial point x_0 is, provided x_0 is different from $\frac{\beta}{2}$, the trajectory is composed alternatively by x_0 and $\beta - x_0$; in this case, one speaks of a *2-cycle* solution (for $x_0 = \frac{\beta}{2}$ the trajectory reduces to the point x_0). An example of a k-cycle is obtained taking the mapping corresponding to f^k, which means to compose k times the mapping f.

1.2 Linear stability

Let us consider the dynamical system described by the equation (1.1) with ℓ degrees of freedom and let J be the Jacobian matrix defined as

$$J = J(\underline{x}) \equiv D\underline{f}(\underline{x}) = \begin{pmatrix} \frac{\partial f_1}{\partial x_1} & \cdots & \frac{\partial f_1}{\partial x_\ell} \\ \vdots & & \vdots \\ \frac{\partial f_\ell}{\partial x_1} & \cdots & \frac{\partial f_\ell}{\partial x_\ell} \end{pmatrix} .$$

In the proximity of an equilibrium position $\underline{x} = \underline{x}_0$ we refer to

$$\dot{\underline{y}} = J(\underline{x}_0)\underline{y} , \qquad \underline{y} \in \mathbf{R}^\ell , \tag{1.6}$$

as the *linearized system*. Let ℓ_0 be the number of eigenvalues of $J(\underline{x}_0)$ with zero real part, ℓ_u be the number of eigenvalues with positive real part and ℓ_s those with negative real part.

Definition. The equilibrium position $\underline{x} = \underline{x}_0$ is called *hyperbolic* if $\ell_0 = 0$; it is called an *attractor* if $\ell_s = \ell$, a *repeller* if $\ell_u = \ell$ and a *saddle* if ℓ_u and ℓ_s are both strictly positive.

Near a hyperbolic equilibrium the dynamics of (1.1) is equivalent to that of the linearized system (1.6) as stated by the following result [84, 93]:

Hartman–Grobman theorem. *Consider the dynamical system described by* (1.1) *and assume that it admits a hyperbolic equilibrium position* $\underline{x}_0$. *In a suitable neighborhood of* $\underline{x}_0$ *the system* (1.1) *is topologically equivalent*[1] *to the linearized system* (1.6).

Let us now discuss the stability of an equilibrium position; we first introduce the following definition of stability according to Lyapunov for a generic solution of the system (1.1). To fix the notations, let $\|\cdot\|$ denote the distance function, e.g. the Euclidean norm in $\mathbf{R}^\ell$.

Definition. The solution $\underline{x}_* = \underline{x}_*(t)$ is said to be *stable according to Lyapunov,* if for any $\varepsilon > 0$ there exists $\delta = \delta(\varepsilon)$ such that if $\|\underline{x}(t_0) - \underline{x}_*(t_0)\| < \delta$ at the initial time $t = t_0$, then $\|\underline{x}(t) - \underline{x}_*(t)\| < \varepsilon$ for any $t \geq t_0$.
The solution is said to be *asymptotically stable* if, for $\|\underline{x}(t_0) - \underline{x}_*(t_0)\| < \delta$, one has $\lim_{t\to\infty} \|\underline{x}(t) - \underline{x}_*(t)\| = 0$.

When the solution coincides with an equilibrium point, its linear stability can be inferred from the analysis of the eigenvalues of the Jacobian matrix; we denote by $(\lambda_1, \ldots, \lambda_\ell)$ the roots of the characteristic equation $\det(J - \lambda I_\ell) = 0$ (I_ℓ is the $\ell \times \ell$ identity matrix) and we denote by $\underline{v}_j \in \mathbf{R}^\ell$, $j = 1, \ldots, \ell$, the eigenvector associated to λ_j. Then, the solution of the linearized system can be written in the form $\underline{y}(t) = \sum_{j=1}^{\ell} \alpha_j \underline{v}_j e^{\lambda_j t}$ for some real or complex coefficients α_j which are determined by the initial conditions.

Assuming for simplicity that $\ell = 2$, the equilibrium is characterized as follows, according to the nature of the eigenvalues λ_1, λ_2:

[1] i.e., there exists a homeomorphism which conjugates (1.1) and (1.6).

- if λ_1, λ_2 are real negative numbers, then the equilibrium is a *stable node;*
- if λ_1, λ_2 are real positive numbers, then the equilibrium is an *unstable node;*
- if λ_1, λ_2 are real numbers with opposite signs, then the equilibrium is a *saddle;*
- if λ_1, λ_2 are complex numbers with negative real part, then the equilibrium is a *stable focus;*
- if λ_1, λ_2 are complex numbers with positive real part, then the equilibrium is an *unstable focus;*
- if λ_1, λ_2 are purely imaginary and non–zero, then the equilibrium is a *center.*

We next introduce the definitions of stable and unstable manifolds as well as those of homoclinic and heteroclinic intersections (see, e.g., [80] for applications to Celestial Mechanics). We denote by $\underline{\Phi}(t;\tilde{\underline{x}})$ the *flow* at time t with initial condition $\tilde{\underline{x}}$, namely the solution at time t of (1.1) starting from $\underline{x} = \tilde{\underline{x}}$.

Definition. Let $\underline{x}_0$ be a saddle point and let $\underline{\Phi}(t;\underline{x})$ be the flow at time t with initial condition $\underline{x}$. The *stable manifold* $W^s(\underline{x}_0)$ is the set of points which end–up at $\underline{x}_0$, i.e.

$$W^s(\underline{x}_0) \equiv \{\underline{x} \in \mathbf{R}^\ell \ : \ \lim_{t\to\infty} \underline{\Phi}(t;\underline{x}) = \underline{x}_0\} \ .$$

Analogously, we define the *unstable manifold* $W^u(\underline{x}_0)$ as the set of points which tend to $\underline{x}_0$ for negative times, i.e.

$$W^u(\underline{x}_0) \equiv \{\underline{x} \in \mathbf{R}^\ell \ : \ \lim_{t\to -\infty} \underline{\Phi}(t;\underline{x}) = \underline{x}_0\} \ .$$

The intersection between W^s and W^u is called a *homoclinic* point; if W^s and W^u belong to different equilibrium positions, then the intersection is called a *hetero-clinic* point (see Figure 1.2).

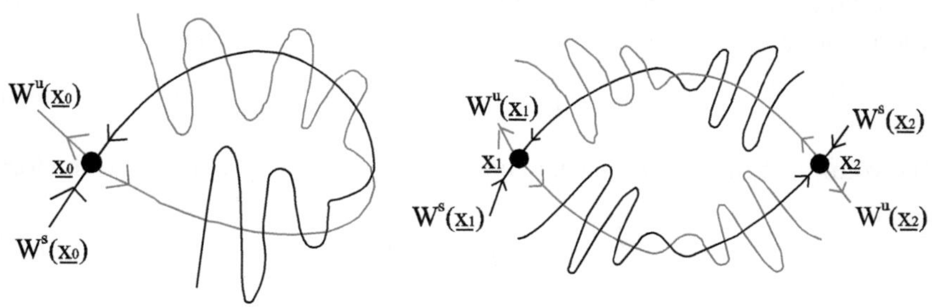

Fig. 1.2. Left: Stable and unstable manifolds of a homoclinic point. Right: An example of heteroclinic intersections.

Let us briefly transpose the main definitions and results to discrete systems. Consider the ℓ–dimensional mapping

$$\underline{x}_{n+1} = \underline{f}(\underline{x}_n) \ , \qquad n \in \mathbf{N} \ , \tag{1.7}$$

where $\underline{x}_n \in \mathbf{R}^\ell$ and $\underline{f} = (f_1, \dots, f_\ell)$; let $\underline{x}_0$ be a fixed point and denote by $J \equiv J(\underline{x}_0)$ the Jacobian matrix of $\underline{f}$ computed at $\underline{x}_0$. The linearized system is written as

$$\underline{y}_{n+1} = J \underline{y}_n \ , \qquad n \in \mathbf{N} \ ,$$

for $\underline{y}_n \in \mathbf{R}^\ell$.

Definition. A fixed point $\underline{x}_0$ of (1.7) is called *hyperbolic,* if the eigenvalues of $J = J(\underline{x}_0)$ are real and one of them has magnitude greater than one. A fixed point $\underline{x}_0$ is called *elliptic,* if the eigenvalues are complex with absolute value equal to one. A fixed point $\underline{x}_0$ is an *attractor* if the eigenvalues have magnitude less than unity and it is a *repellor* if the magnitude is greater than one.

If $\underline{f}$ in (1.7) is defined on a space X with (X, d) being a metric space with metric $d : X \times X \to \mathbf{R}$, then the definition of stability according to Lyapunov of a fixed point $\underline{x}_0$ transposes as follows.

Definition. For any $\varepsilon > 0$, there exists $\delta = \delta(\varepsilon)$ such that for any $\underline{x} \in X$ with $d(\underline{x}_0, \underline{x}) < \delta$, then $d(\underline{f}^n(\underline{x}_0), \underline{f}^n(\underline{x})) < \varepsilon$ for any $n \in \mathbf{N}$. The fixed point is asymptotically stable if $\lim_{n\to\infty} d(\underline{f}^n(\underline{x}_0), \underline{f}^n(\underline{x})) = 0$ whenever $d(\underline{x}_0, \underline{x}) < \delta$.

Concerning the linear stability we proceed as follows. The general solution is a linear combination of the form $\underline{x}_n = \sum_{j=1}^{\ell} \alpha_j \underline{v}_j \lambda_j^n$ for suitable coefficients α_j, while $\underline{v}_j$ denote the eigenvectors. If λ_j, $j = 1, \ldots, \ell$, are the eigenvalues associated to the Jacobian J, then the fixed point is linearly stable if $|\lambda_j| \leq 1$ for all $j = 1, \ldots, \ell$ and it is linearly unstable if for some j it is $|\lambda_j| > 1$.

1.3 Conservative and dissipative systems

Let us consider a continuous system described by the equations (1.1); qualitatively we can say that the system is *conservative,* if the volume of the phase space is preserved. The system is said to be *dissipative* if the volume contracts or expands along the flow.

When the flow associated to the evolution of the system is interpreted as a transformation from $\underline{x}$ to $\underline{y}$ representing the solutions at times t_1 and t_2, say

$$\underline{x} \equiv \underline{x}(t_1) \to \underline{y} \equiv \underline{x}(t_2) , \tag{1.8}$$

then the volume of the phase space, denoted by V, changes according to the formula

$$\int_V d\underline{y} = \int_V |\det(J)| \, d\underline{x} , \tag{1.9}$$

where J is the Jacobian of the transformation (1.8) and $\det(J)$ is its determinant. Writing (1.8) as $\underline{y} = \underline{h}(\underline{x})$ for a suitable vector function $\underline{h}$, one obtains that J coincides with the Jacobian of $\underline{h}$, say $J = D\underline{h}(\underline{x})$. The dynamical system is conservative if $|\det(J)| = 1$, while it is contractive if $|\det(J)| < 1$ and expansive whenever $|\det(J)| > 1$.

With reference to equation (1.1), let $\underline{F}_t(\underline{x}) \equiv \underline{\Phi}(t; \underline{x})$ be the flow at time t with initial datum $\underline{x}$; since $\{\underline{F}_t(\underline{x})\}_{t\in\mathbf{R}}$ is a solution of (1.1), by definition one has

$$\frac{d}{dt}\underline{F}_t(\underline{x}) = \underline{f}(\underline{F}_t(\underline{x})) ;$$

differentiating with respect to $\underline{x}$ one obtains the *variational equations*

$$\frac{d}{dt}D\underline{F}_t(\underline{x}) = D\underline{f}(\underline{F}_t(\underline{x})) \cdot D\underline{F}_t(\underline{x}) \ . \qquad (1.10)$$

Let $J_t \equiv D\underline{F}_t(\underline{x})$ and $A(t) \equiv D\underline{f}(\underline{F}_t(\underline{x}))$; then, $\dot{J}_t = A(t)\ J_t$. According to Liouville's formula [93], denoting by $\mathrm{Tr}(A)$ the trace of the matrix A, one has

$$\frac{d}{dt}(\det(J_t)) = \mathrm{Tr}(A(t)) \cdot \det(J_t) \ , \qquad \text{where} \quad \det J_0 = 1 \ . \qquad (1.11)$$

From (1.11) one finds $\det(J_t) = e^{\int_0^t \mathrm{Tr}(A(s))\ ds}$ and the system is said to be dissipative and contractive if the phase space volume decreases with time, namely if $\mathrm{Tr}(A(t)) < 0$ for any $t > 0$. If we set $\underline{y} = \underline{F}_t(\underline{x})$ in (1.9), we obtain $J_t = J$, so that $\det(J_t) = \det(J)$. In particular, the system is conservative if $\det(J_t) = 1$.

An example of a dissipative system is provided by the damped pendulum described by the equation

$$\ddot{x} + c\dot{x} + \sin x = b \cos t \ ,$$

where c is a positive real constant and b is a real constant. Let us write such equation as the following system of three first–order differential equations:

$$\begin{aligned} \dot{x} &= y \\ \dot{y} &= -cy - \sin x + b \cos t \\ \dot{t} &= 1 \ . \end{aligned}$$

One readily obtains that

$$A(t) = \begin{pmatrix} 0 & 1 & 0 \\ -\cos x(t) & -c & -b \sin t \\ 0 & 0 & 0 \end{pmatrix}$$

with $\mathrm{Tr}(A(t)) = -c < 0$; the contraction rate is therefore represented by the quantity

$$\det(J_t) = e^{\int_0^t \mathrm{Tr}(A(s)) ds} = e^{-ct} \ .$$

1.4 The attractors and basins of attraction

In the framework of dynamical systems theory a relevant concept is provided by that of the *attractor;* qualitatively it is defined as the set toward which the dynamical system evolves. A more precise definition is given as follows.

Definition. For a continuous flow $\underline{\Phi} : \mathbf{R} \times \mathbf{R}^\ell \to \mathbf{R}^\ell$ associated to (1.1), the *ω–limit* set corresponding to the initial condition $\underline{x}_0$ is defined as the set

$$\begin{aligned} \omega(\underline{x}_0) \ = \ &\{\underline{x} \in \mathbf{R}^\ell : \ \text{there exists a sequence } \{t_n\} \to \infty \text{ such that} \\ &\underline{\Phi}(t_n; \underline{x}_0) \to \underline{x} \text{ as } n \to \infty\} \ . \end{aligned}$$

For a mapping of the form (1.4), let $\{\underline{x}, \underline{f}(\underline{x}), \underline{f}^2(\underline{x}), \dots\}$ be the orbit associated to the initial position $\underline{x}$ according to the dynamics generated by $\underline{f}$. The *ω–limit* set corresponding to the initial condition $\underline{x}_0$ is given by the accumulation points $\{\underline{f}^n(\underline{x}_0)\}$, namely

$$\begin{aligned}\omega(\underline{x}_0) &= \{\underline{x} \in \mathbf{R}^\ell : \text{ there exists a sequence } \{n_k\} \in \mathbf{Z} \\ &\quad \text{(strictly increasing) such that } \underline{f}^{n_k}(\underline{x}_0) \to \underline{x} \text{ as } k \to \infty\} \ .\end{aligned}$$

An *attractor* is an ω–limit set which has the property to attract a set with non–zero measure of initial values [146]. We can now give the following qualitative definition of *basin of attraction.*

Definition. Consider a continuous or discrete system which admits an attractor; the corresponding *basin of attraction* is composed by the set of initial conditions which tend to the attractor as time goes on (for continuous systems) or as the number of iterations increases (for discrete systems).

In other words, the basin of attraction is the closure of the set of initial points whose trajectory approaches the attractor asymptotically. An attractor is said to be *chaotic,* when it shows sensitivity to the choice of the initial conditions. More precisely, let us suppose that we observe two trajectories, initially very close to each other. Following their evolution with time, we speak of extreme sensitivity to the initial conditions, if the distance between the two trajectories increases exponentially with time. In such a case it is impossible to make a long–term prediction, since small uncertainties in the initial conditions are greatly amplified in a relatively short time.

There exist different techniques for computing the basins of attraction. The most intuitive is to select a sample of random initial conditions and to let them evolve until they reach the attractor. It may happen that such a procedure requires a very long computational time; henceforth, alternative techniques have been developed in order to compute the basins of attraction using faster algorithms. To provide an example, we mention a simple method, which is based on the following recipe [81]. Select a pair of points, say (P_1, P_2), where P_1 is captured by the attractor, while P_2 is observed to pass it. Having fixed a precision δ, let P_1 and P_2 evolve, checking that their distance remains smaller than δ. Denote by $P_1^{(j)}$, $P_2^{(j)}$ the iterated points at step j; if their distance becomes greater than δ, select a new position coinciding with the point on the middle between $P_1^{(j)}$ and $P_2^{(j)}$, and let it replace $P_1^{(j)}$ or $P_2^{(j)}$ according to whether it evolves, or not, to the attractor. This procedure leads to the delimiting of the boundary of the basin of attraction within the precision δ.

We remark that basin boundaries can be subject to bifurcations, called *metamorphoses,* as a system parameter passes through a critical value; in some cases, such a transition can also take place between a smooth curve and a fractal boundary [146].

1.5 The logistic map

The logistic map is described by the non–linear, one–dimensional equation

$$x_{j+1} = rx_j(1 - x_j) \qquad \text{for } x_j \in [0, 1] \;, \tag{1.12}$$

where r is a real positive parameter. This equation is associated to the logistic continuous equation originally introduced by P.–F. Verhulst [170] as a model of population growth:

$$\frac{dx}{dt} = rx(1 - x) \; .$$

In the mapping (1.12), if x_0 is the initial population, then x_j represents the population at the year j; the parameter r denotes a combined rate for starvation and reproduction. The logistic map was widely investigated in [133]; despite its relative simplicity, it provides a huge variety of complex behaviors. In particular, two effects will be determined which correspond to starvation and reproduction, with a growth rate respectively decreasing and increasing. In order to provide a graphical visualization of the iterations of the logistic map, one can proceed as described in Figure 1.3: first, one draws the parabolic curve $y = rx(1 - x)$ on the diagram x versus y as well as the diagonal line $x = y$. From a given abscissa x_0 one computes the point on the parabola, proceed to meet horizontally the diagonal line, compute the point with the same abscissa on the parabola and continue iterating this procedure. The behavior of the population strongly depends on the value of the parameter r and on the initial condition x_0; for example, in Figure 1.3(a) the dynamics is attracted toward a fixed point, while in Figure 1.3(b) the final trajectory is a periodic orbit of period 2.

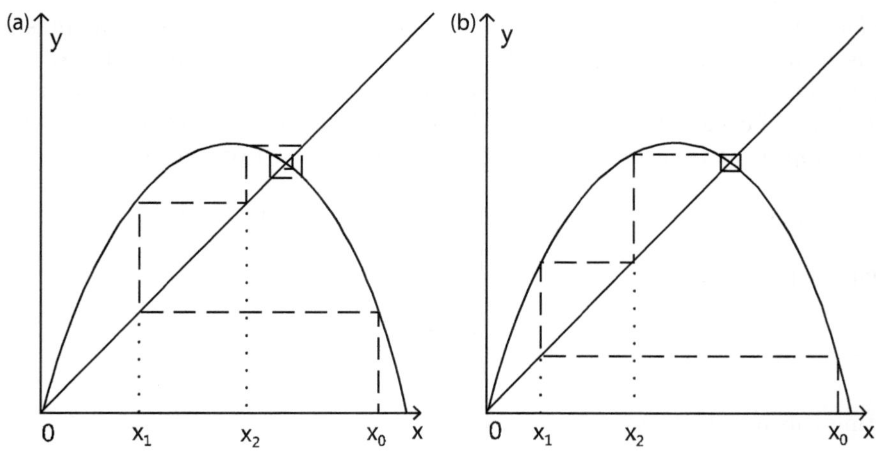

Fig. 1.3. Graphical iteration of the logistic map (1.12). (a) A stable fixed point; (b) a periodic orbit of period 2.

We start by analyzing the logistic map for the specific value $r = 4$, namely

$$x_{j+1} = 4x_j(1 - x_j) \qquad \text{with } x_j \in [0,1] \ .$$

For $x_n \in [0,1]$ define $\xi_n \in [0,1]$ through the change of variables

$$x_n = \sin^2\left(\frac{\pi\xi_n}{2}\right) \ ;$$

one obtains the equality $\sin^2(\frac{\pi\xi_{n+1}}{2}) = \sin^2(\pi\xi_n)$ which implies that $\xi_{n+1} = \pm 2\xi_n + k$ for $k \geq 0$ integer. Since $\xi_n \in [0,1]$, there is a univocal choice for k which leads to the definition of the following map:

$$\begin{aligned} \xi_{n+1} &= 2\xi_n && \text{for } 0 \leq \xi_n < \frac{1}{2} \\ \xi_{n+1} &= 2 - 2\xi_n && \text{for } \frac{1}{2} \leq \xi_n \leq 1 \ . \end{aligned} \tag{1.13}$$

The mapping (1.13) is known as the *tent map* which can be written also as $\xi_{n+1} = 1 - 2|\xi_n - \frac{1}{2}|$. Direct iterations of the mapping show that any initial condition is subject to a stretching and folding process, thus providing an exponential divergence of nearby orbits with the folding keeping the trajectory bounded. Due to the extreme sensitivity to the choice of the initial conditions, the tent mapping is shown to be chaotic.

Let us proceed with the analysis of the logistic map within the interval $0 < r < 4$. The equilibrium points are the solutions of the equation $rx_j(1-x_j) = x_j$, which provides the two points $x^{(0)} = 0$ and $x^{(1)} = \frac{r-1}{r}$. Denoting by $f(x) \equiv rx(1-x)$, since $f_x(0) = r$ one obtains that the origin is stable for $0 < r < 1$ and it is unstable when $1 < r < 4$. The point $x^{(1)}$ does not exist for $0 < r < 1$ (recall that x belongs to the interval $[0,1]$); due to the fact that $f_x(x^{(1)}) = 2 - r$ one finds that $x^{(1)}$ is stable for $|2 - r| < 1$, namely $1 < r < 3$, and it is unstable for $3 < r < 4$. Within the interval $3 < r < 4$ the map shows a sequence of bifurcations with the appearance of cycles of higher length. Two–cycles (namely oscillations between two values) appear as far as $r > 3$. In fact, the condition for two–cycles is

$$x_{n+2} = rx_{n+1}(1 - x_{n+1}) = r[rx_n(1 - x_n)][1 - rx_n(1 - x_n)] = x_n \ ,$$

which is equivalent to

$$-r^3\left(x_n^3 - 2x_n^2 + \frac{1+r}{r}x_n + \frac{1-r^2}{r^3}\right) = 0 \ .$$

Writing this equation as

$$-r^3\left(x_n - \frac{r-1}{r}\right)\left(x_n^2 - \frac{r+1}{r}x_n + \frac{r+1}{r^2}\right) = 0 \ ,$$

we recognize that the quadratic equation admits real solutions provided $r \geq 3$, from which two–cycles appear as fixed points of the square of the mapping. This behavior is described by the *bifurcation diagram* of Figure 1.4, which provides the change of structure of the orbit, through a period–doubling bifurcation, as the parameter r is varied.

The qualitative description of the results is the following (compare also with Figure 1.4):

- if $0 < r < 1$ the origin is stable;
- if $1 < r < 3$ the dynamics admits the stable solution $\frac{r-1}{r}$;
- for $3 \leq r < 3.45$ the dynamics may show an oscillation between 2 values;
- for $3.45 \leq r < 3.54$ the dynamics may show an oscillation between 4 values;
- for $3.54 \leq r < 3.57$ the dynamics may exhibit an oscillation between 8, 16, 32... values, showing a so–called *period–doubling cascade;*
- for $3.57 \leq r < 4$ there is the onset of chaos with high sensitivity to changes in the initial conditions; nevertheless for isolated values of the parameter, there might be islands of stability.

If $\{r_j\}$ denotes the sequence of parameters at which successive period doublings take place, then it can be shown that the sequence formed by the quantities

$$\mu_j \equiv \frac{r_{j+1} - r_j}{r_{j+2} - r_{j+1}}$$

admits the limit

$$\lim_{j \to \infty} \mu_j = 4.669201\ldots$$

The last number is called the *Feigenbaum constant* [65]; it provides the rate of appearance of successive bifurcations as a universal behavior, since it is observed not only for the logistic map, but also for a class of suitable dissipative one–dimensional mappings undergoing a period–doubling cascade.

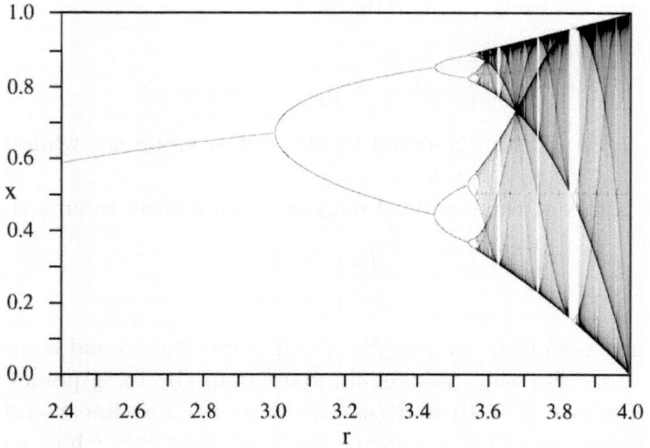

Fig. 1.4. Bifurcation diagram of the logistic map providing the period doubling of the limit cycles as the parameter r is varied
(after http://en.wikipedia.org/wiki/File:LogisticMap_BifurcationDiagram.png).

1.6 The standard map

One of the most popular mappings in the theory of Dynamical Systems is the so–called *standard map,* which was introduced by B.V. Chirikov in [47]. It is defined by the equations

$$\begin{aligned} y' &= y + \varepsilon f(x) \\ x' &= x + y' \ , \end{aligned} \tag{1.14}$$

where $y \in \mathbf{R}$, $x \in \mathbf{T} \equiv \mathbf{R}/(2\pi\mathbf{Z})$, ε is a positive real parameter, called the *perturbing parameter,* and $f = f(x)$ is an analytic, periodic function. Notice that the mapping can also be written using the following notation:

$$\begin{aligned} y_{j+1} &= y_j + \varepsilon f(x_j) \\ x_{j+1} &= x_j + y_{j+1} \qquad \text{for } j \geq 0 \ . \end{aligned} \tag{1.15}$$

The *classical* standard map is obtained setting $f(x) = \sin x$:

$$\begin{aligned} y' &= y + \varepsilon \sin x \\ x' &= x + y' \ . \end{aligned}$$

We list below some properties of the standard map, referring to its formulation (1.15).

(i) The mapping (1.15) is conservative, since the determinant of the corresponding Jacobian is equal to one; in fact, setting $f_x(x_j) \equiv \frac{\partial f(x_j)}{\partial x}$, the determinant of the Jacobian (1.15) is equal to

$$\det \begin{pmatrix} 1 & \varepsilon f_x(x_j) \\ 1 & 1 + \varepsilon f_x(x_j) \end{pmatrix} = 1 \ . \tag{1.16}$$

(ii) For $\varepsilon = 0$ the mapping (1.15) reduces to

$$\begin{aligned} y_{j+1} &= y_j \\ x_{j+1} &= x_j + y_{j+1} \qquad \text{for } j \geq 0 \ . \end{aligned} \tag{1.17}$$

Therefore y_j is constantly equal to the initial value y_0, while we can write $x_j = x_0 + jy_0$ for any $j \geq 0$.

(iii) The fixed points of the standard map are obtained by solving the equations

$$\begin{aligned} y_{j+1} &= y_j \\ x_{j+1} &= x_j \ ; \end{aligned}$$

from the first equation one has $f(x_j) = 0$, while the second equation provides $y_{j+1} = 0 = y_0$. For the classical standard map the fixed points are given by the pairs $(y_0, x_0) = (0, 0)$ and $(y_0, x_0) = (0, \pi)$. The linear stability of such points is investigated by computing the first variation, which can be written as

$$\begin{pmatrix} \delta y_{j+1} \\ \delta x_{j+1} \end{pmatrix} = M \begin{pmatrix} \delta y_j \\ \delta x_j \end{pmatrix} ,$$

where M denotes the matrix appearing in (1.16) computed at the fixed point. The corresponding eigenvalues are determined by solving the characteristic equation

$$\lambda^2 - (2 \pm \varepsilon)\lambda + 1 = 0 \ ,$$

where the positive sign holds for the fixed point $(0, 0)$, while the negative sign must be taken for $(0, \pi)$. Since one eigenvalue associated to $(0, 0)$ is greater than one, the fixed point is unstable. For $\varepsilon < 4$ the eigenvalues associated to the point $(0, \pi)$ are complex conjugate with real part less than one; therefore the position $(0, \pi)$ is stable.

We now describe the behavior of the mapping as ε varies. We have seen (item (ii) above) that for $\varepsilon = 0$ the coordinate y_j is held fixed, while $x_{j+1} = x_j + y_j$. In particular, if the initial value of the y_j variable is equal to a rational multiple of 2π, then the trajectory $\{(x_k, y_k)\}$, $k \in \mathbf{Z}_+$, is a periodic orbit. For example, let us suppose that we start with the initial datum $(y_0, x_0) = (\frac{2}{3}\pi, x_0)$; then, the successive iterations of the x coordinate are given by the following sequence: $x_0 + \frac{2}{3}\pi$, $x_0 + \frac{4}{3}\pi$, $x_0 + 2\pi$. Since x_0 varies on the torus, after 3 iterations one gets back to the initial point; in this case one speaks of a *periodic orbit of period 3.* In general, if $y_0 = 2\pi \frac{p}{q}$ with p, q positive integers ($q \neq 0$), one obtains a periodic orbit of period q. It is readily seen that the quantity p measures how many times the interval $[0, 2\pi)$ is run before coming back to the starting position.

The situation drastically changes when an irrational initial condition y_0 is taken in place of a rational initial point. For $\varepsilon = 0$ the y–value remains constant, while one can show that when the number of iterations of the mapping is increased the x–variable densely fills the line $y = y_0$ (Figure 1.5(a)). Such straight lines are *quasi–periodic invariant curves,* since on these curves a quasi–periodic motion takes place such that the dynamics comes indefinitely close to the initial conditions at regular intervals of time, though never exactly retracing itself (as is the case for the periodic orbits).

In order to distinguish between periodic orbits and quasi–periodic motions, one can introduce the *rotation number* which is defined as the quantity (independent of the initial condition)

$$\omega \equiv \lim_{j \to \infty} \frac{x_j - x_0}{j} \ .$$

In the unperturbed case it is $\omega = y_0$, since for $\varepsilon = 0$ equations (1.15) reduce to

$$\begin{aligned} y_j &= y_0 \\ x_j &= x_0 + j y_0 \ . \end{aligned}$$

According to the value of the rotation number we distinguish between periodic and quasi–periodic motions. Indeed, if $\omega = 2\pi \frac{p}{q}$ with p, q integers ($q \neq 0$), then $y_q = y_0$ and $x_q = x_0 + 2\pi p = x_0$ (modulus 2π) and the motion is periodic. If $\frac{\omega}{2\pi}$ is irrational, the dynamics associated to (1.15) with $\varepsilon = 0$ is quasi–periodic.

In conclusion, for $\varepsilon = 0$ the system reduces to (1.17) and it is said to be *integrable,* since the dynamics can be exactly solved: all motions are recognized as being periodic or quasi–periodic. A *non–integrable* system occurs when it is not possible to find an explicit solution and when chaotic motions appear. For the standard map the transition from an integrable to a non–integrable system occurs whenever ε becomes different from zero. In particular, for ε not zero but sufficiently small, the quasi–periodic invariant curves, also named *rotational invariant curves,* are slightly displaced and deformed with respect to the integrable case (Figure 1.5(b)). The periodic orbits are surrounded by closed trajectories to which we refer as *librational curves.* In addition, there can be chains of islands around periodic orbits with a period multiple of the order of the resonance.

As ε increases the rotational curves are more and more deformed and distorted, while the librational curves increase their amplitude (Figure 1.5(c)); chaotic motions start to appear and they fill an increasing region as ε grows (compare Figures 1.5(c) and (d)).

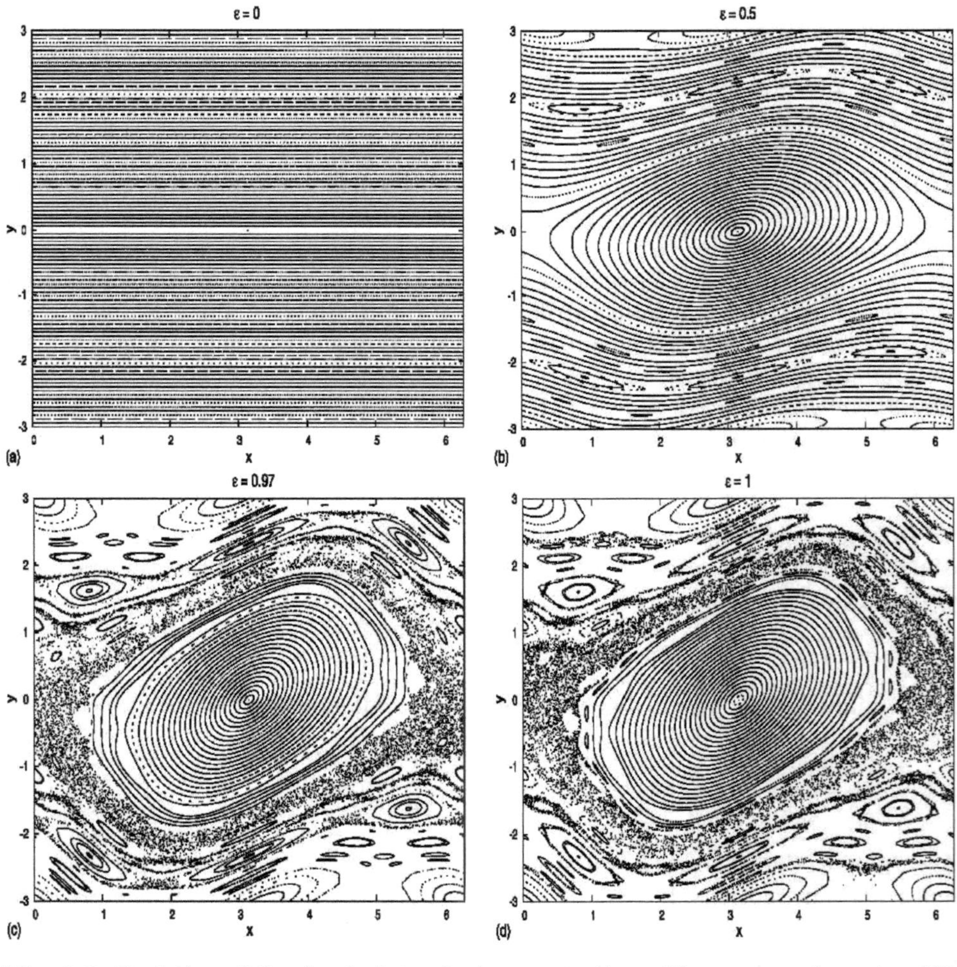

Fig. 1.5. Evolution of the classical standard map, starting with $x_0 = \pi$ and varying 100 initial conditions y_0 within the interval $[0, 3]$. (a) Case $\varepsilon = 0$; (b) case $\varepsilon = 0.5$; (c) case $\varepsilon = 0.97$; (d) case $\varepsilon = 1$.

We remark that a generalization of the classical standard map to more dimensions is provided by the four–dimensional standard map (see [70]) described by the equations

$$\begin{aligned} y' &= y + \varepsilon\big(\sin(x) + \delta \sin(x-t)\big) \\ x' &= x + y' \\ z' &= z + \varepsilon\big(\sin(t) - \delta \sin(x-t)\big) \\ t' &= t + z' \ , \end{aligned}$$

where y, $z \in \mathbf{R}$ and x, $t \in \mathbf{T}$, $\varepsilon \in \mathbf{R}_+$ is the perturbing parameter, while $\delta \in \mathbf{R}_+$ is the coupling parameter. If $\delta = 0$ one obtains two uncoupled standard maps, while if $\varepsilon = 0$ the mapping reduces to two uncoupled circle mappings.

1.7 The dissipative standard map

The standard map can be modified in order to encompass the dissipative case by introducing slight changes with respect to (1.14) [20, 36]. More precisely, we define the *dissipative standard map* through the equations

$$\begin{aligned} y' &= by + c + \varepsilon\ f(x) \\ x' &= x + y' \ , \end{aligned} \tag{1.18}$$

where $y \in \mathbf{R}$, $x \in \mathbf{T}$, $b \in \mathbf{R}_+$, $c \in \mathbf{R}$, $\varepsilon \in \mathbf{R}_+$ and $f(x)$ is an analytic, periodic function. The quantity b is called the *dissipative parameter,* since the determinant of the Jacobian associated to (1.18) amounts to b. We shall be concerned with values of b within the interval $[0, 1]$. As the parameters are varied one obtains the following situations.

- If $b = 1$ and $c = 0$ one recovers the conservative standard mapping (1.14).
- If $b = 0$ one obtains the one–dimensional mapping $x' = x + c + \varepsilon f(x)$.
- If $b = 0$ and $\varepsilon = 0$ one obtains the circle map $x' = x + c$.
- If $0 < b < 1$ the mapping is dissipative (contractive).

In the dissipative case let us define the quantity

$$\alpha \equiv \frac{c}{1-b} \ ;$$

we immediately recognize that for $\varepsilon = 0$ the trajectory $\{y = \alpha\} \times \mathbf{T}$ is invariant. In fact, the condition $y' = y = by + c$ implies $\alpha = b\alpha + c$, namely $c = \alpha(1 - b)$. The latter relation shows that the parameter c becomes zero in the conservative case $b = 1$.

The dynamics associated to the dissipative standard mapping admits attracting periodic orbits and invariant curve attractors as well as strange attractors. In Chapter 2 we shall give a precise definition of strange attractors; for the moment we just mention that these objects have an intricate geometrical structure and that introducing a suitable definition of dimension, the strange attractors are shown

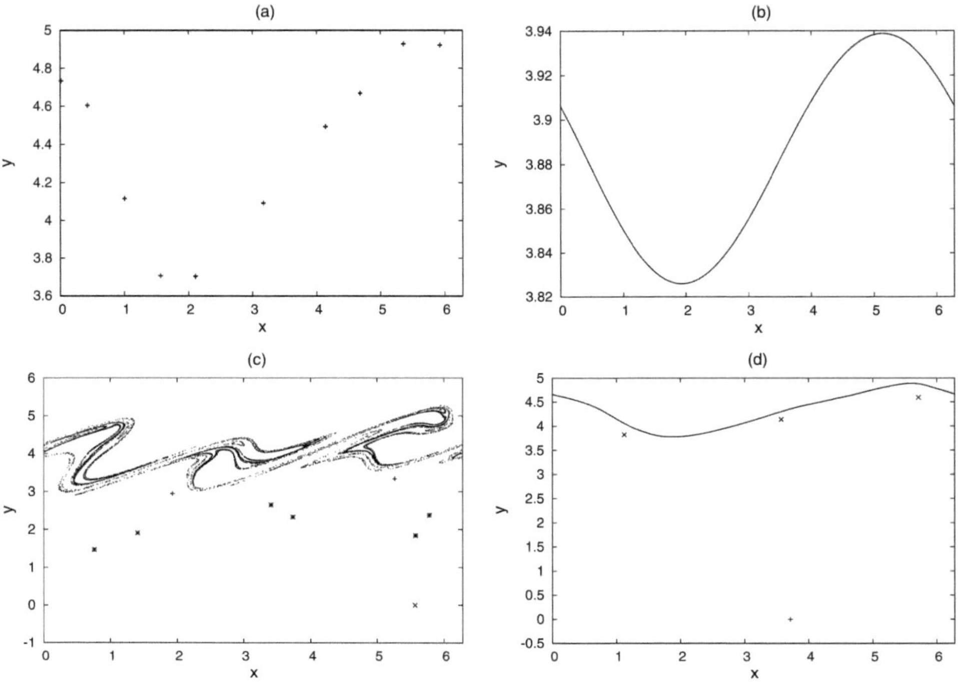

Fig. 1.6. Different attractors associated to the dissipative standard map. (*a*) A periodic orbit for $f(x) = \sin x$ with $\varepsilon = 0.9$, $b = 0.8$, $c = 2\pi \cdot 1.3825$; (*b*) an invariant attractor for $f(x) = \sin x$ with $\varepsilon = 0.1$, $b = 0.9$, $c = 2\pi \cdot 0.0618$; (*c*) a strange attractor and periodic orbits for $f(x) = \sin x + \sin 3x$ with $\varepsilon = 0.5$, $b = 0.8$, $c = 2\pi \cdot 0.1199$; (*d*) an invariant attractor with a periodic orbit of period 0/1 and a periodic orbit of period 1/3 for $f(x) = \sin x$ with $\varepsilon = 0.8$, $b = 0.9$, $c = 2\pi \cdot 0.0618$.

to have a non–integer dimension (namely a *fractal* dimension). Different kinds of attractors are displayed in Figure 1.6; in particular, Figure 1.6(*d*) provides an example of an invariant curve attractor coexisting with two periodic orbits of periods 0/1, 1/3.

The basins of attraction of the attractors shown in Figure 1.6(*d*) are presented in Figure 1.7, which is computed by taking 500×500 random initial data in the (x, y)–plane, evolving the mapping from each of these points and marking a dot when the solution reaches the attractor.

Figure 1.8 shows the evolution of the attractor as the perturbing parameter is varied. An invariant curve is found for $\varepsilon = 0.2$; the curve is distorted as ε grows and for $\varepsilon = 0.32$ one obtains only some remnants of the invariant curve.

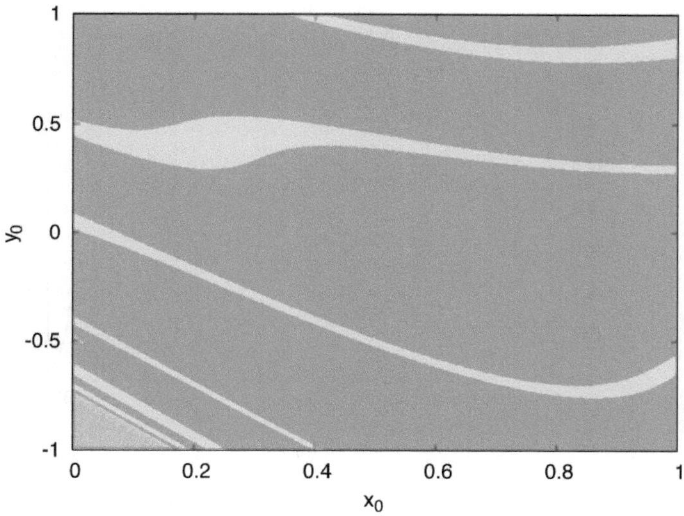

Fig. 1.7. Basins of attraction of the case (*d*) of Figure 1.6. The small portion at the bottom left refers to the 0/1 periodic orbit; the largest region (dark grey) is the basin of attraction of the invariant attractor; the remaining part (light grey) refers to the 1/3 periodic orbit.

1.8 Hénon's mapping

A two–dimensional mapping known as *Hénon's mapping* shows very interesting dynamical behaviors; it is described by the equations

$$\begin{aligned} x' &= y + 1 - ax^2 \\ y' &= bx \ , \end{aligned} \tag{1.19}$$

where a and b are real parameters. The Jacobian of the mapping is given by:

$$J = \begin{pmatrix} -2ax & 1 \\ b & 0 \end{pmatrix} .$$

Therefore the mapping is contractive as far as $-1 < b < 1$, since the determinant of the Jacobian matrix amounts to $-b$. In particular, each region of the (x, y)–plane is shrunk by a constant factor $|b|$ at each iteration of the mapping. In the original paper [95] M. Hénon showed that a good choice of the parameters is the following: $a = 1.4$, $b = 0.3$; for these values of the parameters the attractor of Hénon's mapping is presented in the left panel of Figure 1.9. The other panels provide an enlargement of regions, ever small in size, which make evident the self–similar property of Hénon's attractor; such behavior is typically found in the so–called *fractal* structures (see, e.g., [130]).

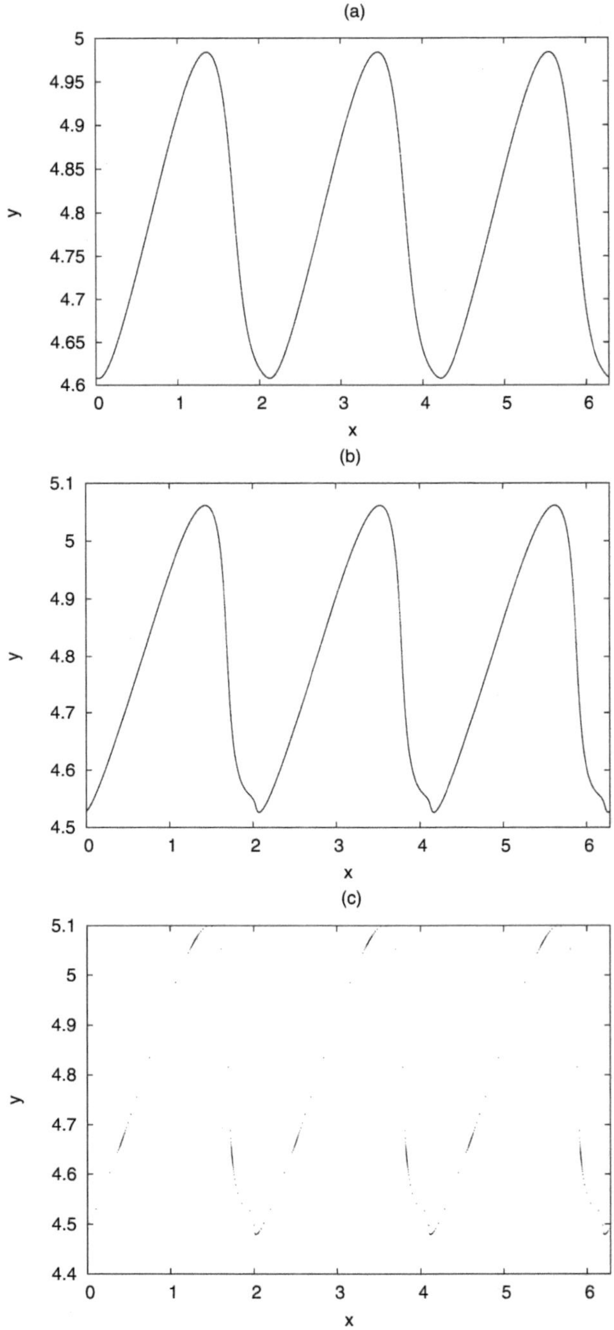

Fig. 1.8. Evolution of the mapping (1.18) with $f(x) = \sin(3x)$, $b = 0.2$, $c = 2\pi \cdot 0.61024$. The initial conditions are $(y_0, x_0) = (5, 0)$. (a) $\varepsilon = 0.2$; (b) $\varepsilon = 0.28$; (c) $\varepsilon = 0.32$.

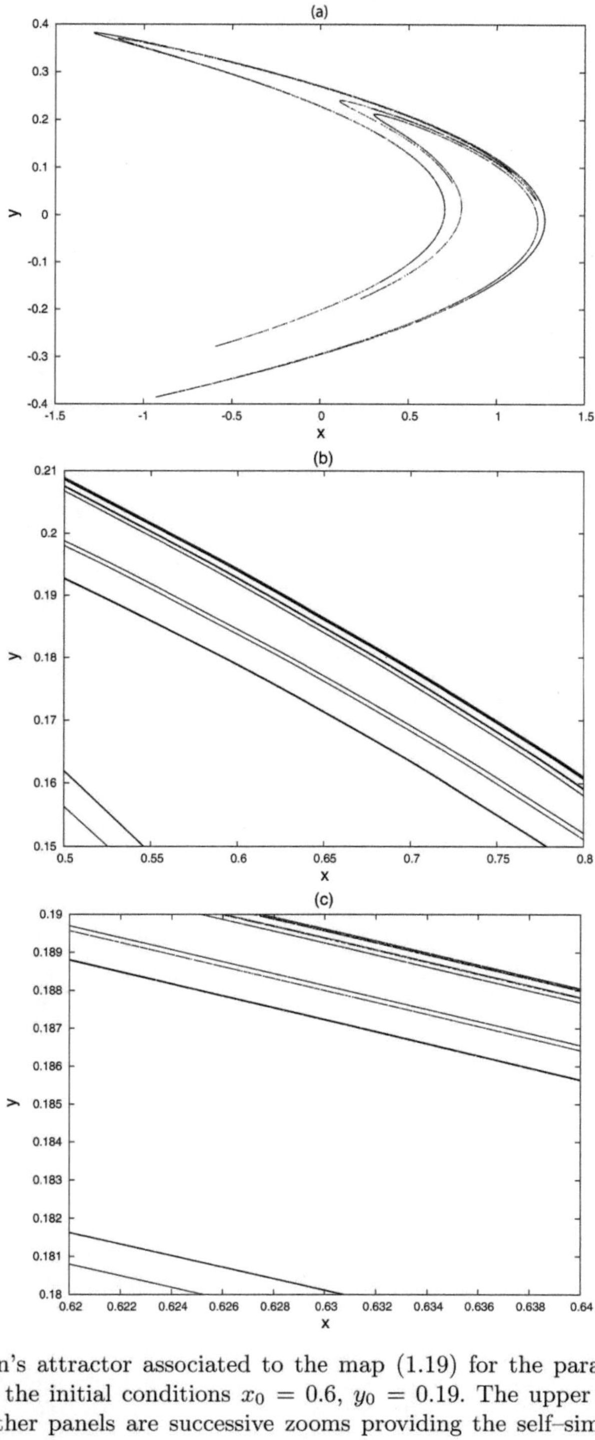

Fig. 1.9. Hénon's attractor associated to the map (1.19) for the parameters $a = 1.4$, $b = 0.3$ and for the initial conditions $x_0 = 0.6$, $y_0 = 0.19$. The upper panel shows the attractor; the other panels are successive zooms providing the self–similar structure of Hénon's attractor.

2 Numerical dynamical methods

A qualitative analysis of dynamical systems can be based on numerical investigations which provide the description of the phase space. Nowadays there exists a large number of numerical tools, some of which we are going to describe in this chapter. The Poincarè mapping (Section 2.1) allows us to reduce the analysis of a continuous system to that of a discrete mapping. The stable or chaotic character of the motion can be investigated through the computation of the Lyapunov exponents (Section 2.2). Whenever an attractor exists, it is useful to evaluate its dimension (Section 2.3). To estimate the attractor's dimension one can implement the Grassberger and Procaccia method, which can be used also for time series analysis (Section 2.4). A qualitative description of the motion can also be inferred from a Fourier analysis (Section 2.5), by looking at the behavior of the frequency of the motion as the parameters are varied (Section 2.6 and Section 2.7) or by determining finite–time Lyapunov exponents known as Fast Lyapunov Indicators (Section 2.8).

2.1 Poincaré map

An effective tool for investigating the dynamical properties of a mechanical system is obtained by introducing the so–called *Poincaré map* [8], which reduces the study of a continuous system to that of a discrete mapping. More precisely, consider the n–dimensional differential system described by the equations

$$\dot{\underline{z}} = \underline{f}(\underline{z}) \ , \qquad\qquad \underline{z} \in \mathbf{R}^n \ , \tag{2.1}$$

where $\underline{f} = \underline{f}(\underline{z})$ is a generic regular vector field. Let $\underline{\Phi}(t;\underline{z}_0)$ be the flow at time t with initial condition $\underline{z}_0$ and let Σ be an $(n-1)$–dimensional hypersurface transverse to the flow[1], which we shall refer to as the *Poincaré section.* For a periodic orbit associated to (2.1), let $\underline{z}_p$ be the intersection of the periodic orbit with Σ and let U be a neighborhood of $\underline{z}_p$ on Σ. Then, for any $\underline{z} \in U$ we define the Poincaré map as $\underline{\Phi}' = \underline{\Phi}(T;\underline{z})$, where T is the first return time of the flow on Σ.

Let us specify the Poincaré mapping in the case of a non–autonomous system with two independent variables. Let us suppose that such a system is described by the equations

$$\begin{aligned} \dot{y} &= f_1(y,x,t) \\ \dot{x} &= f_2(y,x,t) \ , \end{aligned} \tag{2.2}$$

[1] Namely, if $\underline{\nu}(\underline{z})$ denotes the unit normal to Σ at $\underline{z}$, then $\underline{f}(\underline{z}) \cdot \underline{\nu}(\underline{z}) \neq 0$ for any $\underline{z}$ in Σ.

where $y \in \mathbf{R}$, $x \in \mathbf{T}$ and f_1, f_2 are periodic continuous functions in x and t. In the three–dimensional space $\{(y, x, t) \in \mathbf{R} \times \mathbf{T}^2\}$ let us choose a plane, for example $t = 0$; we assume that the system is periodic in time with period 2π. Having fixed an initial condition, we compute the evolution of the trajectory and we mark its crossing with the plane $t = 0$ at intervals 2π. More precisely, let us write the solution of (2.2) as

$$\begin{aligned} y(t) &= y(0) + \int_0^t f_1(y(s), x(s), s) ds \\ x(t) &= x(0) + \int_0^t f_2(y(s), x(s), s) ds \ . \end{aligned}$$

The Poincaré map is defined as the solution at time intervals 2π, as described by the equations

$$\begin{aligned} y' &= y + \int_0^{2\pi} f_1(y(s), x(s), s) ds \\ x' &= x + \int_0^{2\pi} f_2(y(s), x(s), s) ds \ . \end{aligned}$$

This mapping provides a remarkable simplification of the trajectory in the phase space, though still retaining the main dynamical features of the continuous system.

As a concrete example of the computation of the Poincaré mapping, let us consider the equation of a forced pendulum:

$$\ddot{x} + \varepsilon \big[\sin x + \sin(x - t) \big] = 0 \ ,$$

where ε is a positive parameter, while $(x, t) \in \mathbf{T}^2$. This equation can be written as a system of two differential equations of the first order:

$$\begin{aligned} \dot{y} &= -\varepsilon \big[\sin x + \sin(x - t) \big] \\ \dot{x} &= y \ , \end{aligned} \tag{2.3}$$

where $y \in \mathbf{R}$. The Poincaré map associated to (2.3) is provided by the solution at time 2π with initial conditions $(y_0, x_0) \equiv (y, x)$ at time $t = 0$:

$$\begin{aligned} y' &= y - \varepsilon \int_0^{2\pi} \big[\sin x(s) + \sin(x(s) - s) \big] ds \\ x' &= x + \int_0^{2\pi} y(s) ds \ . \end{aligned} \tag{2.4}$$

The iteration of (2.4) for different values of the initial conditions and of the parameter ε is provided in Figure 2.1.

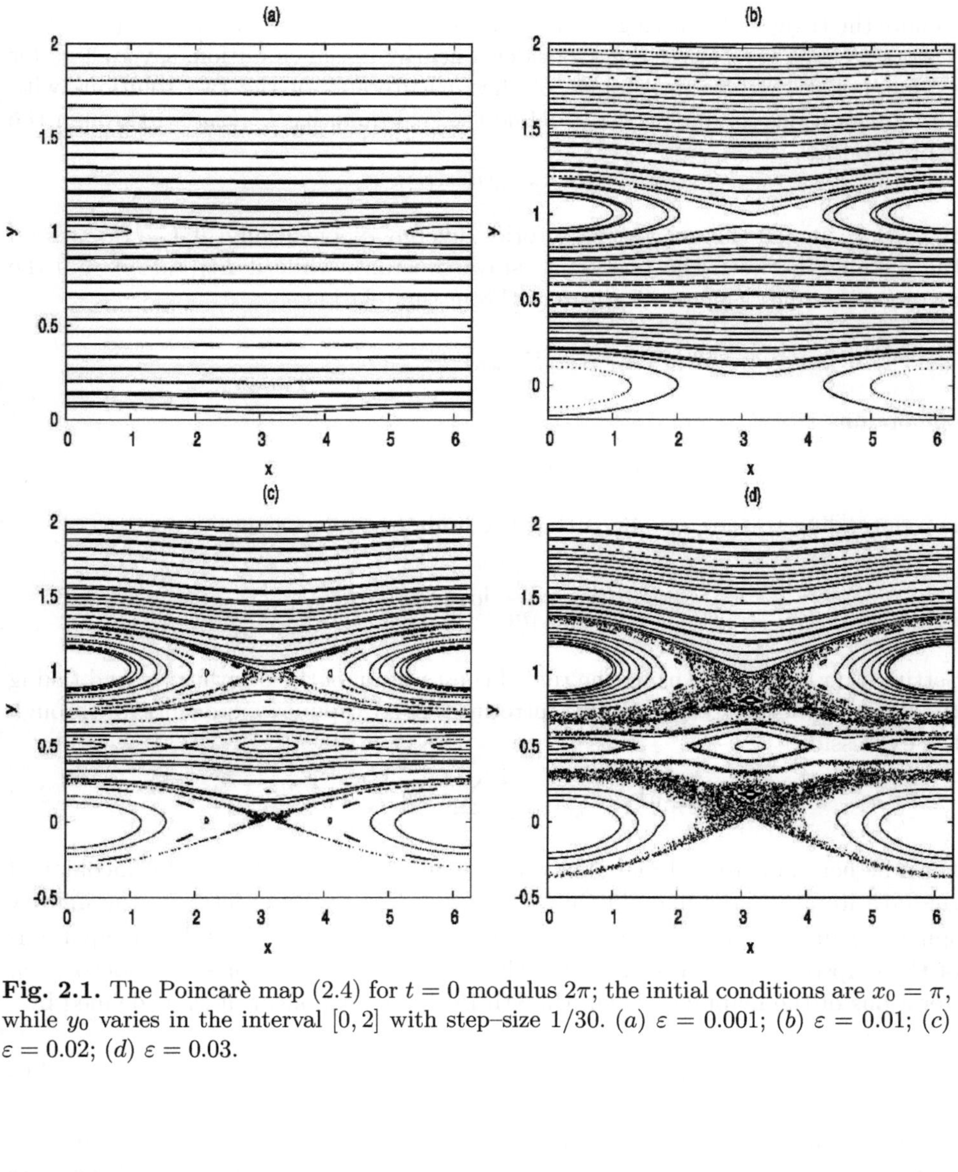

Fig. 2.1. The Poincarè map (2.4) for $t = 0$ modulus 2π; the initial conditions are $x_0 = \pi$, while y_0 varies in the interval $[0, 2]$ with step–size $1/30$. (*a*) $\varepsilon = 0.001$; (*b*) $\varepsilon = 0.01$; (*c*) $\varepsilon = 0.02$; (*d*) $\varepsilon = 0.03$.

2.2 Lyapunov exponents

A classical tool to estimate the rate of divergence of initially close trajectories is provided by the computation of the so–called *Lyapunov exponents* [14, 146, 173]. The number of such quantities equals the dimension of the system, though typically one computes only the largest Lyapunov exponent which provides the maximal excursion associated to nearby orbits. The mathematical definition of the Lyapunov exponents is provided as follows. Let us start from the simple case of a one–dimensional discrete mapping, say $x' = f(x)$ with $f : \mathbf{R} \to \mathbf{R}$ being a continuous function. Set $f^m(x_0) = f(f(\ldots f(x_0)))$, where f is composed m times.

Denote the trajectory starting from the initial condition x_0 as $\{x_0, x_1 = f(x_0), x_2 = f^2(x_0), \dots, x_m = f^m(x_0), \dots\}$. Consider a nearby initial condition, say $x_0 + \varepsilon_0$ for ε_0 small. Let ε_m be the distance of the mth iterates of the two solutions with initial conditions x_0 and $x_0 + \varepsilon_0$; define the real quantity $\tilde{L}_m(x_0; \varepsilon_0)$ through the expression

$$\varepsilon_m = \varepsilon_0 \ e^{m\tilde{L}_m(x_0;\varepsilon_0)} \ .$$

If $\tilde{L}_m(x_0; \varepsilon_0) > 0$ the nearby trajectories diverge exponentially; if $\tilde{L}_m(x_0; \varepsilon_0) = 0$ the distance between the two orbits stays constant, while if $\tilde{L}_m(x_0; \varepsilon_0) < 0$ the distance becomes asymptotically zero. From the relation

$$f^m(x_0 + \varepsilon_0) - f^m(x_0) = \varepsilon_0 \ e^{m\tilde{L}_m(x_0;\varepsilon_0)} \ ,$$

one obtains

$$\tilde{L}_m(x_0; \varepsilon_0) = \frac{1}{m} \log \left| \frac{f^m(x_0 + \varepsilon_0) - f^m(x_0)}{\varepsilon_0} \right| \ .$$

For ε_0 tending to zero, one finds that the limit $L_m(x_0)$ is given by

$$L_m(x_0) = \frac{1}{m} \log \left| \frac{df^m(x_0)}{dx} \right| \ .$$

Setting $f_x(x_j) \equiv \frac{df(x_j)}{dx}$, using the rule of composition of the derivatives[2] and taking the limit as m goes to infinity, one introduces the Lyapunov exponents [8] through the expression

$$L(x_0) \ \equiv \ \lim_{m\to\infty} \frac{1}{m} \sum_{j=1}^{m} \log |f_x(x_{j-1})| \ . \tag{2.5}$$

For a higher–dimensional mapping with $\underline{f}$ defined over $\mathbf{R}^n$, denote the Jacobian of the mth iterate of the mapping at $\underline{x}_0$ as $J_m \equiv D\underline{f}^m(\underline{x}_0)$ and let S be the unitary sphere around $\underline{x}_0$; let $r_k^{(m)}$ $(k = 1, \dots, n)$ be the length of the kth principal axis of the ellipsoid $J_m S$, thus providing the rate of contraction or expansion over m iterations of the mapping. Then, the kth Lyapunov exponent at $\underline{x}_0$ is defined by

$$L_k = L_k(\underline{x}_0) \equiv \lim_{m\to\infty} \frac{1}{m} \ \log(r_k^{(m)}) \ . \tag{2.6}$$

For a continuous system of the form $\dot{\underline{x}} = \underline{f}(\underline{x})$, $\underline{x} \in \mathbf{R}^n$, let $\underline{F}_t(\underline{x}_0) \equiv \underline{\Phi}(t; \underline{x}_0)$ be the flow at time t with initial condition $\underline{x}_0$; the Lyapunov exponents $L_k(\underline{x}_0)$ are defined as the Lyapunov exponents associated to the time–1 map which is defined as $\underline{F}' = \underline{F}_1(\underline{x}_0)$. The computation of the Lyapunov exponents requires the knowledge of $D\underline{F}_1(\underline{x}_0)$, which can be obtained through a simultaneous integration of the equations of motion and of the variational equations (1.10). Given the Lyapunov exponent $L_k(\underline{x}_0)$, the *Lyapunov characteristic number* is defined as

$$\Lambda_k(\underline{x}_0) \equiv e^{L_k(\underline{x}_0)} \ .$$

---

[2] For example: $\frac{df^2(x_0)}{dx} = \frac{df(f(x_0))}{dx} = f_x(x_0) f_x(f(x_0)) = f_x(x_0) f_x(x_1)$, being $x_1 = f(x_0)$.

In some cases the limit introduced in (2.5) or (2.6) cannot exist; nevertheless the *Oseledec theorem* guarantees the existence of the limit for a large class of systems (see, e.g., [158] for an exhaustive discussion).

The behavior of a dynamical system can be studied through the analysis of the Lyapunov exponents: chaotic attractors are characterized by at least one finite positive Lyapunov exponent, while they are zero for stable and periodic orbits. When the dynamical system admits a positive Lyapunov exponent, there exists a time horizon beyond which all predictions fail, thus leading to the introduction of the *chaos* concept as a long–term *aperiodic* behavior, exhibiting an extreme sensitivity to initial conditions (see, e.g., [115]). Notice that the chaotic behavior does not imply the instability of the solution, but rather the impossibility of forecasting the evolution over long time scales. This behavior is widely known as the paradigmatic *butterfly effect* introduced by E. Lorenz [122]: a small change of the initial conditions of a dynamical system can provoke large variations in the long–term behavior.

2.3 The attractor's dimension

Let $L_1 \geq \cdots \geq L_n$ denote the Lyapunov exponents of an n–dimensional system; the *Lyapunov dimension* is introduced as the quantity

$$D_L \equiv j + \frac{\sum_{i=1}^{j} L_i}{|L_{j+1}|} ,$$

where j is the maximum index such that $\sum_{i=1}^{j} L_i \geq 0$. If the dynamical system admits attractors, then we shall refer to *chaotic attractors* whenever the associated exponents are positive; attractors with a fractal structure are called *strange attractors*. Typically these sets are characterized by a non–integer *box–counting* dimension defined as follows.

Let A be an n–dimensional attractor; cover A with a grid of n–dimensional cubes with length r and let $N_c(r)$ be the number of cubes necessary to cover A. In the limit as the length of the cubes goes to zero, the *box–counting dimension* of A is defined as the quantity

$$D_0 \equiv \lim_{r\to 0} \frac{\log\ N_c(r)}{\log(\frac{1}{r})} . \tag{2.7}$$

While studying the geometry of the attractor, one is interested in the cubes in which the trajectory spends more time; to this end one can introduce the *natural measure* as the amount of time that the orbit spends in a given region of the phase space. Within the dynamical system described by (2.1) let $\underline{z}_0$ be an initial condition in the basin of attraction of the attractor A and let $\underline{z}(t;\underline{z}_0)$ be the trajectory at time t originating from $\underline{z}_0$. For a given cube C of the phase space, we define $\mu(C;\underline{z}_0,\tau)$ as the fraction of time that the orbit $\underline{z}(t;\underline{z}_0)$ spends in the cube C during the time interval $[0,\tau]$. If there exists the limit

$$\mu(C;\underline{z}_0) = \lim_{\tau\to\infty} \mu(C;\underline{z}_0,\tau)$$

and if such a quantity is the same for almost all initial conditions (with respect to the Lebesgue measure), say $\mu(C; \underline{z}_0) = \mu(C)$, then the quantity $\mu(C)$ is called the *natural measure* of the cube C. Suppose that $N_c(r)$ cubes C_i ($i = 1, \ldots, N_c(r)$) of unit size r are needed to cover the attractor; then, the *information dimension* is defined as

$$D_1 \equiv \lim_{r \to 0} \frac{\sum_{i=1}^{N_c(r)} \mu(C_i) \log \mu(C_i)}{\log r} \ . \tag{2.8}$$

The *generalized dimension* D_q computes the frequency of visits to the cubes by giving them a weight according to their natural measure; the positive index q measures the strength of their weighting. More precisely, the generalized dimension of order q is defined as

$$D_q \equiv \frac{1}{1-q} \lim_{r \to 0} \frac{\log \ I(q,r)}{\log(\frac{1}{r})} \ , \tag{2.9}$$

where $I(q,r) = \sum_{i=1}^{N_c(r)} \mu(C_i)^q$. One finds that $D_q \geq D_p$ for $p < q$. For $q = 0$ one recovers the box–counting dimension (2.7); for $q = 1$ one obtains the information dimension (2.8). The quantity corresponding to $q = 2$ is called the *correlation dimension,* which coincides with the box–counting dimension if all boxes are equally occupied. According to a conjecture due to J.L. Kaplan and J.A. Yorke (see, e.g., [69]), the Lyapunov dimension coincides in most cases with the information dimension, thus relating the Lyapunov exponents to the attractor's geometry.

2.4 Time series analysis

Discrete time series are formed by a sequence of data of the type $\{\underline{x}_1, \underline{x}_2, \underline{x}_3, \ldots\}$, where the quantities $\underline{x}_j$ are n–dimensional real vectors with $n \geq 1$. Dynamical system theory provides some techniques for analyzing the discrete time series; for example, one can detect their deterministic or noisy character by computing the correlation dimension or the Lyapunov exponents. We remark that a serious limitation of most methods is the necessity of dealing with time series long enough to avoid unreliable results, due to poor statistics. We start by presenting the *Grassberger and Procaccia method,* which allows us to distinguish between deterministic and stochastic time series and eventually to evaluate the dimension of the embedding space [1].

For simplicity, let us consider a one–dimensional time series formed by K data, $\{x_1, \ldots, x_K\}$, $x_j \in \mathbf{R}$. A set of *delay coordinates* in a d–dimensional embedding space is defined by setting

$$\begin{aligned} \underline{y}_1 &= (x_1, \ldots, x_d) \\ \underline{y}_2 &= (x_2, \ldots, x_{d+1}) \\ &\ldots \\ \underline{y}_N &= (x_N, \ldots, x_K) \ , \end{aligned}$$

where $N = K - d + 1$. We denote by $Y \equiv \{\underline{y}_1, \ldots, \underline{y}_N\}$ the set of delay vectors.

For $r > 0$ and for any $\underline{y}_j \in Y$, let $n_j(r;d)$ be the number of points $\underline{y}_i \in \mathbf{R}^d$ $(i \neq j)$, which are contained in the d–dimensional hypersphere of radius r around $\underline{y}_j$:

$$n_j(r;d) = \sum_{i=1, i\neq j}^{N} \Theta(r - \|\underline{y}_i - \underline{y}_j\|_d) ,$$

where Θ is the Heaviside function[3] and $\|\cdot\|_d$ denotes the Euclidean norm in $\mathbf{R}^d$. We define the *correlation integral* functions as

$$C_{N,d}(r) \equiv \frac{1}{N^2} \sum_{j=1}^{N} n_j(r;d) , \tag{2.10}$$

which counts the number of pairs of points within distance r from the attractor, normalized by the total number of pairs of points. According to (2.9) the correlation dimension is defined as

$$D_2 = \lim_{r\to 0} \frac{\log \sum_{i=1}^{N_c(r)} \mu(C_i)^2}{\log r} , \tag{2.11}$$

where $N_c(r)$ has been defined as in (2.8). For simplicity of notation, let $N = N_c(r)$; one can approximate the sum appearing in (2.11) as

$$\sum_{i=1}^{N} \mu(C_i)^2 \simeq \frac{1}{N^2} \sum_{i=1}^{N} \sum_{j=1, j\neq i}^{N} \Theta(r - \|\underline{y}_i - \underline{y}_j\|_d) .$$

Taking into account the definition (2.10), one can compute the correlation dimension as

$$D_2 \equiv \lim_{r\to 0} \lim_{N\to\infty} \frac{\log C_{N,d}(r)}{\log r} , \tag{2.12}$$

for d sufficiently large. In general one has $D_2 \leq D_1 \leq D_0$, while $D_2 = D_1 = D_0$ if the points on the attractor are uniformly distributed. According to (2.12) the correlation dimension corresponds to the slope of the graph of the function $\log C_{N,d}(r)$ versus $\log r$, whenever its value is nearly constant as the embedding dimension d is varied. The minimal value of d at which the slopes are convergent determines the dimension of the embedding space. A stochastic behavior is recognized by a constant increase of the slopes with d.

In practical applications, the slope of the curves $\log C_{N,d}(r)$ versus $\log r$ must be evaluated in a meaningful range of values of the radius, say (r_0, r_1), denoted as the *scaling region*. Particular care must be devoted to the choice of such a region (see [59]): below r_0 the curves might be distorted since few points are counted in the hypersphere of radius r_0, while above r_1 the curves might tend to flatten since the attractor has finite size.

Time–delay coordinates are often used to reconstruct the attractor. More precisely, consider the time series $\{x(t_1), x(t_2), \ldots\}$; for $j \geq 0$ and $T > 0$ let

[3] The Heaviside function is defined as $\Theta(x) = 1$ if $x \geq 0$, $\Theta(x) = 0$ if $x < 0$.

$\{x(t_1 + jT), x(t_2 + jT), \dots\}$ be the set of time–delay coordinates. Let $\underline{y}_j$ be the d–dimensional time–delay vectors defined as

$$\begin{aligned} \underline{y}_1 &= (x(t_1), x(t_1 + T), \dots, x(t_1 + (d-1)T)) \\ \underline{y}_2 &= (x(t_2), x(t_2 + T), \dots, x(t_2 + (d-1)T)) \\ &\dots \end{aligned}$$

Such vectors can be used to reconstruct the attractor; however, particular care must be taken with the choice of the time delay T. Indeed, if T is too small, the coordinates of the vector $\underline{y}_j$ are almost equal to each other and the reconstruction fails, since the coordinates are too close to provide useful information. On the other hand, if T is too large, the coordinates might be too distant and therefore uncorrelated. Notice that if the system shows a rough periodicity, then T can be chosen of the order of the period, while more sophisticated criteria must be adopted if there is no dominant period. For example, one can examine the correlation between pairs of points as a function of their time separation. To this end, let a correlation function be defined as

$$\varphi(T) = \frac{\langle x(t)x(t+T)\rangle}{\langle x(t)^2\rangle} ,$$

where $\langle\cdot\rangle$ denotes the average over all points of the time series. Then, determine the time T_0 of the first zero crossing of $\varphi(T)$ as a function of T; since we look for a value of T which yields high correlation and still provides a time development, a modest fraction of T_0 often represents a reasonable choice.

Let us now briefly review some methods for the computation of the Lyapunov exponents for discrete time series. As in [173] we select two points P_0 and P_0' and follow their evolution. Let P_1, P_1' be the evolved points; if the distance between P_1 and P_1' exceeds a given threshold, replace P_1' with a point P_1'' closer to P_1 and such that the vector with endpoints P_1, P_1'' has the same orientation as the vector with endpoints P_1, P_1' (see Figure 2.2). Let $\{t_k\}$ be the sequence of times at which the replacements take place; denote by $d(t_k)$ the distance between P_k and P_k', and by $d'(t_k)$ the distance between P_k and P_k''. Then, the largest Lyapunov exponent is computed as

$$L_1 = \frac{1}{t_\ell - t_0} \sum_{k=1}^{\ell} \log \frac{d(t_k)}{d'(t_k)} ,$$

where ℓ is the total number of replacements.

An alternative method was developed in [59]; it allows us to compute all Lyapunov exponents and not just the largest one. Suppose that the dynamics is ruled by the mapping

$$\underline{x}_{j+1} = \underline{f}(\underline{x}_j) , \qquad \underline{x}_j \in \mathbf{R}^n ,$$

for a suitable vector function $\underline{f} : \mathbf{R}^n \to \mathbf{R}^n$ and let $D_{\underline{x}_j}$ be the Jacobian matrix at the point $\underline{x}_j$. We look for an approximation of $D_{\underline{x}_j}$ using the discrete time series as follows. Consider the evolution of all points P_i', P_i'', etc., whose distance from a preassigned point P_i is less than r. Consider those points whose images P_{i+m}', P_{i+m}'', etc. of P_i', P_i'', etc. after m iterations are still at a distance less than r from

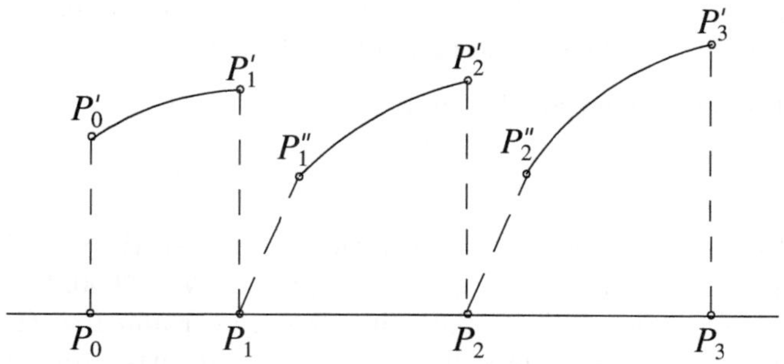

Fig. 2.2. Computation of Lyapunov's exponents following [173].

P_{i+m}, which denotes the image of P_i after m iterations (see Figure 2.3). Compute $D_{\underline{x}_j}$ with a least squares approximation over the points P'_i through the expression

$$D_{\underline{x}_j}(P'_i - P_i) \simeq P'_{i+m} - P_{i+m} \ .$$

In a similar way determine the matrices $D_{\underline{x}_{i+m}}$, $D_{\underline{x}_{i+2m}}$, etc. Furthermore, decompose the matrix $D_{\underline{x}_1}$ as $D_{\underline{x}_1} = Q_1 R_1$, where Q_1 is an orthogonal matrix and R_1 is an upper triangular matrix with non–zero diagonal elements. Analogously let $D_{\underline{x}_{1+m}} Q_1 = Q_2 R_2$, ..., $D_{\underline{x}_{1+jm}} Q_j = Q_{j+1} R_{j+1}$. The Lyapunov exponents are finally computed through the formula

$$L_k = \frac{1}{\tau m M} \sum_{j=1}^{M-1} \log(R_j)_{kk} \ ,$$

where $(R_j)_{kk}$ is the kth diagonal element of R_j, M is the total number of available matrices and τ is the sampling time–step, having assumed that $\underline{x}_j \equiv \underline{x}(j\tau)$ for some vector function $\underline{x} = \underline{x}(t)$; notice that $L_k > L_{k+1}$.

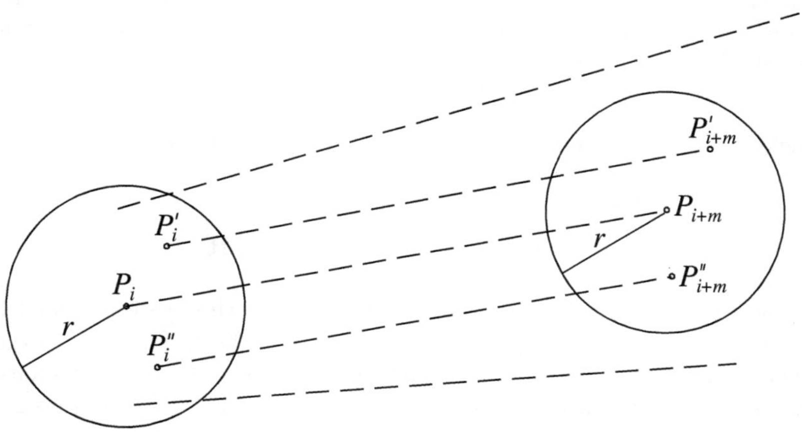

Fig. 2.3. Computation of Lyapunov's exponents following [59].

Let us now consider two concrete applications of the Grassberger and Procaccia method, respectively, to the logistic map and to Hénon's mapping.

(1) Consider the one–dimensional logistic map:

$$x_{j+1} = rx_j\,(1 - x_j)\ , \qquad x_j \in [0,1]\ ,$$

where r is a positive real parameter. Let us define a time series as being composed by the iterates of the x variable, say $\{x_1, \ldots, x_N\}$ with $N = 2000$. Fix $r = 3.57$ and select the initial condition as $x_0 = 0.3$. The Grassberger and Procaccia algorithm allows us to determine that the system evolves on an attractor embedded in a 1–dimensional space with a correlation dimension $D_2 = 0.5$ (see Figure 2.4). In this example the correlation integral functions are displayed for $d = 1, \ldots, 6$; notice that the scaling region can be restricted to the interval $-7 \le \log(r) \le -3$.

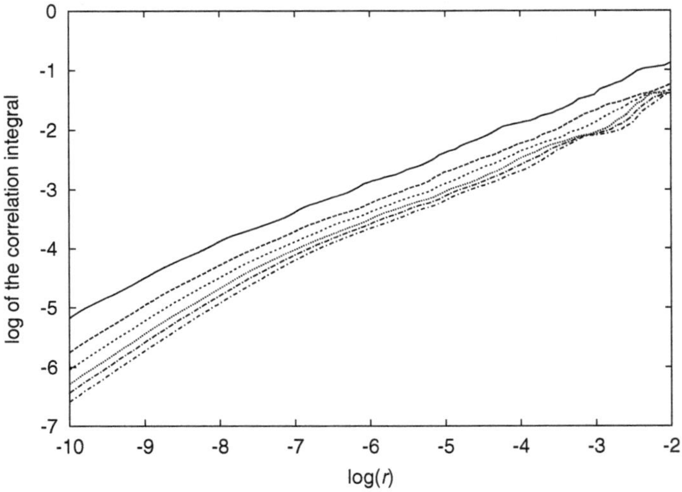

Fig. 2.4. The Grassberger and Procaccia method is implemented on the logistic map over 2000 iterations with initial condition $x_0 = 0.3$ and for $r = 3.57$. The correlation integral functions are displayed for $d = 1, \ldots, 6$.

(2) Consider the two–dimensional dissipative Hènon mapping:

$$\begin{cases} x_{j+1} = -ax_j^2 + y_j + 1 \\ y_{j+1} = bx_j\ , \qquad\qquad x_j, y_j \in \mathbf{R}\ , \end{cases}$$

where a, b are real parameters. Consider the discrete time series formed by the iterates of the x variable, say $\{x_1, \ldots, x_N\}$ with $N = 2000$. The Grassberger and Procaccia method provides an embedding dimension $d = 2$ and a correlation dimension $D_2 = 1.17$ within the scaling region $-6 \le \log(r) \le 0$ (see Figure 2.5).

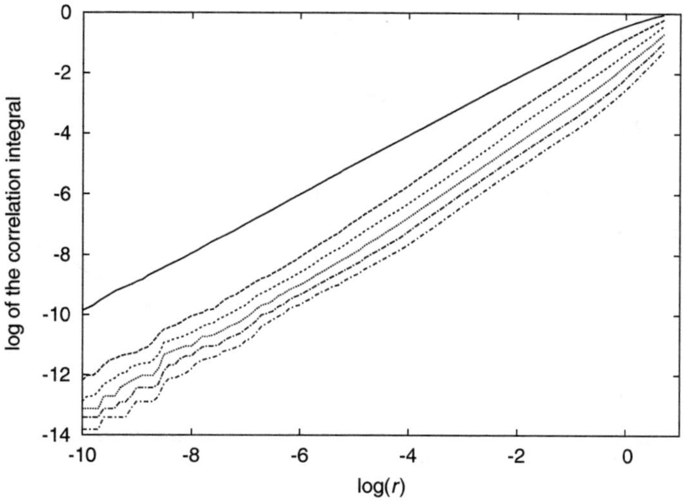

Fig. 2.5. The Grassberger and Procaccia method is implemented on the Hènon mapping over 2000 iterations with initial conditions $x_0 = 0.6$, $y_0 = 0.19$ and parameters $a = 1.4$, $b = 0.3$. The correlation integral functions are displayed for $d = 1, \ldots, 6$.

2.5 Fourier analysis

Fourier analysis is a widely used tool for studying the behavior of a signal associated to a time series [8]. The Fourier series approximation of a periodic function $f = f(t)$ of period 2π is obtained by computing the coefficients

$$\begin{aligned} a_0 &= \frac{1}{\pi}\int_{-\pi}^{\pi} f(\tau)d\tau \\ a_k &= \frac{1}{\pi}\int_{-\pi}^{\pi} f(\tau)\cos(k\tau)d\tau \\ b_k &= \frac{1}{\pi}\int_{-\pi}^{\pi} f(\tau)\sin(k\tau)d\tau \ , \end{aligned}$$

so that an approximation $f_N(t)$ to $f(t)$ can be written as

$$f_N(t) = \frac{a_0}{2} + \sum_{k=1}^{N} \left(a_k \cos kt + b_k \sin kt\right) \ . \tag{2.13}$$

For example, the Fourier coefficients associated to the *sawtooth wave function* defined as

$$f(t) = \begin{cases} t \ , & 0 \le t \le \pi \\ t - 2\pi \ , & \pi < t \le 2\pi \end{cases}$$

are given by $a_k = 0$ for all $k \ge 0$, $b_k = (-1)^{k+1}\frac{2}{k}$, so that the Fourier series representation is equal to $f_N(t) = 2\sum_{k=1}^{N} \frac{(-1)^{k+1}}{k} \sin kt$ (Figure 2.6).

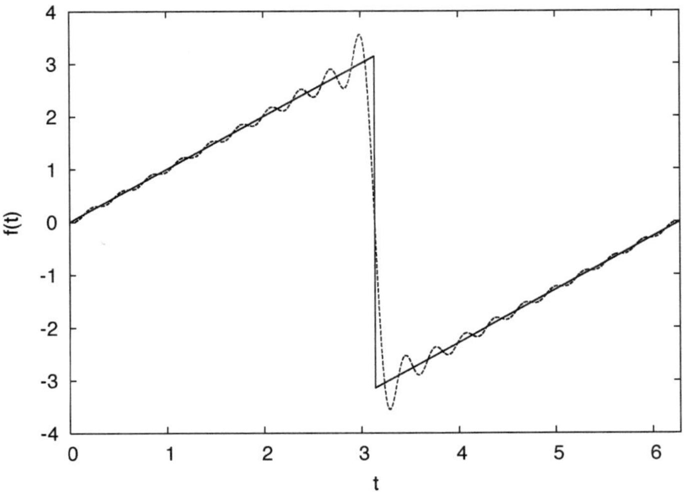

Fig. 2.6. The sawtooth map (thick line) and its Fourier approximation (dashed line) of order $N = 20$.

The *power spectrum* associated to (2.13) is defined by the expression

$$\sigma(k) \equiv \sqrt{a_k^2 + b_k^2} \ ;$$

the behavior of $\sigma(k)$ versus k provides the main frequencies of the motion as this simple example shows. Consider the function

$$f(t) = \frac{1}{2} \cos 2t + \cos 5t \ ;$$

then

$$a_k = -\frac{k(k^2+17)\sin(\pi k)}{\pi(100 - 29k^2 + k^4)} \ ,$$

while $b_k = 0$ for all values of k. Finally, $\sigma(k) = |a_k|$ and it is easy to see that the graph of $\sigma(k)$ versus k shows that the main frequencies are rightly given by 2 and 5.

2.6 Frequency analysis

Frequency analysis is a powerful numerical technique to use when looking for a quasi–periodic approximation of the solution of a dynamical system over a finite time interval [107–109]. It is based on the computation of the fundamental frequencies and on the investigation of their behavior with respect to the parameters. Given a complex function $f = f(t)$ on a finite time interval, say $[-T, T]$, in analogy to (2.13), but using complex notation, we can write an approximation $f_N(t)$ of $f(t)$ as

$$f_N(t) = \sum_{k=-N}^{N} \alpha_k \ e^{i\omega_k t} \ ,$$

for complex coefficients α_k and real frequencies ω_k. The maximal frequency ω_1 is computed as the maximum of the function

$$F(\omega) = \frac{1}{2T} \int_{-T}^{T} f(\tau)\, e^{-i\omega_k \tau} \chi(\tau)\, d\tau \ ,$$

where $\chi(\tau)$ is a positive *weight function,* with average equal to 1 over $[0, 2\pi]$; for example, one can use the Hanning filter $\chi(\tau) = 1 + \cos\frac{\pi\tau}{T}$. Once ω_1 is known, one can determine the coefficient α_1 by orthogonal projection as

$$\alpha_1 = \frac{1}{2T} \int_{-T}^{T} f(\tau)\, e^{-i\omega_1 \tau} \chi(\tau)\, d\tau \ ;$$

then one can repeat the same process starting from the function $f_1(t) = f(t) - \alpha_1 e^{i\omega_1 t}$. Notice that the quantities $e^{i\omega_k t}$ ($k \in \mathbf{Z}$) may not be orthogonal; in that case a Graham–Schmidt orthonormalization procedure must be implemented[4]. Frequency analysis provides an effective tool for distinguishing different kinds of motion, since the frequencies ω_k associated to quasi–periodic motions will be constants with respect to time, while their variation with time is an indication of chaotic dynamics.

Frequency analysis can provide many interesting results about the dynamical behavior of the system. For example, let us consider the two–dimensional standard map (1.14) (see [109]). Having fixed an initial condition x_0 we can compute the frequency as a function of y_0, thus yielding a frequency map $\omega = \omega(y_0)$. The behavior of the frequency map provides a qualitative investigation of the dynamics: a smooth change of $\omega = \omega(y_0)$ indicates a region of regular motion, while sudden jumps of the frequency map denote regions of chaotic motion; in fact, invariant tori intersect only once a vertical line, since they are graphs of continuous functions. Let us consider two initial conditions, say y_A and y_B with $y_A < y_B$. In a regular regime the corresponding rotation numbers satisfy $\omega_A < \omega_B$. On the other hand, if $\omega_A > \omega_B$, then there cannot exist an invariant curve whose frequency belongs to the interval $[\omega_A, \omega_B]$ (compare with [109]).

2.7 Hénon's method

An efficient method of computing the frequency of motion associated to a given conservative two–dimensional mapping M has been devised by M. Hénon (see, e.g., [35]). Assume that a point P_0 belongs to an invariant curve with frequency ω and let $P_n = M^n(P_0) \equiv (x_n, y_n)$ be the nth iterate point under the mapping M. Perform N iterations from P_0 providing the set of points $(P_1, \dots, P_N)$; within this set, determine the nearest neighbor to P_0 and denote by n_1 its index, say P_{n_1}. Next, define the integer p_1 through the relation:

$$n_1 \omega = p_1 + \varepsilon_1 \ , \tag{2.14}$$

[4] If $e_k \equiv e^{i\omega_k t}$, define an orthogonal basis as $e'_k \equiv \frac{\tilde{e}_k}{\|\tilde{e}_k\|}$, where $\tilde{e}_k \equiv \sum_{j=1}^{k-1} [e_k, e'_j] e'_j$ with $[e_k, e'_j] \equiv \frac{1}{2T} \int_{-T}^{T} e_k(t) \bar{e}'_j(t) \chi(t) dt$, the bar denoting complex conjugacy.

where ε_1 is a given small threshold. Equation (2.14) shows that p_1 counts the number of revolutions around the invariant curve, so that ω can be approximated by the ratio $\frac{p_1}{n_1}$. By increasing the number N of iterations we get better approximations; furthermore, the errors ε_i satisfy the following sequence of inequalities, $\varepsilon_1 > \varepsilon_2 > \ldots. > \varepsilon_k > \ldots$, implying that

$$|n_1\omega - p_1| > |n_2\omega - p_2| > \cdots > |n_k\omega - p_k| > \ldots ,$$

where the indexes n_i's are the smallest integers satisfying the above inequalities. We also remark that

$$\varepsilon_k < 1/(n_k n_{k+1}) \qquad \text{with} \quad n_{k+1} \geq a_k n_k + n_{k-1} ,$$

where a_k is the kth term of the *continued fraction* representation of ω, namely

$$\omega = \frac{1}{a_1 + \frac{1}{a_2 + \frac{1}{a_3 + \ldots}}}$$

These relations allow us to estimate the precision with which one obtains the next nearest neighbor to P_0. The rotation number is finally provided by the limit

$$\omega = \lim_{k\to\infty} \frac{p_k}{n_k} .$$

As an example consider the conservative standard mapping; Figure 2.7 shows the evolution of the frequency ω computed implementing Hénon's method for $\varepsilon = 0.9$ and for different values of the initial conditions with $x_0 = 0$ and $0 \leq y_0 \leq 3$. A monotone behavior denotes regular motion, a constant value corresponds to periodic orbits, while an irregular behavior is associated to chaotic motions.

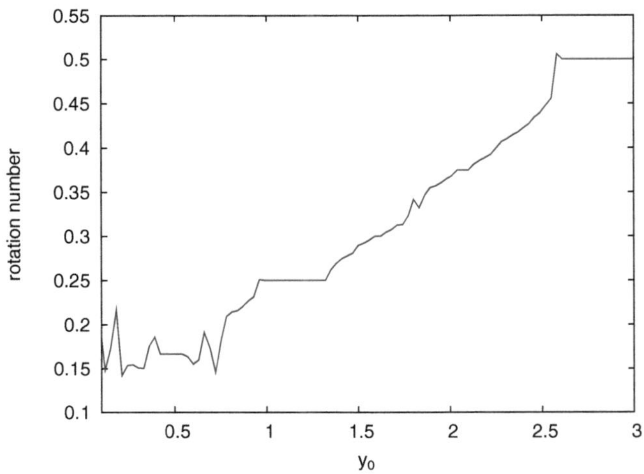

Fig. 2.7. The rotation number of the classical standard map is computed using Hènon's method for $\varepsilon = 0.9$ and for a set of 100 initial conditions with $x_0 = 0$ and $0 \leq y_0 \leq 3$.

2.8 Fast Lyapunov Indicators

The Fast Lyapunov Indicator (hereafter FLI) introduced in [71] is obtained as the value of the largest Lyapunov characteristic exponent at a fixed time, say T. A comparison of the FLIs as the initial conditions are varied allows one to distinguish between different kinds of motion (regular, resonant or chaotic). For a dynamical system described by the equations $\dot{\underline{x}} = \underline{f}(\underline{x})$, $\underline{x} \in \mathbf{R}^n$, let the variational equations be

$$\dot{\underline{v}} = \left(\frac{\partial \underline{f}(\underline{x})}{\partial \underline{x}} \right) \underline{v} \ ,$$

where $\underline{v}$ is an n–dimensional vector[5].

The FLI can be introduced as follows: given the initial conditions $\underline{x}(0) \in \mathbf{R}^n$, $\underline{v}(0) \in \mathbf{R}^n$, the FLI at time $T \geq 0$ is provided by the expression [71]

$$FLI(\underline{x}(0), \underline{v}(0), T) \equiv \sup_{0<t\leq T} \log ||\underline{v}(t)|| \ .$$

As an example of the application of the FLIs, we consider the two–dimensional standard map (1.14). With reference to Figure 2.8, we consider three different orbits: a librational region about $(\pi, 0)$, the surrounding chaotic separatrix and an invariant curve above the librational region. Moderate values of the FLI denote regular regimes, as in the case of the librational region and of the invariant curve; diverging values of the FLI correspond to a chaotic behavior.

Let us now proceed to extend the definition of the FLI to dissipative systems, the so–called *differential* FLI [36]. Consider a dissipative system described by the equations $\dot{\underline{x}} = \underline{f}(\underline{x})$ for $\underline{x} \in \mathbf{R}^n$. Denote by $\underline{v}(t)$ the solution of the differential system

$$\begin{aligned} \dot{\underline{x}} &= \underline{f}(\underline{x}) \\ \dot{\underline{v}} &= \left(\frac{\partial \underline{f}}{\partial \underline{x}}(\underline{x}(t)) \right) \underline{v}(t) \end{aligned}$$

with initial data $\underline{x}(0)$, $\underline{v}(0)$, where $\underline{v}(0)$ has unitary norm. As in [36] introduce the quantity

$$DFLI_0(\underline{x}(0), \underline{v}(0), t) \equiv F(\underline{x}(0), \underline{v}(0), 2t) - F(\underline{x}(0), \underline{v}(0), t) \ ,$$

where $F(\underline{x}(0), \underline{v}(0), t) = F(t) \equiv \log \|\underline{v}(t)\|$. As for the Lyapunov exponents the quantity $DFLI_0$ is zero for invariant attractors, negative for periodic orbits and positive for chaotic attractors. In order to take care of the oscillations of the norm of the vector $\underline{v}$, we introduce the definition of the *differential FLI* as

$$DFLI(T) \equiv G_{2T}(F(t)) - G_T(F(t)) \ ,$$

where

$$\begin{cases} G_\tau(F(t)) = \sup_{0\leq t\leq \tau} F(t) & \text{if } \ F(\tau) \geq 0 \\ G_\tau(F(t)) = \inf_{0\leq t\leq \tau} F(t) & \text{if } \ F(\tau) < 0 \ . \end{cases}$$

The DFLI can be conveniently applied to dissipative systems providing a qualitative description of the dynamics as the FLI do in the conservative context.

---

[5] For a mapping $\underline{x}' = M(\underline{x})$ the variational equation is given by $\underline{v}' = \left(\frac{\partial M(\underline{x})}{\partial \underline{x}} \right) \underline{v}$.

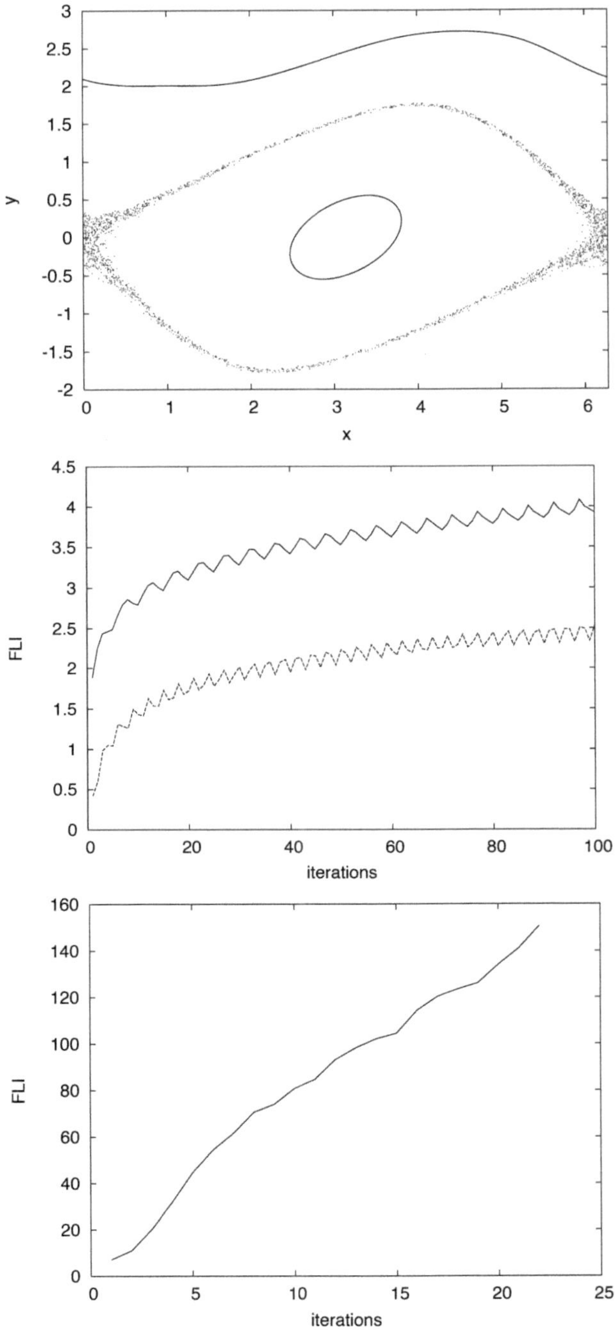

Fig. 2.8. The computation of the FLI for the classical standard map. The upper panel shows a librational island, a chaotic separatrix and an invariant curve. The middle panel shows the FLIs for the librational island (bottom line) and for the invariant curve (upper line). The bottom panel shows the FLI for the chaotic region.

Frequency analysis and FLI (or DFLI) can be viewed as complementary tools for investigating the dynamics as it appears by implementing both techniques on concrete examples. To this end, consider the dissipative standard mapping (1.18) with $f(x) = \sin x + \frac{1}{3} \sin 3x$ and initial conditions $y_0 = 5$, $x_0 = 0$. The frequency of the motion varies with the perturbing parameter ε; according to [36] we study its variation by selecting a grid of 1000 values equally spaced in the interval $0 < \varepsilon < 1$. A regular behavior of $\omega = \omega(\varepsilon)$ (see Figure 2.9(a)) shows a region of invariant tori; a plateau corresponds to periodic orbits, while an irregular behavior is associated to chaotic motions. The computation of the corresponding DFLI is reported in Figure 2.9(b) on the same grid of values of the perturbing parameter. Invariant curve attractors exist up to $\varepsilon = 0.3$ (the DFLI is zero); then we observe periodic orbit attractors (plateaus in the frequency map and negative values of the DFLI) and strange attractors (noisy variation of both quantities).

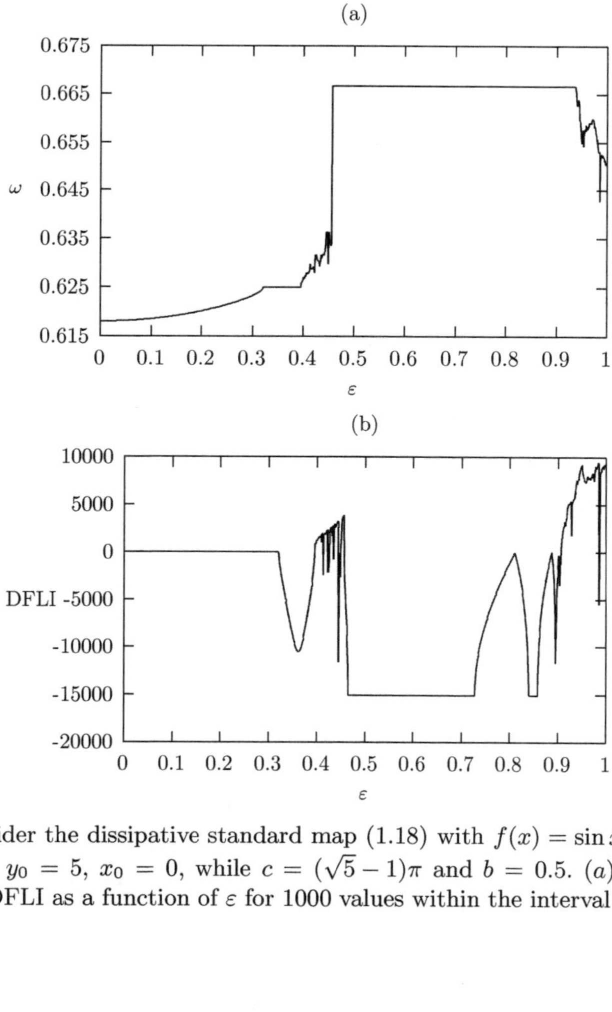

Fig. 2.9. Consider the dissipative standard map (1.18) with $f(x) = \sin x + \frac{1}{3} \sin 3x$; the initial data are $y_0 = 5$, $x_0 = 0$, while $c = (\sqrt{5} - 1)\pi$ and $b = 0.5$. (a) Variation of ω and (b) of the DFLI as a function of ε for 1000 values within the interval $[0, 1]$ (reprinted from [36]).

3 Kepler's problem

The *two–body problem* is the study of the motion of two material points $\mathcal{P}_1$ and $\mathcal{P}_2$, with masses respectively m_1 and m_2; when the two bodies are subject to the mutual gravitational attraction one speaks of *Kepler's problem,* whose dynamics is described by the three so–called *Kepler's laws* (see, e.g., [157]). In this chapter we concentrate on the mathematical description of the two–body problem. The starting point is the gravitational law and Newton's three laws of dynamics. The gravitational law states that two bodies attract each other through a force which is directly proportional to the product of the masses and inversely proportional to the square of their distance r:

$$\underline{F} = -\mathcal{G}\frac{m_1 m_2}{r^2}\ \underline{e}_{12}\ ,$$

where $\mathcal{G}$ is the gravitational constant, amounting to $\mathcal{G} = 6.67 \cdot 10^{-11}\ \mathrm{m}^3\,\mathrm{kg}^{-1}\,\mathrm{s}^{-2}$, and $\underline{e}_{12}$ is the unit vector joining the two bodies. Newton's laws of dynamics can be stated as follows:

(i) First law (law of inertia): without external forces every body remains at rest or moves uniformly on a straight line.
(ii) Second law: the net force experienced by a body is equal to the rate of change of its momentum.
(iii) Third law (action and reaction principle): for every action, there is an equal and opposite reaction.

As a consequence of the second law, if the mass of the body is constant, one gets the fundamental principle of classical mechanics according to which the net force is equal to the product of the mass of the particle times its acceleration:

$$\underline{F} \;=\; m\underline{a}\ . \tag{3.1}$$

After the investigation of the motion of the barycenter (Section 3.1), the solution of the two–body problem (Section 3.2) will be provided in terms of the three Kepler's laws, whose solution can also be given as a time series (Section 3.3); elliptic (Section 3.4), parabolic (Section 3.5) and hyperbolic (Section 3.6) motions will be analyzed and classified according to the value of the total mechanical energy (Section 3.7). We briefly remark that the Keplerian solution is also used for mission design as for the Hohmann transfer maneuvers (Section 3.8). Action–angle variables for the two–body problem are the so–called Delaunay variables (Section 3.9), which are also used to formulate Gylden's problem concerning a two–body model with variable mass (Section 3.10).

3.1 The motion of the barycenter

We start by introducing the following notation. In an inertial reference frame $(\mathcal{O}, X, Y, Z)$ with origin in $\mathcal{O}$, let $\underline{r}_1$ and $\underline{r}_2$ be the distance vectors of $\mathcal{P}_1$ and $\mathcal{P}_2$ from $\mathcal{O}$. Let $\underline{r} = \underline{r}_2 - \underline{r}_1$ be the relative distance between $\mathcal{P}_1$ and $\mathcal{P}_2$. Denote by $\mathcal{B}$ the barycenter of the two bodies and let $\underline{R}$ be the distance vector of $\mathcal{B}$ from $\mathcal{O}$ (Figure 3.1). Let $\underline{F}_1$ be the force exerted by $\mathcal{P}_2$ on $\mathcal{P}_1$ and let $\underline{F}_2$ be the force exerted by $\mathcal{P}_1$ on $\mathcal{P}_2$.

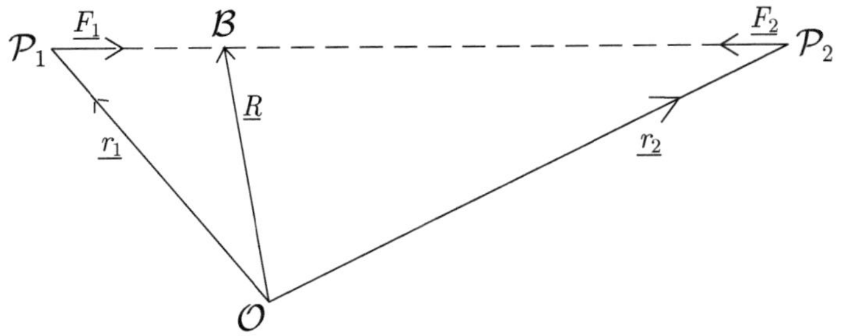

Fig. 3.1. Distance vectors in an inertial reference frame with origin in $\mathcal{O}$.

By the action and reaction principle one has

$$\underline{F}_1 = -\underline{F}_2 \ , \qquad \text{where} \qquad \underline{F}_1 = \mathcal{G}\frac{m_1 m_2}{r^2}\frac{\underline{r}}{r} \ .$$

Using (3.1), the equations of motion are given by the expressions

$$\begin{aligned} m_1 \frac{d^2 \underline{r}_1}{dt^2} &= \mathcal{G}\frac{m_1 m_2}{r^2}\frac{\underline{r}}{r} \\ m_2 \frac{d^2 \underline{r}_2}{dt^2} &= -\mathcal{G}\frac{m_1 m_2}{r^2}\frac{\underline{r}}{r} \ . \end{aligned} \qquad (3.2)$$

Adding the above equations one obtains

$$m_1 \frac{d^2 \underline{r}_1}{dt^2} + m_2 \frac{d^2 \underline{r}_2}{dt^2} = \underline{0} \ ,$$

whose integration provides the relations:

$$m_1 \frac{d\underline{r}_1}{dt} + m_2 \frac{d\underline{r}_2}{dt} = \underline{C}_1 \ , \qquad m_1 \underline{r}_1 + m_2 \underline{r}_2 = \underline{C}_1 t + \underline{C}_2 \ ,$$

with $\underline{C}_1$, $\underline{C}_2$ being constant vectors.

Let M be the total mass, namely $M = m_1 + m_2$; the location of the barycenter is given by

$$M\underline{R} = m_1 \underline{r}_1 + m_2 \underline{r}_2 \ .$$

Therefore we obtain the equations

$$M\frac{d\underline{R}}{dt} = \underline{C}_1 \ , \qquad M\underline{R} = \underline{C}_1 t + \underline{C}_2 \ ,$$

which express that the barycenter moves with constant velocity.

3.2 The solution of Kepler's problem

Let us divide the first of (3.2) by m_1 and the second by m_2; subtracting the two resulting equations one obtains:

$$\frac{d^2}{dt^2}(\underline{r}_1 - \underline{r}_2) = \mathcal{G}(m_1 + m_2)\frac{\underline{r}}{r^3} \ ;$$

being $\underline{r} = \underline{r}_2 - \underline{r}_1$ one finds:

$$\frac{d^2\underline{r}}{dt^2} + \mu\frac{\underline{r}}{r^3} = \underline{0} \ , \tag{3.3}$$

where $\mu \equiv \mathcal{G}(m_1 + m_2)$. Multiplying (3.3) by the vector $\underline{r}$ one gets

$$\underline{r} \wedge \frac{d^2\underline{r}}{dt^2} = \underline{0} \ ,$$

namely

$$\underline{r} \wedge \frac{d\underline{r}}{dt} = \underline{h} \ , \tag{3.4}$$

where $\underline{h}$ is a constant vector which represents the total angular momentum; such a vector turns out to be perpendicular to the orbital plane. From (3.4) one obtains that the two bodies move at any instant on the same plane.

On such a plane of motion we introduce a reference frame $(\mathcal{P}_1, x, y, z)$ with axes parallel to the inertial frame, the z–axis being orthogonal to the plane of motion and with the origin centered in $\mathcal{P}_1$ (Figure 3.2); let us denote by $\underline{i}$, $\underline{j}$, $\underline{k}$ the unit vectors corresponding to the reference axes. Let (r, ϑ) be the polar coordinates of $\mathcal{P}_2$ with respect to $\mathcal{P}_1$. Since the coordinates of $\mathcal{P}_2$ are $(x, y, z) = (r\cos\vartheta, r\sin\vartheta, 0)$, one obtains

$$\det\left(\underline{r} \wedge \frac{d\underline{r}}{dt}\right) = \det\begin{pmatrix} \underline{i} & \underline{j} & \underline{k} \\ r\cos\vartheta & r\sin\vartheta & 0 \\ \dot{r}\cos\vartheta - r\dot{\vartheta}\sin\vartheta & \dot{r}\sin\vartheta + r\dot{\vartheta}\cos\vartheta & 0 \end{pmatrix} = r^2\dot{\vartheta}\ \underline{k} \ .$$

Denoting by h the absolute value of $\underline{h}$ one has

$$r^2\dot{\vartheta} = h \ , \tag{3.5}$$

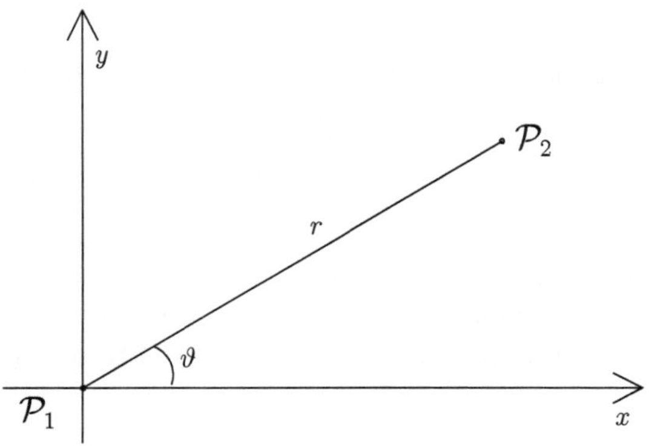

Fig. 3.2. Geometrical configuration of Kepler's problem.

which provides Kepler's second law, whose physical interpretation is the following: the areal velocity spanned by the radius vector is constant. In fact, let us evaluate the area $\delta\mathcal{A}$ spanned by the radius vector $r(t)$ at time t and by the vector $r(t+\delta t)$ at time $t+\delta t$:

$$\delta\mathcal{A} = \frac{1}{2} r(t) r(t+\delta t) \sin\delta\vartheta \ ,$$

where $\delta\vartheta$ represents the angle spanned from $r(t)$ to $r(t+\delta t)$. The variation of $\mathcal{A}$ with respect to the time is given by

$$\frac{\delta\mathcal{A}}{\delta t} = \frac{1}{2} r(t) r(t+\delta t) \frac{\sin\delta\vartheta}{\delta\vartheta} \frac{\delta\vartheta}{\delta t} \ ;$$

in the limit of δt tending to zero the areal velocity is given by

$$\dot{\mathcal{A}} = \frac{1}{2} r^2 \dot{\vartheta} \ . \tag{3.6}$$

We next consider the scalar product of (3.3) with $\dot{\underline{r}}$:

$$\dot{\underline{r}} \cdot \frac{d^2 \underline{r}}{dt^2} + \mu \frac{\dot{\underline{r}} \cdot \underline{r}}{r^3} = 0 \ ,$$

which provides

$$\frac{1}{2} \left(\frac{d\underline{r}}{dt} \right)^2 - \frac{\mu}{r} = E \ , \tag{3.7}$$

where E is a suitable real constant. Equation (3.7) provides the preservation of the total energy.

In order to solve Kepler's problem, we need to compute the radial and orthogonal components of the acceleration. In cartesian coordinates one finds

$$\begin{aligned} \ddot{x} &= \ddot{r}\cos\vartheta - 2\dot{r}\dot{\vartheta}\sin\vartheta - r\ddot{\vartheta}\sin\vartheta - r\dot{\vartheta}^2\cos\vartheta \\ \ddot{y} &= \ddot{r}\sin\vartheta + 2\dot{r}\dot{\vartheta}\cos\vartheta + r\ddot{\vartheta}\cos\vartheta - r\dot{\vartheta}^2\sin\vartheta \\ \ddot{z} &= 0 \ . \end{aligned} \tag{3.8}$$

Multiplying the first equation by $\cos\vartheta$, the second by $\sin\vartheta$ and adding the results, the radial component of the acceleration is given by

$$\ddot{r} - r\dot{\vartheta}^2 = -\frac{\mu}{r^2} \ . \tag{3.9}$$

Moreover, multiplying the second of (3.8) by $\cos\vartheta$, the first by $\sin\vartheta$ and subtracting the results, the orthogonal component is equal to

$$r\ddot{\vartheta} + 2\dot{r}\dot{\vartheta} = 0 \ .$$

Such an equation can be written in the form $\frac{d}{dt}(r^2\dot{\vartheta}) = 0$, which provides the constancy of the angular momentum h as in (3.5).

Let us introduce the quantity $\rho \equiv \frac{1}{r}$; using the constancy of the angular momentum, the radial component (3.9) can be written in terms of ρ as

$$\frac{d^2\rho}{d\vartheta^2} + \rho = \frac{\mu}{h^2} \ . \tag{3.10}$$

In fact, from

$$\frac{d\rho}{d\vartheta} = \frac{d(1/r)}{dt}\frac{r^2}{h} = -\frac{\dot{r}}{h}\ , \qquad \frac{d^2\rho}{d\vartheta^2} = -\ddot{r}\frac{r^2}{h^2}\ ,$$

one obtains the equation

$$\ddot{r} - \frac{h^2}{r^3} = -\frac{\mu}{r^2}\ ,$$

while using $\dot{\vartheta} = \frac{h}{r^2}$ one gets (3.9).

The equation (3.10) can be solved to provide the variation of ρ as a function of ϑ as

$$\rho(\vartheta) = \frac{\mu}{h^2} + A\cos(\vartheta - g_0)\ ,$$

where A, g_0 are suitable constants. Recalling that $\rho = \frac{1}{r}$ and introducing the quantities $p \equiv \frac{h^2}{\mu}$, called the ellipse *parameter,* and $\mathrm{e} \equiv \frac{Ah^2}{\mu}$, called the *eccentricity,* one obtains the expression providing the radius vector as a function of the angle ϑ:

$$r = \frac{p}{1 + \mathrm{e}\cos(\vartheta - g_0)}\ . \tag{3.11}$$

The quantity g_0, usually called the *argument of perihelion,* represents the angle between the x–axis of the reference frame and the direction of the semimajor axis of the ellipse, called the *perihelion axis* (compare with Figure 3.3). Introducing the *true anomaly* f as

$$f \equiv \vartheta - g_0\ ,$$

equation (3.11) can be equivalently written as

$$r = \frac{p}{1 + \mathrm{e}\cos f}\ . \tag{3.12}$$

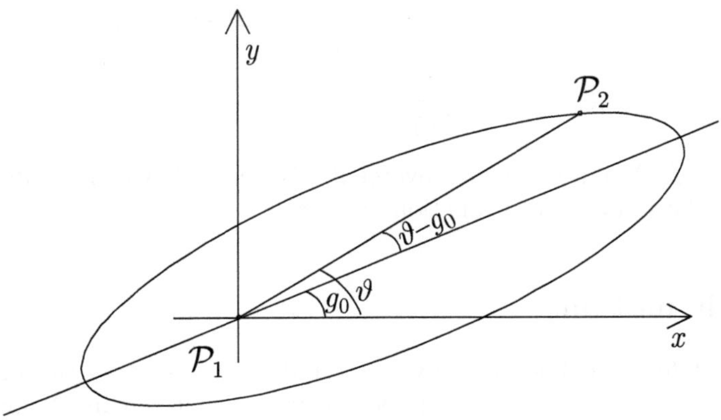

Fig. 3.3. The argument of perihelion g_0.

3.3 $\tilde{f}$ and $\tilde{g}$ series

Kepler's problem admits a solution in the form of a time series, the coefficients of such a series being functions of the mass parameter μ, of the initial values of the radius vector $\underline{r}$ and of its derivative. Inserting in (3.3) the change of time given by

$$\tau = \sqrt{\mu}\, t \ ,$$

one obtains the equation

$$\frac{d^2\underline{r}}{d\tau^2} + \frac{\underline{r}}{r^3} = \underline{0} \ . \tag{3.13}$$

For short we denote by $\underline{r}' = \frac{d\underline{r}}{d\tau}$, $\underline{r}'' = \frac{d^2\underline{r}}{d\tau^2}$ and so on. With this notation, the equation (3.13) can be written as $\underline{r}'' = -\frac{\underline{r}}{r^3}$. Differentiating (3.13) we obtain

$$\begin{aligned} \underline{r}''' &= -\left(\frac{\underline{r}'}{r^3} - \frac{3\underline{r}}{r^4}\frac{\underline{r}'\cdot\underline{r}}{r}\right) \\ \underline{r}'''' &= -\left\{\left[\frac{2\mu}{r^6} - \frac{3\underline{r}'\cdot\underline{r}'}{r^5} + \frac{15(\underline{r}'\cdot\underline{r})^2}{r^7}\right]\underline{r} - 6\frac{\underline{r}'\cdot\underline{r}}{r^5}\underline{r}'\right\} \\ & \ldots \end{aligned} \tag{3.14}$$

Expanding $\underline{r}$ in Taylor series around $\tau = 0$ and setting $\underline{r}_0 = \underline{r}(0)$ (similarly for the derivatives) we obtain

$$\underline{r} = \underline{r}_0 + \underline{r}_0'\tau + \frac{1}{2}\underline{r}_0''\tau^2 + \frac{1}{3!}\underline{r}_0'''\tau^3 + \ldots$$

Using (3.14) and rearranging the terms we can write

$$\underline{r} = \tilde{f}\underline{r}_0 + \tilde{g}\underline{r}_0' \ ,$$

where $\tilde{f}$ and $\tilde{g}$ are the following series in τ:

$$\begin{aligned} \tilde{f}(\tau) &= 1 - \frac{1}{2r_0^3}\tau^2 + \frac{1}{2r_0^3}\frac{\underline{r}_0'\cdot\underline{r}_0}{r_0^2}\tau^3 + \ldots \\ \tilde{g}(\tau) &= \tau - \frac{1}{6r_0^3}\tau^3 + \ldots \end{aligned}$$

The series $\tilde{f} = \tilde{f}(\tau)$ and $\tilde{g} = \tilde{g}(\tau)$ converge if τ is small; they can be efficiently used to determine the solution as a function of time.

3.4 Elliptic motion

We prove that for eccentricities between 0 and 1 ($0 \leq \mathrm{e} < 1$) the motion takes place on an ellipse. We consider a reference frame centered in $\mathcal{P}_1$ and with the abscissa coinciding with the perihelion line. Having denoted by r the size of the radius vector joining $\mathcal{P}_1$ and $\mathcal{P}_2$, and by f the angle spanned by the radius vector with respect to the perihelion axis, the coordinates of $\mathcal{P}_2$ are given by

$$(x, y) = (r\cos f, r\sin f) \ .$$

Therefore from (3.12) we obtain $p = r + \mathrm{e}x$; taking the square of such equation and recalling that $r = \sqrt{x^2 + y^2}$, one obtains

$$x^2(1 - \mathrm{e}^2) + 2p\mathrm{e}\, x + y^2 = p^2 \ .$$

This equation can be written as

$$\frac{(x - x_0)^2}{a^2} + \frac{y^2}{b^2} = 1 \ , \tag{3.15}$$

where

$$x_0 = -\frac{p\mathrm{e}}{1 - \mathrm{e}^2} \ , \qquad a = \frac{p}{1 - \mathrm{e}^2} \ , \qquad b = \frac{p}{\sqrt{1 - \mathrm{e}^2}} \ . \tag{3.16}$$

Notice that (3.15) describes an ellipse with semimajor axes a and b oriented according to the x and y axes. Moreover we find that the quantity $\mathrm{e} = \sqrt{1 - \frac{b^2}{a^2}}$ coincides with the *eccentricity* of the ellipse.

We have thus proved Kepler's first law, which states the following: assuming that $\mathcal{P}_1$ coincides with the Sun and $\mathcal{P}_2$ with a planet, then the motion of the planet takes place on an ellipse with the Sun located at one of the two foci.

From the second of (3.16) and from $p = \frac{h^2}{\mu}$, one obtains $h = \sqrt{\mu a(1 - \mathrm{e}^2)}$. From (3.5) and (3.6) the angular momentum h is equal to twice the areal velocity; denoting by T the period of revolution, being πab the area of the ellipse, one obtains that $h = \frac{2}{T}\pi ab$. Using the relation $b = a\sqrt{1 - \mathrm{e}^2}$ one gets that the period of revolution and the semimajor axis are linked by the expression:

$$T^2 = \frac{4\pi^2}{\mu}\, a^3 \ . \tag{3.17}$$

Equation (3.17) provides the content of Kepler's third law.

We are finally in the position to summarize Kepler's laws, which were proved in the present and previous sections.

First law. The orbit of each planet around the Sun is an ellipse with the Sun at one focus.

Second law. The radius vector sweeps equal areas in equal intervals of time.

Third law. The square of the period of revolution is proportional to the third power of the semimajor axis.

We remark that, among other consequences, Kepler's third law allows us to estimate the mass of a planet, once the orbital elements of one of its satellites are known. More precisely, let us denote by $m_{\mathcal{S}un}$, $m_{\mathcal{P}}$, $m_{\mathcal{S}}$ the masses of the Sun, of the planet $\mathcal{P}$ and of its satellite $\mathcal{S}$. Let $a_{\mathcal{P}}$, $a_{\mathcal{S}}$ and $T_{\mathcal{P}}$, $T_{\mathcal{S}}$ be, respectively, the semimajor axes and the periods of the planet around the Sun, and of the satellite around the planet; we assume that these quantities can be obtained by direct measurements. Applying Kepler's third law to the pairs Sun–planet and planet–satellite, one obtains

$$\mathcal{G}(m_{\mathcal{S}un} + m_{\mathcal{P}}) = 4\pi^2 \frac{a_{\mathcal{P}}^3}{T_{\mathcal{P}}^2} \ , \qquad \mathcal{G}(m_{\mathcal{P}} + m_{\mathcal{S}}) = 4\pi^2 \frac{a_{\mathcal{S}}^3}{T_{\mathcal{S}}^2} \ .$$

The ratio of the two equations provides

$$\frac{m_{\mathcal{P}} + m_{\mathcal{S}}}{m_{Sun} + m_{\mathcal{P}}} = \left(\frac{a_{\mathcal{S}}}{a_{\mathcal{P}}}\right)^3 \left(\frac{T_{\mathcal{P}}}{T_{\mathcal{S}}}\right)^2 .$$

Assuming that the mass of the satellite is negligible with respect to that of the planet and that the mass of the Sun is known, the previous equation provides an estimate for the mass of the planet as

$$m_{\mathcal{P}} = m_{Sun} \frac{\left(\frac{a_{\mathcal{S}}}{a_{\mathcal{P}}}\right)^3 \left(\frac{T_{\mathcal{P}}}{T_{\mathcal{S}}}\right)^2}{1 - \left(\frac{a_{\mathcal{S}}}{a_{\mathcal{P}}}\right)^3 \left(\frac{T_{\mathcal{P}}}{T_{\mathcal{S}}}\right)^2} . \tag{3.18}$$

For example, let us take $\mathcal{P}$ as Jupiter and $\mathcal{S}$ as its satellite Io; their elements are $a_{\mathcal{P}} = 7.78 \cdot 10^8$ km, $T_{\mathcal{P}} = 4331.87$ days, $a_{\mathcal{S}} = 421\,800$ km, $T_{\mathcal{S}} = 1.769$ days, while $m_{Sun} = 2 \cdot 10^{30}$ kg. The expression (3.18) provides an estimate for the mass of Jupiter equal to $1.9 \cdot 10^{27}$ kg in full agreement with the experimental data.

To conclude the description of the elliptic motion, we provide the formula for the squared velocity which, expressed in terms of the polar coordinates, takes the form,

$$v^2 = \dot{r}^2 + r^2 \dot{f}^2 .$$

From (3.12) and (3.5) one finds

$$\dot{r} = \frac{h}{p} \mathrm{e} \sin f , \qquad r\dot{f} = \frac{h}{p}(1 + \mathrm{e} \cos f) .$$

Computing the square, adding the two equations and using $\frac{h^2}{\mu} = p = a(1-\mathrm{e}^2)$ one obtains

$$v^2 = \mu \left(\frac{2}{r} - \frac{1}{a}\right) .$$

We remark that at perihelion $r = a(1-\mathrm{e})$ so that the velocity $v^2 = \frac{\mu}{a}\frac{1+\mathrm{e}}{1-\mathrm{e}}$ is maximum, while at aphelion $r = a(1+\mathrm{e})$ so that $v^2 = \frac{\mu}{a}\frac{1-\mathrm{e}}{1+\mathrm{e}}$ and the velocity is minimum. We also remark that for $\mathrm{e} = 0$ the orbit reduces to a circle.

3.4.1 Mean and eccentric anomaly

If T denotes the period of revolution of $\mathcal{P}_2$ around $\mathcal{P}_1$, we introduce the *mean motion* as

$$n \equiv \frac{2\pi}{T} . \tag{3.19}$$

The angular momentum can be expressed in terms of the mean motion as $h = \frac{2}{T}\pi a^2 \sqrt{1-\mathrm{e}^2} = na^2\sqrt{1-\mathrm{e}^2}$. Let t_0 be the time of passage at perihelion; we define the *mean anomaly* ℓ_0 as the angle described by the radius vector rotating around the focus with mean angular velocity n during the interval $t - t_0$:

$$\ell_0 \equiv n(t - t_0) .$$

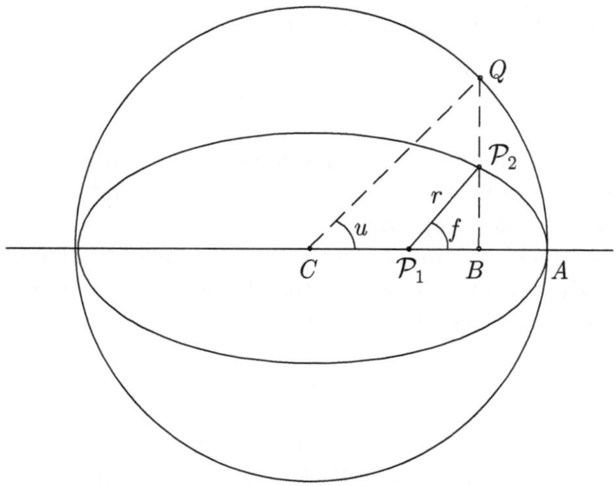

Fig. 3.4. The eccentric anomaly.

We next introduce a quantity u called the *eccentric anomaly:* draw the circle with radius equal to the semimajor axis of the ellipse (see Figure 3.4); from the instantaneous position of $\mathcal{P}_2$ on the ellipse, draw the perpendicular to the semimajor axis until it meets the circle and let u be the angle QCA formed by the direction to the center and the direction corresponding to the semimajor axis.

The mathematical relations within the true, mean and eccentric anomalies can be easily derived from the geometry of the problem. With reference to Figure 3.4 one has: $\mathcal{P}_1B = CB - C\mathcal{P}_1 = a\cos u - a\mathrm{e}$ and, since $\mathcal{P}_1B = r\cos f$, it follows that

$$r\cos f = a(\cos u - \mathrm{e}) \ . \tag{3.20}$$

By elementary properties of the ellipse one has $\frac{\mathcal{P}_2B}{QB} = \frac{b}{a}$, namely $\frac{r\sin f}{a\sin u} = \frac{b}{a}$; by this relation one has:

$$r\sin f = a\sqrt{1-\mathrm{e}^2}\sin u \ . \tag{3.21}$$

Computing the square of (3.20), (3.21) and adding the two equations one obtains

$$r^2 = a^2 + a^2\mathrm{e}^2\cos^2 u - 2a^2\mathrm{e}\cos u \ ,$$

from which it follows that

$$r = a(1 - \mathrm{e}\cos u) \ ; \tag{3.22}$$

this relation provides the radius vector as a function of the eccentric anomaly.

Taking into account that $2r\sin^2\frac{f}{2} = r(1-\cos f)$ and using (3.20), (3.21), one obtains

$$\begin{aligned} 2r\sin^2\frac{f}{2} &= a(1+\mathrm{e})(1-\cos u) \\ 2r\cos^2\frac{f}{2} &= a(1-\mathrm{e})(1+\cos u) \ ; \end{aligned}$$

computing the ratio of the two equations one gets

$$\tan\frac{f}{2} = \sqrt{\frac{1+\mathrm{e}}{1-\mathrm{e}}}\ \tan\frac{u}{2}\ , \tag{3.23}$$

which provides the true anomaly as a function of the eccentric anomaly.

Let us now derive the relation between the eccentric and mean anomalies; this formula is known as *Kepler's equation.*

From Kepler's second law we can state that the ratio bewteen the area of the region defined by $\mathcal{P}_1\mathcal{P}_2A$ and the area of the ellipse amounts to $\frac{t-t_0}{T}$; recalling the definition of the mean anomaly one has that

$$area(\mathcal{P}_1\mathcal{P}_2A) = \frac{1}{2}ab\ell_0\ .$$

On the other hand this area can be obtained as the sum of the area $\mathcal{P}_1\mathcal{P}_2B$ and of the area $B\mathcal{P}_2A$, where the area $B\mathcal{P}_2A$ is equal to $\frac{b}{a}$ times the area of QBA; therefore one has the sequence of relations:

$$\begin{aligned}
area(\mathcal{P}_1\mathcal{P}_2A) &= area(\mathcal{P}_1\mathcal{P}_2B) + \frac{b}{a}area(QBA) \\
&= area(\mathcal{P}_1\mathcal{P}_2B) + \frac{b}{a}(area(QCA) - area(QCB)) \\
&= \frac{1}{2}r^2\sin f\cos f + \frac{b}{a}\left(\frac{1}{2}a^2u - \frac{1}{2}a^2\sin u\cos u\right) \\
&= \frac{1}{2}ab(u - \mathrm{e}\sin u)\ .
\end{aligned}$$

One thus obtains that the relation between ℓ_0 and u is given by

$$\ell_0 = u - \mathrm{e}\sin u\ , \tag{3.24}$$

which is known as *Kepler's equation.* It is now necessary to solve this equation to get u as a function of the time, being $\ell_0 = n(t-t_0)$. Once such equation is solved, and therefore $u = u(t)$ is obtained, one inserts the resulting expression in (3.22) and (3.23) to obtain the variation with time of the radius vector and the true anomaly, thus providing the solution of the equation of motion.

3.4.2 Solution of Kepler's equation

In order to find the eccentric anomaly as a function of the time, it is necessary to solve the implicit Kepler's equation (3.24). An approximate solution can be computed as long as the eccentricity e is small. Indeed, the inversion of (3.24) provides u as a function of ℓ_0 as a series in the eccentricity:

$$\begin{aligned}
u &= \ell_0 + \mathrm{e}\sin u \\
&= \ell_0 + \mathrm{e}\sin(\ell_0 + \mathrm{e}\sin u) \\
&= \ell_0 + \mathrm{e}\sin(\ell_0 + \mathrm{e}\sin(\ell_0 + \mathrm{e}\sin u)) \\
&= \ell_0 + \left(\mathrm{e} - \frac{\mathrm{e}^3}{8}\right)\sin\ell_0 + \frac{1}{2}\mathrm{e}^2\sin(2\ell_0) + \frac{3}{8}\mathrm{e}^3\sin(3\ell_0) + O(\mathrm{e}^4)\ ,
\end{aligned}$$

where $O(e^4)$ denotes a quantity of order e^4. The complete solution can be expressed as

$$u = \ell_0 + e \sum_{k=1}^{\infty} \frac{1}{k} \Big[J_{k-1}(ke) + J_{k+1}(ke) \Big] \sin(k\ell_0) \ , \tag{3.25}$$

where $J_k(x)$ are the *Bessel's functions* of order k and argument x; they are defined by the relation

$$J_k(x) \equiv \frac{1}{2\pi} \int_0^{2\pi} \cos(kt - x \sin t) dt \ .$$

The functions $J_k(x)$ can be developed as follows:

$$\begin{aligned} J_0(x) &\equiv \sum_{m=0}^{\infty} \frac{(-1)^m}{(m!)^2} \left(\frac{x}{2}\right)^{2m} \\ J_k(x) &\equiv \left(\frac{x}{2}\right)^k \frac{1}{k!} \sum_{m=0}^{\infty} \frac{(-1)^m}{m! \prod_{j=1}^{m} (k+j)} \left(\frac{x}{2}\right)^{2m} \ . \end{aligned} \tag{3.26}$$

Notice that equations (3.25), (3.26) provide the solution of Kepler's equation with arbitrary precision.

3.5 Parabolic motion

When $e = 1$ one gets the open trajectory described by the equation

$$r = \frac{p}{1 + \cos f} \ . \tag{3.27}$$

From (3.27) it follows that $r + r\cos f = p$, namely $r + x = p$; using $r = \sqrt{x^2 + y^2}$ one obtains $y^2 = -2px + p^2$, namely

$$x = -\frac{y^2}{2p} + \frac{p}{2} \ ,$$

which describes a parabola in the plane (y, x) with vertex coinciding with $(\frac{p}{2}, 0)$ (see Figure 3.5).

Notice that equation (3.27) can be written in the form

$$r = \frac{p}{2} \left(1 + \tan^2 \frac{f}{2} \right) \ .$$

Using (3.5), (3.27), one has:

$$\left(\frac{p}{2}\right)^2 \frac{1}{\cos^4 \frac{f}{2}} \dot{f} = \sqrt{p\mu} \ ,$$

whose integration yields

$$2 \left(\frac{\mu}{p^3} \right)^{1/2} (t - t_0) = \tan \frac{f}{2} + \frac{1}{3} \tan^3 \frac{f}{2} \ , \tag{3.28}$$

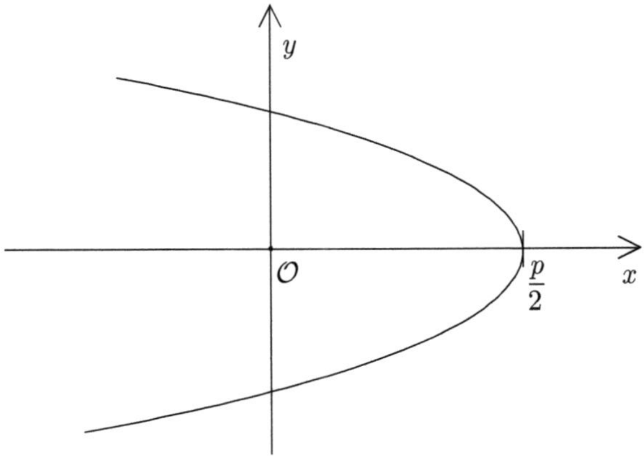

Fig. 3.5. The parabolic solution of Kepler's problem.

where t_0 is the time of passage at perihelion. The solution of equation (3.28), known as *Barker's equation,* provides the variation of the true anomaly as a function of time in the case of a parabolic orbit.

3.6 Hyperbolic motion

For $\mathrm{e} > 1$, we write the polar equation as

$$r = \frac{a(\mathrm{e}^2 - 1)}{1 + \mathrm{e}\cos f} \ . \tag{3.29}$$

Using the relations

$$x = r\cos f \ , \qquad r = \sqrt{x^2 + y^2} \ , \qquad b = a\sqrt{\mathrm{e}^2 - 1} \ ,$$

we obtain

$$x^2(\mathrm{e}^2 - 1) - y^2 + a^2(\mathrm{e}^2 - 1)^2 - 2a\mathrm{e}(\mathrm{e}^2 - 1)x = 0 \ ; \tag{3.30}$$

since $b = a\sqrt{\mathrm{e}^2 - 1}$, the equation (3.30) becomes

$$\frac{(x - x_0)^2}{a^2} - \frac{y^2}{b^2} = 1 \ ,$$

where we have introduced $x_0 = a\mathrm{e}$. From the angular momentum integral the velocity can be written as

$$v^2 = \mu\left(\frac{2}{r} + \frac{1}{a}\right) \ .$$

Notice that r tends to infinity with a non–zero velocity given by $v^2 = \frac{\mu}{a}$.

From (3.29) one has

$$\cos f = \frac{a(\mathrm{e}^2 - 1)}{\mathrm{e}r} - \frac{1}{\mathrm{e}} \ . \tag{3.31}$$

From (3.17) and (3.19) one has $n^2 a^3 = \mu$; computing the derivative of (3.31) with respect to r and using the angular momentum integral in the form $h = \sqrt{\mu p} = r^2 \dot{f}$ as well as $p = a(\mathrm{e}^2 - 1)$, one finds

$$n \frac{dt}{dr} = \frac{r}{a\sqrt{(a+r)^2 - a^2 \mathrm{e}^2}} \ .$$

Introducing the hyperbolic eccentric anomaly u_h such that

$$r \equiv a(\mathrm{e} \cosh u_h - 1) \ , \qquad (3.32)$$

one obtains

$$n \frac{dt}{du} = \mathrm{e} \cosh u_h - 1$$

whose integration provides the hyperbolic Kepler's equation

$$\ell_0 = \mathrm{e} \sinh u_h - u_h \ ,$$

where ℓ_0 is the mean anomaly. Notice that such equation is not periodic and that the solution tends quickly to infinity. From (3.29) and (3.32) one gets

$$\frac{\mathrm{e}^2 - 1}{1 + \mathrm{e} \cos f} = \mathrm{e} \cosh u_h - 1 \ ;$$

using the formulae

$$\cos f = \frac{1 - \tan^2 \frac{f}{2}}{1 + \tan^2 \frac{f}{2}} \ , \qquad \cosh u_h = \frac{1 + \tanh^2 \frac{u_h}{2}}{1 - \tanh^2 \frac{u_h}{2}} \ ,$$

one obtains the relation between the true and eccentric anomaly in the case of hyperbolic motion:

$$\tan \frac{f}{2} = \sqrt{\frac{\mathrm{e}+1}{\mathrm{e}-1}} \ \tanh \frac{u_h}{2} \ .$$

Numerical methods for solving Kepler's equation in the hyperbolic case were developed for example in [72, 135].

3.7 Classification of the orbits

According to the value of the parameter e (the eccentricity) the trajectory coincides with the following conic sections:

(i) $\mathrm{e} = 0$: the trajectory is a circle;
(ii) $0 < \mathrm{e} < 1$: the trajectory is an ellipse;
(iii) $\mathrm{e} = 1$: the trajectory is a parabola;
(iv) $\mathrm{e} > 1$: the trajectory is a hyperbola.

The same classification of the orbits can be inferred as a function of the energy. In polar coordinates the energy is given by (see (3.7))

$$E = \frac{1}{2}(\dot{r}^2 + r^2\dot{\vartheta}^2) - \frac{\mu}{r} ,$$

where $\mu \equiv \mathcal{G}(m_1 + m_2)$; using the angular momentum integral we can write

$$E = \frac{1}{2}\dot{r}^2 + V_e(r) ,$$

where $V_e(r)$ is the effective potential given by

$$V_e(r) = \frac{h^2}{2r^2} - \frac{\mu}{r} .$$

Then, we obtain

$$\frac{dr}{dt} = \sqrt{2(E - V_e(r))} .$$

Through the angular momentum integral one gets

$$\vartheta - \vartheta_0 = h \int \frac{dr}{r^2\sqrt{2(E - V_e(r))}} = \arccos \frac{\frac{r_0}{r} - 1}{\sqrt{1 - \frac{E}{E_0}}} ,$$

where r_0 is such that $V_e(r_0)$ is minimum and $E_0 = E(r_0)$, namely

$$r_0 = \frac{h^2}{\mu} , \qquad E_0 = -\frac{\mu^2}{2h^2} .$$

Recalling (3.11) we find

$$p = r_0 , \qquad \mathrm{e} = \sqrt{1 - \frac{E}{E_0}} , \qquad \vartheta_0 = g_0 .$$

In summary we obtain that the parameter e is related to the energy E by

$$\mathrm{e} = \sqrt{1 + \frac{2h^2 E}{\mu^2}} .$$

According to the classification of the orbits in terms of the eccentricity we obtain the following classification of the trajectories in terms of the energy:

(i) $E = -\frac{\mu^2}{2h^2}$ (i.e. $\mathrm{e} = 0$): the trajectory is a circle;
(ii) $E < 0$ (i.e. $0 < \mathrm{e} < 1$): the trajectory is an ellipse;
(iii) $E = 0$ (i.e. $\mathrm{e} = 1$): the trajectory is a parabola;
(iv) $E > 0$ (i.e. $\mathrm{e} > 1$): the trajectory is a hyperbola.

3.8 Spacecraft transfers

As a practical implementation of Keplerian orbits, we consider the problem of spacecraft transfers. The transfer of a spacecraft from one orbit to another is obtained by implementing proper orbital maneuvers (see, e.g., [51]). The classical ones are the so–called *Hohmann transfer* and *bi–elliptic Hohmann transfer* maneuvers, which are based on a careful combination of suitable Keplerian elliptic orbits. Impulse maneuvers require a short firing of the on–board engines, so to allow for a change of sign and direction of the velocity vector. A Hohmann transfer requires two impulse maneuvers for transferring the spacecraft from one circular orbit of radius r_A to another coplanar circular orbit of radius r_B, through an elliptic orbit which is tangent to both circles at their periapses (see Figure 3.6(a)). The changes of velocities required at the periapses can be easily computed using the angular momentum integral. Bi–elliptic Hohmann transfers between the circles of radii r_A and r_B are constructed using two semi–ellipses as in Figure 3.6(b). The first semi–ellipse allows us to reach a point C outside the external circle (see Figure 3.6(b)), while the second semi–ellipse joins with the target point B on the external circle.

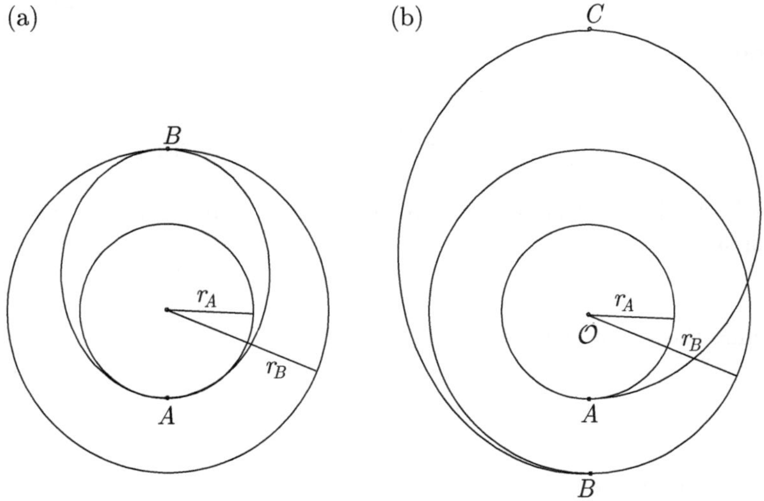

Fig. 3.6. (a) A Hohmann transfer from the circular orbit of radius r_A to the circular orbit of radius r_B. (b) A bi–elliptic Hohmann transfer from the point A on the circle of radius r_A to the point B on the circle of radius r_B.

3.9 Delaunay variables

Classical action–angle variables (see [73] and Appendix A) for the two–body problem are known as Delaunay variables [18, 169]. We present their detailed derivation for the planar motion and we provide the results for the spatial case. Let (r, ϑ) be the polar coordinates as in Figure 3.2 and let (p_r, p_ϑ) be the conjugated momenta;

it is readily seen that $p_\vartheta = h$. The Hamiltonian function[1] governing the two–body motion is given by

$$\mathcal{H}(p_r, p_\vartheta, r, \vartheta) = \frac{1}{2}\left(p_r^2 + \frac{p_\vartheta^2}{r^2}\right) - \frac{\mu}{r} .$$

Being ϑ a cyclic variable, we introduce the effective potential (see Figure 3.7) as

$$V_e(r) = \frac{p_\vartheta^2}{2r^2} - \frac{\mu}{r} , \tag{3.33}$$

so that the Hamiltonian can be written as a one–dimensional Hamiltonian of the form

$$\mathcal{H}(p_r, r) = \frac{p_r^2}{2} + V_e(r) . \tag{3.34}$$

For a fixed value E of the energy, let $r_\pm = r_\pm(E)$ be the roots of $V_e(r) = E$, so that

$$E - V_e(r) = -\frac{E}{r^2}(r_+ - r)(r - r_-) \qquad \text{with} \qquad r_\pm(E) = \frac{\mu \pm \sqrt{\mu^2 + 2Ep_\vartheta^2}}{-2E} .$$

The period of the motion can be expressed as

$$T(E) = 2\int_{r_-(E)}^{r_+(E)} \frac{dr}{\sqrt{2(E - V_e(r))}} = 2\pi\mu \left(\frac{1}{-2E}\right)^{3/2} .$$

By Kepler's third law (3.17) one obtains the following relation between the semi-major axis and the energy:

$$a = -\frac{\mu}{2E} . \tag{3.35}$$

Let us define the action variable L_0 as

$$L_0 \equiv \sqrt{-\frac{\mu^2}{2E}}$$

which in view of (3.35) provides

$$L_0 = \sqrt{\mu a} .$$

On the other hand, since (3.34) does not depend on ϑ, we can define another action variable as

$$G_0 \equiv p_\vartheta ;$$

using the expression for the angular momentum $h = \sqrt{\mu a(1 - \mathrm{e}^2)}$ and being $p_\vartheta = h$, one gets

$$G_0 = L_0\sqrt{1 - \mathrm{e}^2} .$$

[1] See Appendix A for a basic introduction to Hamiltonian dynamics.

Notice that one can express the elliptic elements, namely semimajor axis and eccentricity, in terms of the Delaunay action variables as

$$a = \frac{L_0^2}{\mu} , \qquad \mathrm{e} = \sqrt{1 - \frac{G_0^2}{L_0^2}} .$$

The Hamiltonian function expressed in terms of the action variables becomes

$$\mathcal{H} = \mathcal{H}(L_0) = -\frac{\mu^2}{2L_0^2} . \tag{3.36}$$

As for the angle variables, we proceed as follows. Using the relations

$$p_r = p_r(L_0, G_0, r) = \sqrt{-\frac{\mu^2}{L_0^2} + \frac{2\mu}{r} - \frac{G_0^2}{r^2}} , \qquad p_\vartheta = G_0 ,$$

we introduce the generating function defining the Delaunay variables as

$$\Phi(L_0, G_0, r, \vartheta) = \int p_r dr + \int p_\vartheta d\vartheta = \int \sqrt{-\frac{\mu^2}{L_0^2} + \frac{2\mu}{r} - \frac{G_0^2}{r^2}} dr + G_0 \vartheta .$$

The angle variable conjugated to L_0 is defined as

$$\ell_0 = \frac{\partial \Phi}{\partial L_0} = \int \frac{\mu^2}{L_0^3 \sqrt{-\frac{\mu^2}{L_0^2} + \frac{2\mu}{r} - \frac{G_0^2}{r^2}}} dr .$$

Using (3.22) it follows that ℓ_0 coincides with the mean anomaly, namely

$$\ell_0 = u - \mathrm{e} \sin u .$$

The angle variable conjugated to G_0 is computed as

$$g_0 = \frac{\partial \Phi}{\partial G_0} = \vartheta - \int \frac{G_0}{r^2 \sqrt{-\frac{\mu^2}{L_0^2} + \frac{2\mu}{r} - \frac{G_0^2}{r^2}}} dr .$$

Using (3.22) one finds that $g_0 = \vartheta - f$, which coincides with the argument of perihelion.

In the spatial case, namely when the three bodies are not constrained to move on the same plane, one needs to add a third pair of action–angle variables. Indeed, in polar coordinates (r, ϑ, φ) the spatial two–body Hamiltonian is given by

$$\mathcal{H}(p_r, p_\vartheta, p_\varphi, r, \vartheta, \varphi) = \frac{1}{2} \left(p_r^2 + \frac{p_\vartheta^2}{r^2} + \frac{p_\varphi^2}{r^2 \sin^2 \vartheta} \right) - \frac{\mu}{r} ,$$

where $(p_r, p_\vartheta, p_\varphi)$ are conjugated to (r, ϑ, φ). Define

$$G_0 = \sqrt{p_\vartheta^2 + \frac{p_\varphi^2}{\sin^2 \vartheta}} , \qquad H_0 = p_\varphi$$

and let the energy be

$$E = \frac{1}{2}\left(p_r^2 + \frac{G_0^2}{r^2}\right) - \frac{\mu}{r} .$$

One easily finds that

$$p_\vartheta = \pm\sqrt{G_0^2 - \frac{H_0^2}{\sin^2\vartheta}} , \qquad p_r = \pm\sqrt{2\left(E + \frac{\mu}{r}\right) - \frac{G_0^2}{r^2}} .$$

Having set

$$\vartheta_- = \arcsin\frac{H_0}{G_0} , \qquad \vartheta_+ = 2\pi - \vartheta_-$$

and

$$r_\pm = -\frac{1}{2E}\left(\mu \pm \sqrt{\mu^2 + 2EG_0^2}\right) ,$$

the action variables can be defined as

$$\begin{aligned} A_1 &\equiv \frac{1}{2\pi}\int_0^{2\pi} p_\varphi d\varphi = H_0 \\ A_2 &\equiv \frac{1}{2\pi}\int_{\vartheta_-}^{\vartheta_+} p_\vartheta d\vartheta = G_0 - |H_0| \\ A_3 &\equiv \frac{1}{2\pi}\int_{r_-}^{r_+} p_r dr = -G_0 + L_0 . \end{aligned}$$

Being $L_0^2 = -\frac{\mu^2}{2E}$, the new Hamiltonian is given by

$$\mathcal{H}(A_1, A_2, A_3) = -\frac{\mu^2}{2(A_1 + A_2 + A_3)^2} .$$

Let α_1, α_2, α_3 be the conjugated angle variables. The relation with the Delaunay variables is obtained through the symplectic change of coordinates

$$\begin{aligned} L_0 &= |A_1| + A_2 + A_3 & \qquad \ell_0 &= \alpha_3 \\ G_0 &= |A_1| + A_2 & g_0 &= \alpha_2 - \alpha_3 \\ H_0 &= |A_1| & h_0 &= \alpha_1 - \alpha_2 , \end{aligned}$$

where it can be shown (see, e.g., [53]) that H_0 is related to G_0 by

$$H_0 = G_0 \cos i ,$$

being i the inclination of the orbital plane with respect to a fixed inertial reference frame. The angle variable h_0 conjugated to H_0 coincides with the *longitude of the ascending node,* namely the angle formed by the x–axis of the reference frame with the line of nodes given by the intersection of the orbital plane with the xy–reference plane. The Hamiltonian function of the spatial case becomes

$$\mathcal{H} = \mathcal{H}(L_0) = -\frac{\mu^2}{2L_0^2} .$$

We remark that if the eccentricity is zero, then $G_0 = L_0$ and the argument of perihelion is not defined; similarly, when the inclination is zero, then $H_0 = G_0$ and the longitude of the ascending node is not defined. In these cases it is convenient to introduce the *modified* Delaunay variables defined as

$$\begin{aligned} \Lambda &= L_0 & \lambda &= \ell_0 + g_0 + h_0 \\ \Gamma &= L_0 - G_0 = L_0\left(1 - \sqrt{1-e^2}\right) & \gamma &= -g_0 - h_0 \\ \Phi &= G_0 - H_0 = 2G_0 \sin^2 \frac{i}{2} & \varphi &= -h_0 \ . \end{aligned}$$

The singularities are now represented by $\Gamma = 0$ and $\Phi = 0$, for which γ and φ are not defined. We remark that for small values of the eccentricity and of the inclination, it is often convenient to introduce the so–called *Poincaré variables* defined as

$$\begin{aligned} p_1 &= \sqrt{2\Gamma}\cos\gamma & p_2 &= \sqrt{2\Phi}\cos\varphi \\ q_1 &= \sqrt{2\Gamma}\sin\gamma & q_2 &= \sqrt{2\Phi}\sin\varphi \ . \end{aligned}$$

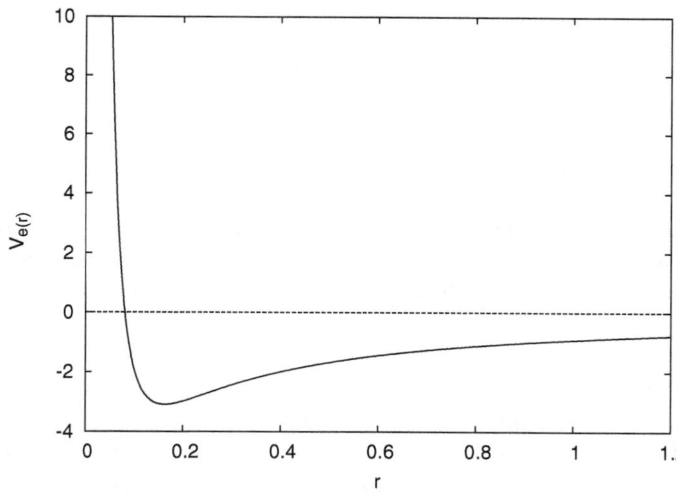

Fig. 3.7. Graph of the effective potential $V_e(r)$ given in (3.33) for $p_\vartheta = 0.4025$ and $\mu = 1$.

3.10 The two–body problem with variable mass

3.10.1 The rocket equation

In this section we study the two–body problem formed by a planet and a satellite and we assume that the mass of the satellite is not constant, but varies with time. For example, we can imagine dealing with an artificial satellite, whose mass variation is due to the loss of fuel. We assume that the decrease of the mass of the rocket is constant, namely

$$\frac{dm}{dt} = -b \tag{3.37}$$

for some positive constant b. Let us denote by $\underline{v}_p$ the exhaust velocity of the expelled particles with respect to the spacecraft; let $\Delta t = t - t_0$ with t_0 being the initial time and let $\Delta \underline{v} = \underline{v}(t) - \underline{v}(t_0)$ be the variation of the velocity. Let $m\underline{v}$ be the momentum at time t, and let $(m - b\Delta t)(\underline{v} + \Delta \underline{v})$ be the momentum at time $t + \Delta t$. Without external forces acting on the rocket (as in the case of high–thrust engines), the total change of linear momentum is given by (see [152])

$$(m - b\Delta t)(\underline{v} + \Delta \underline{v}) + b\Delta t(\underline{v} + \underline{v}_p) - m\underline{v} = \underline{0} \ .$$

In the limit for Δt tending to zero, one gets the *rocket equation* (see, e.g., [152]):

$$m\frac{d\underline{v}}{dt} = -b\underline{v}_p \ . \tag{3.38}$$

Recalling (3.37) and assuming $\underline{v}_p$ constant, the solution of (3.38) is given by

$$\Delta \underline{v} = -\underline{v}_p \log \frac{m(t_0)}{m} \ ,$$

where $m(t_0)$ is the initial mass. Notice that the quantity $\Delta \underline{v}$ is the velocity needed for the maneuver, which depends on the rate of mass loss.

3.10.2 Gylden's problem

A physical sample described by a two–body problem with variable mass is composed by a planet orbiting around a central star which varies its mass [58, 88, 90, 116]. Following classical results by Jeans [98] one can assume a mass variation according to the law

$$\frac{dm}{dt} = -\alpha m^j \ , \tag{3.39}$$

where α is usually small and j varies in the interval $[1.4, 4.4]$. For example, in the case of the Sun, the decrease of mass by radiation implies that α is of the order of 10^{-12} or 10^{-13}, where the units of measure have been assumed as the solar mass, the astronomical unit and the year; a bigger α must be adopted in the case of corpuscolar emission.

Denoting by $\underline{v}_C$ the velocity of the center of mass and by $\underline{F}$ the sum of all external forces, the equation of motion is given by

$$\frac{d}{dt}(m\underline{v}_C) = \underline{F} \ .$$

For a point within the body let $\underline{v}$ be its velocity and let $\underline{v}_m$ be the relative velocity of the escaping or incident mass with respect to the center of mass [91]; then, we can write the equation of motion as

$$m(t)\frac{d\underline{v}}{dt} = \underline{F} + \underline{v}_m \frac{dm(t)}{dt} \ . \tag{3.40}$$

When the mass is ejected isotropically, for example as for the solar wind, the sum of the contributions of the second term of the right–hand side of (3.40) cancels out and the equation of motion reduces to the so–called *Gylden's equation*

$$m(t)\frac{d\underline{v}}{dt} = \underline{F} \ . \tag{3.41}$$

If the body travels within a stationary cloud and accumulates mass, then it is $\underline{v}_m = -\underline{v}$ and the equation of motion (3.40) reduces to the so–called *Levi–Civita's equation*

$$\frac{d}{dt}(m(t)\underline{v}) = \underline{F} \ .$$

In the rest of this section we will concentrate on the analysis of Gylden's equation (3.41). Let us denote by $\underline{x} \in \mathbf{R}^3$ the two–body relative position vector and by $\underline{X} \in \mathbf{R}^3$ the conjugated momentum vector. In suitable units of measure let us write the Hamiltonian function associated to (3.41) as

$$\mathcal{H}_0(\underline{X}, \underline{x}, t) = \frac{1}{2}\underline{X} \cdot \underline{X} - \frac{k(t)}{\|\underline{x}\|} \ , \tag{3.42}$$

where in (3.41) we assumed $\underline{F} = -\frac{k(t)}{\|\underline{x}\|^3}\underline{x}$ ($\|\cdot\|$ denotes the Euclidean norm in $\mathbf{R}^3$) with $k(t)$ taking into account the mass variation (eventually one can assume that the dependence upon time is due to a time variation of the gravitational constant). According to [56] we assume that

$$k(t) \equiv \frac{k_0}{\varepsilon(t)} \ , \tag{3.43}$$

where $\varepsilon(t_0) = 1$ for some initial time t_0; we also assume that at any time $\varepsilon(t)$ is positive and that $\dot{\varepsilon}(t) \neq 0$. From (3.42) the equations of motion read as

$$\begin{aligned} \dot{\underline{x}} &= \underline{X} \\ \dot{\underline{X}} &= -k(t)\frac{\underline{x}}{\|\underline{x}\|^3} \ , \end{aligned}$$

from which we obtain that the angular momentum vector $\underline{h} = \underline{x} \wedge \underline{X}$ is constant (as it follows from $\dot{\underline{h}} = \dot{\underline{x}} \wedge \underline{X} + \underline{x} \wedge \dot{\underline{X}}$). Let us show that a suitable coordinate and time transformation gives (3.42) in the form of a perturbed Kepler's problem. Let $(\underline{y}, \underline{Y})$ denote a new set of variables obtained through the generating function

$$\Phi_1(\underline{X}, \underline{y}, t) \equiv \varepsilon \underline{y} \left(\underline{X} - \frac{1}{2}\dot{\varepsilon}\underline{y} \right) \ ,$$

which provides the change of coordinates:

$$\underline{x} = \varepsilon \underline{y} \ , \qquad \underline{X} = \frac{1}{\varepsilon}\underline{Y} + \dot{\varepsilon}\underline{y} \ .$$

Denoting by $\delta(t) \equiv \varepsilon^3 \ddot{\varepsilon}$, the Hamiltonian in the new variables takes the form

$$\mathcal{H}_1(\underline{Y}, \underline{y}, t) = \frac{1}{\varepsilon^2}\left(\frac{1}{2}\underline{Y} \cdot \underline{Y} - \frac{k_0}{\|\underline{y}\|} + \frac{1}{2}\delta(t)\|\underline{y}\|^2 \right) \ .$$

Next we perform a change of time according to

$$dt = \varepsilon^2 d\tau \ ; \tag{3.44}$$

setting $\delta(\tau) = \delta(t(\tau))$ through the transformation (3.44), the new Hamiltonian becomes

$$\mathcal{H}_2(\underline{Y}, \underline{y}, \tau) = \frac{1}{2}\underline{Y}\cdot\underline{Y} - \frac{k_0}{\|\underline{y}\|} + \frac{1}{2}\delta(\tau)\|\underline{y}\|^2 \ , \tag{3.45}$$

with associated Hamilton's equations

$$\frac{d\underline{y}}{d\tau} = \frac{\partial\mathcal{H}_2}{\partial\underline{Y}} \ , \qquad \frac{d\underline{Y}}{d\tau} = -\frac{\partial\mathcal{H}_2}{\partial\underline{y}} \ .$$

From (3.43) and (3.44) we obtain

$$\frac{\dot{k}}{k} = -\frac{\dot{\varepsilon}}{\varepsilon} \ , \qquad \frac{\ddot{\tau}}{\dot{\tau}} = -\frac{2\dot{\varepsilon}}{\varepsilon} \ ,$$

which yield

$$\frac{\ddot{\tau}}{\dot{\tau}} = \frac{2\dot{k}}{k} \ ,$$

usually referred to as the *law of marginal acceleration in ephemeris time* [56].

Let us express the Hamiltonian (3.45) in terms of the Delaunay variables introduced in Section 3.9. To this end, we set $\underline{y} = (r\cos\vartheta, r\sin\vartheta)$, $\underline{Y} = (p_r\cos\vartheta - \frac{p_\vartheta}{r}\sin\vartheta, p_r\sin\vartheta + \frac{p_\vartheta}{r}\cos\vartheta)$ and we perform a change of variables from $(p_r, p_\vartheta, r, \vartheta)$ to Delaunay variables (L_0, G_0, ℓ_0, g_0) through the generating function

$$\Phi_2(L_0, G_0, r, \vartheta, t) = \int_{r_0}^{r}\sqrt{-\frac{k^2}{L_0^2} + \frac{2k}{r} - \frac{G_0^2}{r^2}}\, dr + G_0\vartheta \ ,$$

where r_0 is a root of the function $A(L_0, G_0, r) \equiv -\frac{k^2}{L_0^2} + \frac{2k}{r} - \frac{G_0^2}{r^2}$. With the present notation the semimajor axis and the eccentricity of the osculating Keplerian orbit are related to the action Delaunay variables by

$$a = \frac{L_0^2}{k} \ , \qquad \mathrm{e} = \sqrt{1 - \frac{G_0^2}{L_0^2}} \ .$$

The time derivative of the generating function is given by

$$\begin{aligned}
\frac{\partial\Phi_2}{\partial t} &= \frac{\dot{k}}{k}\left[k\int_{r_0}^{r}\frac{dr}{r\sqrt{A(L_0, G_0, r)}} - \frac{k}{a}\int_{r_0}^{r}\frac{dr}{\sqrt{A(L_0, G_0, r)}}\right] \\
&= \frac{\dot{k}}{k}\left[\frac{L_0^3}{k^2}\frac{k}{a}u - \frac{L_0^3}{k^2}\frac{k}{a}(u - \mathrm{e}\sin u)\right] \\
&= \frac{\dot{k}}{k}L_0\mathrm{e}\sin u \ .
\end{aligned}$$

Finally, the one–dimensional, time–dependent Hamiltonian function describing Gylden's problem is given by

$$\mathcal{H}_3(L_0, \ell_0, t; G_0) = \mathcal{H}_2 + \frac{\partial \Phi_2}{\partial t} = -\frac{k^2}{2L_0^2} + \frac{\dot{k}}{k} L_0 \mathrm{e} \sin u \ ,$$

where u is related to ℓ_0 by Kepler's equation $\ell_0 = u - \mathrm{e} \sin u$, which can be inverted to provide $\sin u = \sin \ell_0 + \frac{\mathrm{e}}{2} \sin 2\ell_0 - \frac{\mathrm{e}^2}{8}(\sin \ell_0 - 3 \sin 3\ell_0) + O(\mathrm{e}^3)$. Here L_0, G_0, ℓ_0 should be interpreted as the osculating elements of the Keplerian motion having $k = k(t)$ constant.

In the following example we choose $j = 3$ in (3.39), so that the variation of the mass is given by the equation $\dot{m} = -\alpha m^3$, whose integration provides $m(t) = \frac{1}{\sqrt{2\alpha t}}$. We assume that the gravitational constant does not vary with time and we normalize it to one, so that $k(t)$ coincides with $m(t)$. The Hamiltonian function of Gylden's problem, depending parametrically on the eccentricity e and on the perturbing parameter α, turns out to be:

$$\begin{aligned} \mathcal{H}_G(L_0, \ell_0, t; \mathrm{e}, \alpha) &= -\frac{m(t)^2}{2L_0^2} - \alpha m(t)^2 L_0 \mathrm{e} \Big(\sin \ell_0 + \frac{\mathrm{e}}{2} \sin 2\ell_0 \\ &\quad - \frac{\mathrm{e}^2}{8} (\sin \ell_0 - 3 \sin 3\ell_0) \Big) \ . \end{aligned}$$

4 The three–body problem and the Lagrangian solutions

The solution of the two–body problem is provided by Kepler's laws, which state that for negative energies a point–mass moves on an ellipse whose focus coincides with the other point–mass. As shown by Poincaré [149], the dynamics becomes extremely complicated when you add the gravitational influence of a third body. In Section 4.1 we shall focus on a particular three–body problem, known as the *restricted* three–body problem, where it is assumed that the mass of one of the three bodies is so small that its influence on the others can be neglected (see, e.g., [21, 44, 94, 131, 163, 169]). As a consequence the primaries move on Keplerian ellipses around their common barycenter; a simplified model consists in assuming that the primaries move on circular orbits and that the motion takes place on the same plane. Action–angle Delaunay variables are introduced for the restricted three–body problem and the expansion of the perturbing function is provided.
In the framework of the planar, circular, restricted three–body problem we derive the special solutions found by Lagrange, which are given by stationary points in the synodic reference frame (Section 4.2). The existence and stability of such solutions is also discussed in the framework of a model in which the primaries move on elliptic orbits (Section 4.3) as well as in the context of the elliptic, unrestricted three–body problem (Section 4.4).

4.1 The restricted three–body problem

Let $\mathcal{P}_1$, $\mathcal{P}_2$, $\mathcal{P}_3$ be three bodies with masses m_1, m_2, m_3, respectively; throughout this section the three bodies are assumed to be point–masses. In the *restricted* problem one takes m_2 much smaller than m_1 and m_3, so that $\mathcal{P}_2$ does not affect the motion of $\mathcal{P}_1$ and $\mathcal{P}_3$. As a consequence we can assume that the motion of $\mathcal{P}_1$ and $\mathcal{P}_3$, to which we refer as the *primaries,* is Keplerian. Concerning the motion of $\mathcal{P}_2$ around the primaries, the region where the attraction of $\mathcal{P}_1$ or that of $\mathcal{P}_3$ is dominant is called the *sphere of influence;* an estimate of such a domain is provided in Appendix B.

4.1.1 The planar, circular, restricted three–body problem

The simplest non–trivial three–body model assumes that $\mathcal{P}_1$ and $\mathcal{P}_3$ move on a circular orbit around the common barycenter and that the motion of the three bodies takes place on the same plane. We refer to such a model as the *planar, circular, restricted three–body problem.* In an inertial reference frame whose origin coincides with the barycenter of the three bodies, let $\underline{\xi}_1$, $\underline{\xi}_2$, $\underline{\xi}_3 \in \mathbf{R}^2$ be the corresponding coordinates. From Newton's gravitational law one obtains that the motion of $\mathcal{P}_1$ and $\mathcal{P}_2$ is described by the equations

$$\frac{d^2\underline{\xi}_1}{dt^2} = \mathcal{G}\frac{m_2(\underline{\xi}_2-\underline{\xi}_1)}{|\underline{\xi}_2-\underline{\xi}_1|^3}+\mathcal{G}\frac{m_3(\underline{\xi}_3-\underline{\xi}_1)}{|\underline{\xi}_3-\underline{\xi}_1|^3}\ ,$$
$$\frac{d^2\underline{\xi}_2}{dt^2} = -\mathcal{G}\frac{m_1(\underline{\xi}_2-\underline{\xi}_1)}{|\underline{\xi}_2-\underline{\xi}_1|^3}-\mathcal{G}\frac{m_3(\underline{\xi}_2-\underline{\xi}_3)}{|\underline{\xi}_2-\underline{\xi}_3|^3}\ .$$

Next we consider a (heliocentric) reference frame with origin coinciding with $\mathcal{P}_1$; let $\underline{r}_2 \equiv \underline{\xi}_2-\underline{\xi}_1$, $\underline{r}_3 \equiv \underline{\xi}_3-\underline{\xi}_1$ be the relative positions with $\rho_2 \equiv |\underline{r}_2|$, $\rho_3 \equiv |\underline{r}_3|$. Then, one obtains

$$\frac{d^2\underline{r}_2}{dt^2} = -\frac{\mathcal{G}(m_1+m_2)\underline{r}_2}{\rho_2^3}-\frac{\mathcal{G}m_3\underline{r}_3}{\rho_3^3}+\frac{\mathcal{G}m_3(\underline{r}_3-\underline{r}_2)}{|\underline{r}_3-\underline{r}_2|^3}\ .$$

Setting $\mu \equiv \mathcal{G}(m_1+m_2)$ and $\varepsilon = \mathcal{G}m_3$, one has

$$\frac{d^2\underline{r}_2}{dt^2}+\frac{\mu\underline{r}_2}{\rho_2^3} = -\varepsilon\frac{\partial R}{\partial \underline{r}_2}\ ,$$

where the function R takes the form

$$R = \frac{\underline{r}_2\cdot\underline{r}_3}{\rho_3^3}-\frac{1}{|\underline{r}_3-\underline{r}_2|}\ . \tag{4.1}$$

Notice that for $\varepsilon = 0$ the dynamics reduces to the two–body problem of the motion of $\mathcal{P}_2$ around $\mathcal{P}_1$. For this reason we shall refer to ε as the perturbing parameter and to R as the perturbing function of the Keplerian motion. Recalling (3.36) we can write the three–body Hamiltonian as

$$\mathcal{H}_0(L_0,G_0,\ell_0,g_0) = -\frac{\mu^2}{2L_0^2}+\varepsilon R(L_0,G_0,\ell_0,g_0)\ , \tag{4.2}$$

where R is given by (4.1) and the functions $\underline{r}_2$, $\underline{r}_3$ must be expressed in terms of the Delaunay variables. Since the motion of $\mathcal{P}_3$ around $\mathcal{P}_1$ is circular, normalizing the time so that the angular velocity of $\mathcal{P}_3$ is equal to one, one obtains $\underline{r}_3 = (\rho_3\cos t,\ \rho_3\sin t)$. Denoting by ϑ the longitude of $\mathcal{P}_2$ and using $\underline{r}_2\cdot\underline{r}_3 = \rho_2\rho_3\cos(\vartheta-t)$, one obtains $|\underline{r}_3-\underline{r}_2| = \sqrt{\rho_2^2+\rho_3^2-2\rho_2\rho_3\cos(\vartheta-t)}$. As a consequence, the perturbing function takes the form

$$R = \frac{\rho_2\cos(\vartheta-t)}{\rho_3^2}-\frac{1}{\sqrt{\rho_2^2+\rho_3^2-2\rho_2\rho_3\cos(\vartheta-t)}}\ . \tag{4.3}$$

We immediately remark that R depends upon the difference $\vartheta-t$; being $\vartheta = g_0+f$, one obtains that R depends on the difference g_0-t. Therefore we perform the canonical change of variables from the Delaunay coordinates (L_0, G_0, ℓ_0, g_0) introduced in Chapter 3 to a new set of variables (L, G, ℓ, g) defined as

$$\begin{aligned} \ell &= \ell_0 \ , & L &= L_0 \ , \\ g &= g_0 - t \ , & G &= G_0 \ . \end{aligned}$$

The transformed Hamiltonian takes the form

$$\begin{aligned} \mathcal{H}(L, G, \ell, g) &= -\frac{\mu^2}{2L^2} - G \\ &+ \varepsilon \, \frac{\rho_2 \cos(g + f)}{\rho_3^2} - \frac{\varepsilon}{\sqrt{\rho_2^2 + \rho_3^2 - 2\rho_2\rho_3 \cos(g + f)}} \ , \end{aligned}$$

where ρ_2, f are intended to be expressed in terms of the mean anomaly.

4.1.2 Expansion of the perturbing function

The perturbing function (4.3) can be written in terms of the Delaunay variables. Here we compute explicitly the first few coefficients of its Fourier–Taylor series expansion and we refer to Appendix C (see also [61, 67, 68]) for general formulae valid at any order.

Let us introduce the Legendre polynomials $P_j(x)$ defined through the recursive relations

$$\begin{aligned} P_0(x) &= 1 \\ P_1(x) &= x \\ P_{j+1}(x) &= \frac{(2j+1)P_j(x)x - jP_{j-1}(x)}{j+1} \qquad \text{for any } j \geq 1 \ . \end{aligned}$$

Apart from a constant factor, the second term in (4.3) becomes

$$\frac{1}{\sqrt{\rho_2^2 + \rho_3^2 - 2\rho_2\rho_3 \cos(\vartheta - t)}} = \frac{1}{\rho_3} \sum_{j=0}^{\infty} P_j(\cos(\vartheta - t)) \left(\frac{\rho_2}{\rho_3}\right)^j \ ,$$

from which we obtain

$$R = -\frac{1}{\rho_3} \sum_{j=2}^{\infty} P_j(\cos(\vartheta - t)) \left(\frac{\rho_2}{\rho_3}\right)^j \ . \tag{4.4}$$

The inversion of Kepler's equation (3.24) up to the second order in the eccentricity yields

$$u = \ell + \mathrm{e} \sin \ell + \frac{\mathrm{e}^2}{2} \sin(2\ell) + O(\mathrm{e}^3) \ .$$

Using (3.23) one obtains

$$f = \ell + 2\mathrm{e} \sin \ell + \frac{5}{4}\mathrm{e}^2 \sin 2l + O(\mathrm{e}^3) \ ,$$

so that

$$\vartheta - t = g + \ell + 2\mathrm{e}\sin\ell + \frac{5}{4}\mathrm{e}^2\sin 2\ell + O(\mathrm{e}^3)\ . \tag{4.5}$$

In a similar way, from $\rho_2 = a(1 - \mathrm{e}\cos u)$ one obtains:

$$\rho_2 = a\ \left(1 + \frac{1}{2}\mathrm{e}^2 - \mathrm{e}\cos\ell - \frac{1}{2}\mathrm{e}^2\cos 2\ell\right) + O(\mathrm{e}^3)\ . \tag{4.6}$$

Recall that the ecccentricity is a function of the Delaunay variables through the relation $\mathrm{e} = \sqrt{1 - \frac{G^2}{L^2}}$. The powers $(\frac{\rho_2}{a})^j$ for $j = 2, 3, \ldots$ admit the following expansions:

$$\begin{aligned}
\left(\frac{\rho_2}{a}\right)^2 &= 1 + \frac{3}{2}\mathrm{e}^2 - 2\mathrm{e}\cos\ell - \frac{1}{2}\mathrm{e}^2\cos 2\ell + O(\mathrm{e}^3)\\
\left(\frac{\rho_2}{a}\right)^3 &= 1 + 3\mathrm{e}^2 - 3\mathrm{e}\cos\ell + O(\mathrm{e}^3)\\
\left(\frac{\rho_2}{a}\right)^4 &= 1 + 5\mathrm{e}^2 - 4\mathrm{e}\cos\ell + \mathrm{e}^2\cos 2\ell + O(\mathrm{e}^3)\\
\left(\frac{\rho_2}{a}\right)^5 &= 1 - 5\mathrm{e}\cos\ell + \mathrm{e}^2\left(\frac{15}{2} + \frac{5}{2}\cos 2\ell\right) + O(\mathrm{e}^3)\ \ldots\ .
\end{aligned}$$

From (4.4), one gets:

$$\begin{aligned}
R &= -\tfrac{1}{\rho_3}\ [P_2(\cos(\vartheta - t))\ (\tfrac{\rho_2}{a})^2(\tfrac{a}{\rho_3})^2 + P_3(\cos(\vartheta - t))\ (\tfrac{\rho_2}{a})^3(\tfrac{a}{\rho_3})^3\\
&+ P_4(\cos(\vartheta - t))\ (\tfrac{\rho_2}{a})^4(\tfrac{a}{\rho_3})^4 + P_5(\cos(\vartheta - t))\ (\tfrac{\rho_2}{a})^5(\tfrac{a}{\rho_3})^5] + \ldots\ .
\end{aligned}$$

Casting together (4.5), (4.6) and normalizing the unit of length so that $\rho_3 = 1$, one obtains the expansion

$$\begin{aligned}
R &= R_{00}(L,G) + R_{10}(L,G)\cos\ell + R_{11}(L,G)\cos(\ell + g)\\
&+ R_{12}(L,G)\cos(\ell + 2g) + R_{22}(L,G)\cos(2\ell + 2g)\\
&+ R_{32}(L,G)\cos(3\ell + 2g) + R_{33}(L,G)\cos(3\ell + 3g)\\
&+ R_{44}(L,G)\cos(4\ell + 4g) + R_{55}(L,G)\cos(5\ell + 5g) + \ldots\ ,
\end{aligned} \tag{4.7}$$

where the coefficients R_{ij} are given by the following expressions:

$$\begin{aligned}
R_{00} &= -\frac{L^4}{4}\left(1 + \frac{9}{16}L^4 + \frac{3}{2}\mathrm{e}^2\right) + \ldots\ , & R_{10} &= \frac{L^4\mathrm{e}}{2}\left(1 + \frac{9}{8}L^4\right) + \ldots\\
R_{11} &= -\frac{3}{8}L^6\left(1 + \frac{5}{8}L^4\right) + \ldots\ , & R_{12} &= \frac{L^4\mathrm{e}}{4}(9 + 5L^4) + \ldots\\
R_{22} &= -\frac{L^4}{4}\left(3 + \frac{5}{4}L^4\right) + \ldots\ , & R_{32} &= -\frac{3}{4}L^4\mathrm{e} + \ldots\\
R_{33} &= -\frac{5}{8}L^6\left(1 + \frac{7}{16}L^4\right) + \ldots\ , & R_{44} &= -\frac{35}{64}L^8 + \ldots\\
R_{55} &= -\frac{63}{128}L^{10} + \ldots
\end{aligned} \tag{4.8}$$

4.1.3 The planar, elliptic, restricted three–body problem

If we assume that $\mathcal{P}_3$ orbits around $\mathcal{P}_1$ on an elliptic orbit with eccentricity e$'$, the corresponding motion is described by a Hamiltonian function with three degrees of freedom; if ψ denotes the longitude of $\mathcal{P}_3$ and Ψ is the conjugated action variable, the Hamiltonian of the elliptic case is given by

$$\mathcal{H}(L,G,\Psi,\ell,g,\psi) = -\frac{1}{2L^2} + \Psi + \varepsilon R(L,G,\ell,g,\psi;\mathrm{e}') \ ,$$

where $R(L,G,\ell,g,\psi;\mathrm{e}')$ depends parametrically on e$'$ and, in normalized units, ε is the primaries mass–ratio. Up to constants, the first few Fourier coefficients of the series expansion of the perturbing function are the following:

$$\begin{aligned}
R(L,G,\ell,g,\psi) &= \\
&= -\frac{L^4}{4}\Big(\frac{5}{2}+\frac{9}{16}L^4-\frac{3}{2}\frac{G^2}{L^2}+\frac{3}{2}\mathrm{e}'^2\Big)+L^4\frac{\mathrm{e}}{2}\Big(1+\frac{9}{8}L^4\Big)\cos(\ell) \\
&\quad -\frac{3}{8}L^6\Big(1+\frac{5}{8}L^4\Big)\cos(\ell+g-\psi)+\frac{L^4}{4}\mathrm{e}(9+5L^4)\cos(\ell+2g-2\psi) \\
&\quad -\frac{L^4}{4}\Big(3+\frac{5}{4}L^4\Big)\cos(2\ell+2g-2\psi)-\frac{3}{4}L^4\mathrm{e}\cos(3\ell+2g-2\psi) \\
&\quad -\frac{5}{8}L^6\Big(1+\frac{7}{16}L^4\Big)\cos(3\ell+3g-3\psi)-\frac{35}{64}L^8\cos(4\ell+4g-4\psi) \\
&\quad -\frac{63}{128}L^{10}\cos(5\ell+5g-5\psi)-L^4\Big(\frac{3}{4}\mathrm{e}'+\frac{45}{64}L^4\mathrm{e}'\Big)\cos(\psi) \\
&\quad -L^4\Big(\frac{21}{8}\mathrm{e}'+\frac{45}{32}\mathrm{e}'L^4\Big)\cos(2\ell+2g-3\psi) \\
&\quad -L^4\Big(-\frac{3}{8}\mathrm{e}'+\frac{5}{32}\mathrm{e}'L^4\Big)\cos(2\ell+2g-\psi)+\dots
\end{aligned}$$

4.1.4 The inclined, circular, restricted three–body problem

We assume that the motion of $\mathcal{P}_3$ around $\mathcal{P}_1$ is circular, but we let the planes of the orbits of $\mathcal{P}_2$ and $\mathcal{P}_3$ have a non–zero mutual inclination i. Using the spatial Delaunay variables (L,G,H,ℓ,g,h) introduced in Chapter 3, denoting with ψ the longitude of $\mathcal{P}_3$, the Hamiltonian function takes the form:

$$\mathcal{H}(L,G,H,\ell,g,h,\psi) = -\frac{1}{2L^2} - H + \varepsilon R(L,G,H,\ell,g,h,\psi) \ ,$$

where, setting $\gamma = \sqrt{\frac{1}{2}-\frac{H}{2G}}$, up to constants the first few terms of the Fourier expansion of the perturbing function are given by

$$
\begin{aligned}
R(L,G,H,\ell,g,h,\psi) &= -L^4\Big(\frac{1}{4}+\frac{3}{8}\mathrm{e}^2+\frac{9}{64}L^4-\frac{3}{2}\gamma^2\Big) \\
&-\Big(\frac{3}{4}-\frac{3}{2}\gamma^2+\frac{5}{16}L^4\Big)\cos(2\ell+2g+2h-2\psi) \\
&-\Big(-\frac{1}{2}\mathrm{e}+3\gamma^2\mathrm{e}-\frac{9}{16}\mathrm{e}L^4\Big)\cos(\ell)-\Big(\frac{3}{4}\mathrm{e}-\frac{3}{2}\gamma^2\mathrm{e}\Big)\cos(3\ell+2g+2h-2\psi) \\
&-\Big(-\frac{9}{4}\mathrm{e}+\frac{9}{2}\gamma^2\mathrm{e}-\frac{5}{4}\mathrm{e}L^4\Big)\cos(\ell+2g+2h-2\psi)-\frac{3}{2}\gamma^2\cos(2\ell+2g) \\
&-\frac{3}{2}\gamma^2\cos(2h-2\psi)-\frac{3}{2}\gamma^2\mathrm{e}\cos(3\ell+2g) \\
&+\frac{9}{2}\gamma^2\mathrm{e}\cos(\ell+2g)+\frac{3}{2}\gamma^2\mathrm{e}\cos(\ell+2h-2\psi) \\
&+\frac{3}{2}\gamma^2\mathrm{e}\cos(\ell-2h+2\psi)-L^6\Big(\frac{3}{8}+\frac{15}{64}L^4\Big)\cos(\ell+g+h-\psi) \\
&-\Big(\frac{5}{8}+\frac{35}{128}L^4\Big)L^6\cos(3\ell+3g+3h-3\psi) \\
&-\frac{35}{64}L^8\cos(4\ell+4g+4h-4\psi)-\frac{63}{128}L^{10}\cos(5\ell+5g+5h-5\psi)+\dots
\end{aligned}
$$

4.2 The circular, restricted Lagrangian solutions

In the framework of the restricted, planar, circular three–body problem, Euler and Lagrange proved that in a rotating reference frame the equations of motion admit the existence of equilibrium solutions, known as the collinear and triangular equilibrium points. A concrete example is provided by the Trojan and Greek groups of asteroids, which (approximately) form an equilateral triangle with Jupiter and the Sun.

The mathematical derivation of such equilibrium solutions is the following. Consider a *sidereal* reference frame $(\mathcal{O},\xi,\eta,\zeta)$, where $\mathcal{O}$ coincides with the barycenter of the three bodies, the ξ axis lies along the direction joining the bodies with masses m_1 and m_3 at time $t=0$, η is orthogonal to ξ and belongs to the orbital plane, while ζ is perpendicular to the orbital plane. Let (ξ_i,η_i,ζ_i), $i=1,3$, be the coordinates of the primaries $\mathcal{P}_1$ and $\mathcal{P}_3$. We normalize the units of measure so that the distance between the primaries is unity and that $\mathcal{G}(m_1+m_3)=1$. Without loss of generality we assume that $m_1>m_3$ and let

$$
\overline{\mu}\equiv\frac{m_3}{m_1+m_3}\ ,
$$

so that $\mu_1\equiv\mathcal{G}m_1=1-\overline{\mu}$, $\mu_3\equiv\mathcal{G}m_3=\overline{\mu}$. The equations of motion of $\mathcal{P}_2$ with coordinates (ξ,η,ζ) can be written as

$$\begin{aligned}\ddot{\xi} &= \mu_1 \frac{\xi_1 - \xi}{r_1^3} + \mu_3 \frac{\xi_3 - \xi}{r_3^3} \\ \ddot{\eta} &= \mu_1 \frac{\eta_1 - \eta}{r_1^3} + \mu_3 \frac{\eta_3 - \eta}{r_3^3} \\ \ddot{\zeta} &= \mu_1 \frac{\zeta_1 - \zeta}{r_1^3} + \mu_3 \frac{\zeta_3 - \zeta}{r_3^3} \ , \end{aligned} \tag{4.9}$$

where r_1 and r_3 denote the distances from the primaries:

$$\begin{aligned} r_1 &= \sqrt{(\xi_1 - \xi)^2 + (\eta_1 - \eta)^2 + (\zeta_1 - \zeta)^2} \ , \\ r_3 &= \sqrt{(\xi_3 - \xi)^2 + (\eta_3 - \eta)^2 + (\zeta_3 - \zeta)^2} \ . \end{aligned}$$

Let us introduce a synodic reference frame $(\mathcal{O}, x, y, z)$, rotating with the angular velocity n of the primaries, where n has been normalized to one, due to the choice of the units of measure. Let us fix the axes so that the coordinates of the primaries become $(x_1, y_1, z_1) = (-\mu_3, 0, 0)$, $(x_3, y_3, z_3) = (\mu_1, 0, 0)$. The link between the synodic and the sidereal reference frames is

$$\begin{aligned} \xi &= \cos(t)x - \sin(t)y \\ \eta &= \sin(t)x + \cos(t)y \\ \zeta &= z \ , \end{aligned} \tag{4.10}$$

while the distances of $\mathcal{P}_2$ from the primaries are now given by

$$r_1 = \sqrt{(x + \mu_3)^2 + y^2 + z^2} \ , \qquad r_3 = \sqrt{(x - \mu_1)^2 + y^2 + z^2} \ . \tag{4.11}$$

Computing the second derivative of (4.10) with respect to time and inserting the result in (4.9) one obtains the equations of motion in the synodic frame:

$$\begin{aligned} \ddot{x} - 2\dot{y} &= \frac{\partial U}{\partial x} \\ \ddot{y} + 2\dot{x} &= \frac{\partial U}{\partial y} \\ \ddot{z} &= \frac{\partial U}{\partial z} \ , \end{aligned} \tag{4.12}$$

where the function U is defined as

$$U = U(x, y, z) \equiv \frac{1}{2}(x^2 + y^2) + \frac{\mu_1}{r_1} + \frac{\mu_3}{r_3} \ . \tag{4.13}$$

Multiplying (4.12) by $\dot{x}$, $\dot{y}$, $\dot{z}$ and adding the results, one obtains:

$$\dot{x}\ddot{x} + \dot{y}\ddot{y} + \dot{z}\ddot{z} = \frac{\partial U}{\partial x}\dot{x} + \frac{\partial U}{\partial y}\dot{y} + \frac{\partial U}{\partial z}\dot{z} \ ; \tag{4.14}$$

notice that the left–hand side of (4.14) is equal to $\frac{1}{2}\frac{d}{dt}(\dot{x}^2 + \dot{y}^2 + \dot{z}^2)$, while the right–hand side is equal to $\frac{dU}{dt}$. Therefore, integrating with respect to time one gets

$$\dot{x}^2 + \dot{y}^2 + \dot{z}^2 = 2U - C_J \ , \tag{4.15}$$

where C_J is a constant of integration, called the *Jacobi integral.* Using (4.13) one obtains

$$C_J = x^2 + y^2 + 2\frac{\mu_1}{r_1} + 2\frac{\mu_3}{r_3} - (\dot{x}^2 + \dot{y}^2 + \dot{z}^2) \ . \tag{4.16}$$

Notice that (4.15) implies $2U - C_J \geq 0$. The curves of zero velocity are defined through the expression $C_J = 2U$; such a relation defines a boundary, called *Hill's surface,* which separates regions where motion is allowed or forbidden. An example of Hill's region is given in Figure 4.1.

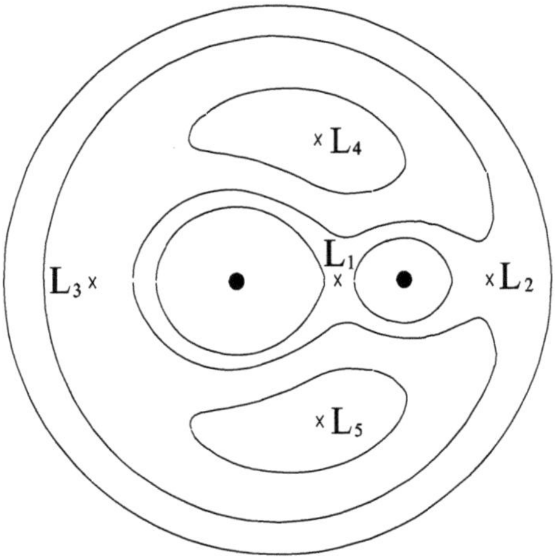

Fig. 4.1. The triangular and collinear equilibrium points with an example of Hill's surfaces.

Let us now turn to the determination of the position of the equilibrium points in the planar case with $z = 0$ [142], since we assumed that the motion of the three bodies takes place on the same plane. Recalling (4.11) and using $\mu_1 + \mu_3 = 1$ one has:

$$\mu_1 r_1^2 + \mu_3 r_3^2 = x^2 + y^2 + \mu_1\mu_3 \ .$$

Inserting such an expression in U one has

$$U = \mu_1 \left(\frac{r_1^2}{2} + \frac{1}{r_1}\right) + \mu_3 \left(\frac{r_3^2}{2} + \frac{1}{r_3}\right) - \frac{1}{2}\mu_1\mu_3 \ .$$

The equilibrium points are the solutions of the system obtained imposing that the partial derivatives of (4.13) with respect to x and y are zero:

$$\begin{aligned}
\frac{\partial U}{\partial x} &= \frac{\partial U}{\partial r_1}\frac{\partial r_1}{\partial x} + \frac{\partial U}{\partial r_3}\frac{\partial r_3}{\partial x} \\
&= \mu_1\left(r_1 - \frac{1}{r_1^2}\right)\frac{x+\mu_3}{r_1} + \mu_3\left(r_3 - \frac{1}{r_3^2}\right)\frac{x-\mu_1}{r_3} = 0 \\
\frac{\partial U}{\partial y} &= \frac{\partial U}{\partial r_1}\frac{\partial r_1}{\partial y} + \frac{\partial U}{\partial r_3}\frac{\partial r_3}{\partial y} \\
&= \mu_1\left(r_1 - \frac{1}{r_1^2}\right)\frac{y}{r_1} + \mu_3\left(r_3 - \frac{1}{r_3^2}\right)\frac{y}{r_3} = 0 \ .
\end{aligned} \tag{4.17}$$

A solution of (4.17) is obtained by solving the equations

$$r_1 - \frac{1}{r_1^2} = 0 \,, \qquad r_3 - \frac{1}{r_3^2} = 0 \,,$$

from which one obtains $r_1 = r_3 = 1$, namely

$$(x + \mu_3)^2 + y^2 = 1 \,, \qquad (x - \mu_1)^2 + y^2 = 1 \,.$$

Solving these equations, one finds the equilibrium solutions

$$\left(\frac{1}{2} - \mu_3, \frac{\sqrt{3}}{2}\right) , \qquad \left(\frac{1}{2} - \mu_3, -\frac{\sqrt{3}}{2}\right) ,$$

which correspond to the *triangular* Lagrangian solutions, usually denoted as L_4 and L_5 (see Figure 4.1).

Other solutions are obtained observing that $y = 0$ solves the second of (4.17); in particular, there exist three *collinear* equilibrium solutions usually denoted as L_1, L_2, L_3, where L_1 is located between the primaries, while L_2 and L_3 are outside the primaries. We derive in detail the location of L_1; the same procedure can be straightforwardly extended to L_2 and L_3.

At L_1 we have $y = 0$ and $r_1 = x + \mu_3$, $r_3 = -x + \mu_1$, so that $r_1 + r_3 = 1$; moreover, $\frac{\partial r_1}{\partial x} = -\frac{\partial r_3}{\partial x} = 1$. Replacing in $\frac{\partial U}{\partial x} = 0$, one obtains

$$\mu_1 \left(1 - r_3 - \frac{1}{(1 - r_3)^2}\right) - \mu_3 \left(r_3 - \frac{1}{r_3^2}\right) = 0 \,,$$

from which one gets

$$\frac{\mu_3}{3\mu_1} = r_3^3 \, \frac{1 - r_3 + \frac{r_3^2}{3}}{(1 + r_3 + r_3^2)(1 - r_3)^3} \,.$$

Define $\alpha \equiv (\frac{\mu_3}{3\mu_1})^{1/3}$; developing α in Taylor series, one finds

$$\alpha = r_3 + \frac{1}{3} r_3^2 + \frac{1}{3} r_3^3 + \frac{53}{81} r_3^4 + \ldots$$

Inverting such relation, for example using the Lagrange inversion method [142], one has

$$r_3 = \alpha - \frac{1}{3}\alpha^2 - \frac{1}{9}\alpha^3 - \frac{23}{81}\alpha^4 + \ldots \qquad (4.18)$$

Since r_3 represents the distance along the x–axis from the body with mass m_3, the solution (4.18) provides the location of the equilibrium point L_1 as a function of the mass ratio α. Similar computations can be performed for L_2 such that $r_1 = x + \mu_3$ and $r_3 = x - \mu_1$ with $r_1 - r_3 = 1$, and for L_3 such that $r_1 = -x - \mu_3$ and $r_3 = -x + \mu_1$ with $r_3 - r_1 = 1$.

To give a concrete example, in the Moon–Earth system the location of the equilibrium points is the following: L_1 lies at $3.26 \cdot 10^5$ km from the Earth, L_2 is at $4.49 \cdot 10^5$ km, L_3 is about $3.82 \cdot 10^5$ km from the Earth, while L_4 and L_5 are the triangular positions at $3.84 \cdot 10^5$ km, being located on the Moon's orbit.

We conclude with a discussion on the linear stability of the equilibrium positions (see [142]). Let us denote by (x_ℓ, y_ℓ) one of the five stationary solutions $(L_1, \ldots, L_5)$; let (δ_x, δ_y) be a small displacement from the equilibrium and let $(x, y) \equiv (x_\ell + \delta_x, y_\ell + \delta_y)$. Let us insert such coordinates in (4.12) and expand the derivatives of U in a neighborhood of the equilibrium solution. Using the notation

$$U_{xx} = \frac{\partial^2 U(x_\ell, y_\ell)}{\partial x^2}\,, \qquad U_{xy} = \frac{\partial^2 U(x_\ell, y_\ell)}{\partial x \partial y}\,, \qquad U_{yy} = \frac{\partial^2 U(x_\ell, y_\ell)}{\partial y^2}\,,$$

the equations for the variations (δ_x, δ_y) can be written as

$$\begin{pmatrix} \dot\delta_x \\ \dot\delta_y \\ \ddot\delta_x \\ \ddot\delta_y \end{pmatrix} = A \begin{pmatrix} \delta_x \\ \delta_y \\ \dot\delta_x \\ \dot\delta_y \end{pmatrix}\,,$$

where

$$A \equiv \begin{pmatrix} 0 & 0 & 1 & 0 \\ 0 & 0 & 0 & 1 \\ U_{xx} & U_{xy} & 0 & 2 \\ U_{xy} & U_{yy} & -2 & 0 \end{pmatrix}\,.$$

The eigenvalues of A are the solutions of the secular equation $\det(A - \lambda I_4) = 0$ (where I_4 is the 4×4 identity matrix), namely

$$\lambda^4 + (4 - U_{xx} - U_{yy})\lambda^2 + (U_{xx}U_{yy} - U_{xy}^2) = 0\,.$$

This equation admits four roots:

$$\begin{aligned}
\lambda_{1,2} &= \pm\left[\frac{1}{2}(U_{xx} + U_{yy} - 4) - \frac{1}{2}[(4 - U_{xx} - U_{yy})^2 - 4(U_{xx}U_{yy} - U_{xy}^2)]^{\frac{1}{2}}\right]^{\frac{1}{2}} \\
\lambda_{3,4} &= \pm\left[\frac{1}{2}(U_{xx} + U_{yy} - 4) + \frac{1}{2}[(4 - U_{xx} - U_{yy})^2 - 4(U_{xx}U_{yy} - U_{xy}^2)]^{\frac{1}{2}}\right]^{\frac{1}{2}}\,.
\end{aligned}$$

The equilibrium solution is stable, if the eigenvalues are purely imaginary.

For the collinear equilibrium position L_1, one has $y_\ell = 0$, $r_1 = x_\ell + \mu_3$, $r_3 = -x_\ell + \mu_1$; defining $M \equiv \frac{\mu_1}{r_1^3} + \frac{\mu_3}{r_3^3}$, the characteristic equation becomes

$$\lambda^4 + (2 - M)\lambda^2 + (1 + M - 2M^2) = 0\,.$$

Therefore the product of the four eigenvalues amounts to $1 + M - 2M^2$, with the constraints $\lambda_1 = -\lambda_2$, $\lambda_3 = -\lambda_4$. The eigenvalues are purely imaginary provided that $\lambda_1^2 = \lambda_2^2 < 0$ and $\lambda_3^2 = \lambda_4^2 < 0$, which imply that $1 + M - 2M^2 > 0$, namely $-\frac{1}{2} < M < 1$. These inequalities would guarantee the stability of the equilibrium point; however, computing M at the collinear point L_1 one finds that $M > 1$. In fact, in the case of L_1 we know that $r_1 < 1$ and $r_3 < 1$, so that $M > \mu_1 + \mu_3 = 1$.

We conclude that the collinear point L_1 is unstable for any value of the masses. The same conclusion holds for L_2 and L_3.

Concerning the triangular equilibrium positions one has $x_\ell = \frac{1}{2} - \mu_3$, $y_\ell = \pm\frac{\sqrt{3}}{2}$, $r_1 = r_3 = 1$. Computing the derivatives of U at the equilibria, one obtains

$$U_{xx} = \frac{3}{4}\,, \qquad U_{yy} = \frac{9}{4}\,, \qquad U_{xy} = \pm\frac{3\sqrt{3}}{4}(1-2\mu_3)\ .$$

The eigenvalues become

$$\begin{aligned}\lambda_{1,2} &= \pm\frac{\sqrt{-1-\sqrt{1-27(1-\mu_3)\mu_3}}}{\sqrt{2}}\,,\\ \lambda_{3,4} &= \pm\frac{\sqrt{-1+\sqrt{1-27(1-\mu_3)\mu_3}}}{\sqrt{2}}\,.\end{aligned}$$

The eigenvalues are purely imaginary provided

$$1-27(1-\mu_3)\mu_3 \geq 0\ ; \tag{4.19}$$

recalling that we assumed $m_1 > m_3$, so that $\mu_1 > \mu_3$ with $\mu_1 + \mu_3 = 1$, taking into account the inequality (4.19) one obtains

$$\mu_3 \leq \frac{27-\sqrt{621}}{54} \simeq 0.0385\ . \tag{4.20}$$

In conclusion, if the masses verify (4.20), then the triangular equilibrium solutions are linearly stable.

4.3 The elliptic, restricted Lagrangian solutions

Consider the planar motion of a body $\mathcal{P}_2(x,y)$ of mass μ_2 in the gravitational field of two primaries, $\mathcal{P}_1(x_1,y_1)$ and $\mathcal{P}_3(x_3,y_3)$ with masses μ_1 and μ_3, which are assumed to move on elliptic orbits around their common center of mass $\mathcal{O}$; let f denote the true anomaly of the common ellipse and let $r = \frac{a(1-e^2)}{1+e\cos f}$ be the distance between $\mathcal{P}_1$ and $\mathcal{P}_3$. In an inertial barycentric reference frame, the cartesian equations of the motion of $\mathcal{P}_2$ are given by

$$\begin{aligned}\ddot{x} &= -\frac{\mu_1(x-x_1)}{r_1^3} - \frac{\mu_3(x-x_3)}{r_3^3}\\ \ddot{y} &= -\frac{\mu_1(y-y_1)}{r_1^3} - \frac{\mu_3(y-y_3)}{r_3^3}\,,\end{aligned}$$

where $r_1 = \sqrt{(x-x_1)^2+(y-y_1)^2}$, $r_3 = \sqrt{(x-x_3)^2+(y-y_3)^2}$; the above equations are associated to the Lagrangian function

$$\mathcal{L}(\dot{x},\dot{y},x,y,r,f) = \frac{1}{2}(\dot{x}^2+\dot{y}^2) + \frac{\mu_1}{r_1} + \frac{\mu_3}{r_3}\,,$$

where the coordinates of the primaries, (x_1, y_1) and (x_3, y_3), depend upon r and f in the following way:

$$\begin{aligned} x_1 &= -\mu_3 r \cos f \ , & x_3 &= \mu_1 r \cos f \ , \\ y_1 &= -\mu_3 r \sin f \ , & y_3 &= \mu_1 r \sin f \ . \end{aligned}$$

Next we move to a barycentric reference frame $(\mathcal{O}, \xi, \eta)$ rotating with variable angular velocity, such that at each instant of time the rotation angle is equal to f with $\dot f = \frac{h}{r^2}$, h being the angular momentum and having assumed $\mathcal{G}(m_1+m_3) = 1$. The transformation equations are given by

$$\begin{aligned} x &= \xi \cos f - \eta \sin f \\ y &= \xi \sin f + \eta \cos f \ . \end{aligned}$$

Thus, the primaries oscillate on the ξ–axis and have coordinates

$$\begin{aligned} \xi_1 &= -\mu_3 r \ , & \xi_3 &= \mu_1 r \ , \\ \eta_1 &= 0 \ , & \eta_3 &= 0 \ . \end{aligned}$$

The new Lagrangian function takes the form:

$$\mathcal{L}(\dot\xi, \dot\eta, \xi, \eta, r, f) = \frac{1}{2}(\dot\xi^2 + \dot\eta^2) + \frac{1}{2}(\xi^2 + \eta^2)\dot f^2 + (\xi\dot\eta - \dot\xi\eta)\dot f + \frac{\mu_1}{r_1} + \frac{\mu_3}{r_3} \ .$$

The transformation to the so–called rotating–pulsating coordinates (X, Y) is achieved through the further change of variables:

$$\begin{aligned} \xi &= rX \\ \eta &= rY \ ; \end{aligned}$$

the primaries are now in a fixed position with coordinates $(X_1, Y_1) = (-\mu_3, 0)$, $(X_3, Y_3) = (\mu_1, 0)$ and the Lagrangian function takes the form

$$\begin{aligned} \mathcal{L}(\dot X, \dot Y, X, Y, r, f) &= \frac{r^2}{2}(\dot X^2 + \dot Y^2) + r\dot r(X\dot X + Y\dot Y) \\ &+(X^2+Y^2)\left(1 + \frac{\dot r^2}{2} + \frac{h^2}{2r^2}\right) + h(X\dot Y - Y\dot X) + \frac{1}{r}\left(\frac{\mu_1}{r_1} + \frac{\mu_3}{r_3}\right) \ . \end{aligned}$$

Finally, we change the time taking the true anomaly as independent variable through the transformation

$$dt = \frac{1}{h} r^2 \, df \ .$$

Denoting by $X' \equiv \frac{dX}{df}$ and $Y' \equiv \frac{dY}{df}$, the new Lagrangian function is given by

$$\mathcal{L}(X', Y', X, Y, r, f) = \frac{1}{2}(X'^2 + Y'^2) + XY' - YX' + \frac{r}{2h^2}(X^2+Y^2) + \frac{r}{h^2}\left(\frac{\mu_1}{r_1^2} + \frac{\mu_3}{r_3^2}\right) \ .$$

The corresponding equations of motion take a form similar to that of the circular case (see (4.12)), being

$$\begin{aligned} X'' - 2Y' &= \Omega_X \\ Y'' + 2X' &= \Omega_Y \ , \end{aligned} \tag{4.21}$$

where we define $\Omega = \Omega(X, Y, f)$ as

$$\Omega = \frac{1}{1 + \mathrm{e}\cos f}\left[\frac{1}{2}(X^2+Y^2) + \frac{\mu_1}{r_1} + \frac{\mu_3}{r_3} + \frac{1}{2}\mu_1\mu_3\right]$$

and Ω_X, Ω_Y denote the derivatives with respect to X, Y, respectively. Let Ω_0 be defined through the relation

$$\Omega_0 \equiv (1 + \mathrm{e}\cos f)\Omega \ .$$

The equivalent of the Jacobi integral is obtained from (4.21) multiplying the first equation by X' and the second by Y'; adding the results one obtains:

$$\left(\frac{dX}{df}\right)^2 + \left(\frac{dY}{df}\right)^2 = 2\int(\Omega_X dX + \Omega_Y dY) \ . \tag{4.22}$$

Let us write the derivative of Ω with respect to the true anomaly as

$$\Omega_f = \frac{\mathrm{e}\sin f}{(1+\mathrm{e}\cos f)^2}\,\Omega_0 \ .$$

Then, (4.22) becomes:

$$\begin{aligned} \left(\frac{dX}{df}\right)^2 + \left(\frac{dY}{df}\right)^2 &= 2\int(d\Omega - \Omega_f df) \\ &= 2\Omega - 2\mathrm{e}\int\frac{\Omega_0 \sin f}{(1+\mathrm{e}\cos f)^2}df - C_\mathrm{e} \ , \end{aligned}$$

where C_e is a constant which reduces to the Jacobi integral in the circular case $\mathrm{e} = 0$.

The stationary solutions of (4.21) are given by

$$\frac{\partial\Omega}{\partial X} = 0 \ , \qquad \frac{\partial\Omega}{\partial Y} = 0$$

or equivalently by

$$\frac{\partial\Omega_0}{\partial X} = 0 \ , \qquad \frac{\partial\Omega_0}{\partial Y} = 0 \ ,$$

which imply that the solutions of the elliptic problem coincide with those of the circular case. In particular, the triangular solutions are located at $(\frac{1}{2} - \mu_3, \pm\frac{\sqrt{3}}{2})$, which pulsate as the coordinates. In order to analyze the stability, one starts by introducing a displacement (δ_X, δ_Y) from the libration points, say $X \equiv X_\ell + \delta_X$, $Y \equiv Y_\ell + \delta_Y$, where (X_ℓ, Y_ℓ) coincides with one of the five stationary solutions; the linearized equations can be written as

$$\begin{aligned}\delta''_X - 2\delta'_Y &= \frac{1}{1+\mathrm{e}\cos f}\,[\Omega^{(\ell)}_{0,XX}\delta_X + \Omega^{(\ell)}_{0,XY}\delta_Y]\\ \delta''_Y + 2\delta'_X &= \frac{1}{1+\mathrm{e}\cos f}\,[\Omega^{(\ell)}_{0,YX}\delta_X + \Omega^{(\ell)}_{0,YY}\delta_Y]\ ,\end{aligned}$$

where $\Omega^{(\ell)}_{0,XX}$ denotes the second derivative of Ω_0 with respect to X computed at the stationary solution (X_ℓ, Y_ℓ) (similarly for the other derivatives). A numerical procedure based on Floquet theory (see Appendix D) and on the computation of the characteristic exponents (see [52]) provides the domain of the linear stability in the parameter plane (μ, e).

Figure 4.2 shows the regions of linear stability of the triangular solutions (compare with [52]): at $\mu \simeq 0.028$ one has linear instability for any value of the eccentricity; at $\mu \simeq 0.038$ one has instability also for $\mathrm{e} = 0$, while the eccentricity can have a stabilizing effect up to $\mu \simeq 0.047$ (notice that the point D in Figure 4.2 has coordinates $D(0.047, 0.314)$). The collinear points are always unstable, as in the circular case, for any value of the eccentricity and of the mass parameter.

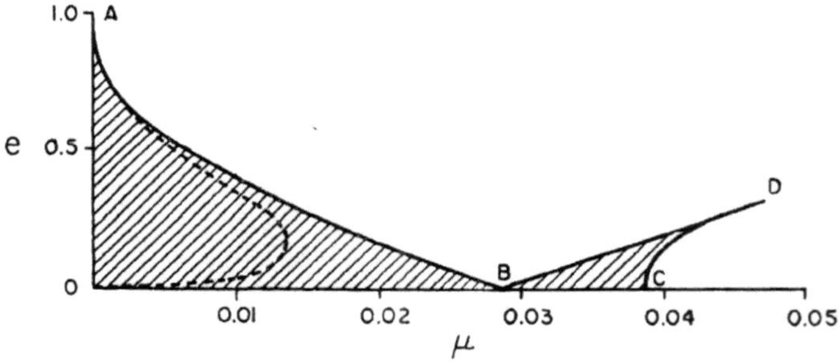

Fig. 4.2. The shaded area denotes a region of equilibrium of the elliptic, restricted triangular solutions as the parameters μ and e are varied (after [52]; reproduced by permission of the AAS).

4.4 The elliptic, unrestricted triangular solutions

Let $\mathcal{P}_1$, $\mathcal{P}_2$, $\mathcal{P}_3$ be three bodies of masses m_1, m_2, m_3 which are subject to the mutual gravitational attraction; we assume that the three bodies move in the same plane and we denote the position vectors in an inertial reference frame by means of the two–dimensional vectors $\underline{q}_1$, $\underline{q}_2$, $\underline{q}_3$. The equations of motion can be written as

$$m_i \ddot{\underline{q}}_i = \frac{\partial U}{\partial \underline{q}_i}\ , \qquad\qquad i = 1, 2, 3\ , \tag{4.23}$$

where

$$U(\underline{q}) = \sum_{1 \le i < j \le 3} \frac{m_i m_j}{\|\underline{q}_j - \underline{q}_i\|}\ .$$

Following [154] (see also [134]) the generalization of the Lagrangian solutions of the restricted case is obtained by looking for a periodic homographic[1] solution of the form

$$\underline{q}_i(t) = \psi(t)\underline{z}_i \ , \qquad i = 1, 2, 3, \tag{4.24}$$

where $\underline{z}_i$ are constant vectors and $\psi(t)$ is an unknown function, which can be found as follows. Inserting (4.24) in (4.23) one obtains

$$m_i \underline{z}_i \ddot{\psi}(t) = \sum_{1 \leq j \leq 3, j \neq i} \frac{m_i m_j \psi(t)(\underline{z}_j - \underline{z}_i)}{|\psi(t)|^3 \|\underline{z}_j - \underline{z}_i\|^3} \ , \qquad i = 1, 2, 3,$$

which can be split as

$$\begin{aligned} \ddot{\psi}(t) &= -\nu \frac{\psi(t)}{|\psi(t)^3|} \\ \sum_{1 \leq j \leq 3, j \neq i} \frac{m_i m_j (\underline{z}_j - \underline{z}_i)}{\|\underline{z}_j - \underline{z}_i\|^3} + \nu m_i \underline{z}_i &= \underline{0} \ , \end{aligned} \tag{4.25}$$

where ν is a real constant. From the first equation we recognize that $\psi(t)$ is a solution of a Keplerian motion; summing the second equation in (4.25) over $i = 1, 2, 3$, one obtains

$$\sum_{i=1}^{3} m_i \underline{z}_i = \underline{0} \ , \tag{4.26}$$

showing that the center of mass is located at the origin of the reference frame. Let d be the length of the sides of the triangular solution; the scaling factor ν can be set to one by a proper choice of d. In fact, the first component of the second equation in (4.25) is given by

$$\nu \underline{z}_1 = \frac{1}{d^3} \left[m_2(\underline{z}_1 - \underline{z}_2) + m_3(\underline{z}_1 - \underline{z}_3) \right] = \frac{M}{d^3} \underline{z}_1 \ ,$$

where $M = m_1 + m_2 + m_3$ denotes the total mass. Setting

$$d = M^{\frac{1}{3}} \ , \tag{4.27}$$

we obtain $\nu = 1$.

If $\underline{p}_i$ $(i = 1, 2, 3)$ denote the momenta conjugated to $\underline{q}_i$, the Hamiltonian governing the three–body problem can be written as

$$\begin{aligned} \mathcal{H}_1(\underline{p}_1, \underline{p}_2, \underline{p}_3, \underline{q}_1, \underline{q}_2, \underline{q}_3) &= \frac{\|\underline{p}_1\|^2}{2m_1} + \frac{\|\underline{p}_2\|^2}{2m_2} + \frac{\|\underline{p}_3\|^2}{2m_3} \\ &- \frac{m_1 m_2}{\|\underline{q}_2 - \underline{q}_1\|} - \frac{m_1 m_3}{\|\underline{q}_3 - \underline{q}_1\|} - \frac{m_2 m_3}{\|\underline{q}_3 - \underline{q}_2\|} \ . \end{aligned} \tag{4.28}$$

[1] A homographic solution is a configuration which remains similar to itself all times.

The center of mass and the total linear momentum can be eliminated through a transformation to *Jacobi coordinates:*

$$
\begin{aligned}
\underline{u}_1 &= \underline{q}_2 - \underline{q}_1 & \underline{v}_1 &= -\frac{m_2}{m_1+m_2}\underline{p}_1 \\
\underline{u}_2 &= \underline{q}_3 - \frac{1}{m_1+m_2}(m_1\underline{q}_1 + m_2\underline{q}_2) & \underline{v}_2 &= -\frac{m_3}{M}(\underline{p}_1+\underline{p}_2+\underline{p}_3)+\underline{p}_3 \\
\underline{u}_3 &= \frac{1}{M}(m_1\underline{q}_1 + m_2\underline{q}_2 + m_3\underline{q}_3) & \underline{v}_3 &= \underline{p}_1+\underline{p}_2+\underline{p}_3 \ . \qquad (4.29)
\end{aligned}
$$

An alternative reduction is obtained through the transformation to heliocentric coordinates as in Appendix E. Recalling (4.24) and (4.26), we obtain

$$
\sum_{i=1}^{3} m_i\underline{q}_i = \psi(t)\sum_{i=1}^{3} m_i\underline{z}_i = \underline{0}\ , \qquad \sum_{i=1}^{3} \underline{p}_i = \dot{\psi}(t)\ \sum_{i=1}^{3} m_i\underline{z}_i = \underline{0}\ ,
$$

which imply the elimination of the center of mass and of the total linear momentum, since the above equations yield that $\underline{u}_3 = \underline{0}$ and $\underline{v}_3 = \underline{0}$. Under the transformation (4.29) the Hamiltonian (4.28) becomes:

$$
\mathcal{H}_2(\underline{v}_1,\underline{v}_2,\underline{u}_1,\underline{u}_2) = \frac{\|\underline{v}_1\|^2}{2M_1} + \frac{\|\underline{v}_2\|^2}{2M_2} - \frac{m_1m_2}{\|\underline{u}_1\|} - \frac{m_1m_3}{\|\underline{u}_2+M_3\underline{u}_1\|} - \frac{m_2m_3}{\|\underline{u}_2+M_4\underline{u}_1\|}\ , \qquad (4.30)
$$

where $M_1 \equiv \frac{m_1m_2}{m_1+m_2}$, $M_2 \equiv \frac{m_3(m_1+m_2)}{M}$, $M_3 \equiv \frac{m_2}{m_1+m_2}$, $M_4 \equiv -\frac{m_1}{m_1+m_2}$. Transforming to polar coordinates and making use of the constancy of the angular momentum allows us to reduce to three degrees of freedom. More precisely, we start by performing a coordinate change from $(\underline{u}_i,\underline{v}_i) \in \mathbf{R}^4$ to $(r_i,\vartheta_i,R_i,\Theta_i) \in \mathbf{R}^4$, defined as

$$
\underline{u}_i = \begin{pmatrix} r_i\cos\vartheta_i \\ r_i\sin\vartheta_i \end{pmatrix}\ , \qquad \underline{v}_i = \begin{pmatrix} R_i\cos\vartheta_i - \frac{\Theta_i}{r_i}\sin\vartheta_i \\ R_i\sin\vartheta_i + \frac{\Theta_i}{r_i}\cos\vartheta_i \end{pmatrix}\ , \qquad i=1,2\ .
$$

The Hamiltonian (4.30) becomes

$$
\begin{aligned}
\mathcal{H}_3(R_1,R_2,\Theta_1,\Theta_2,r_1,r_2,\vartheta_1,\vartheta_2) &= \frac{1}{2M_1}\left(R_1^2+\frac{\Theta_1^2}{r_1^2}\right) + \frac{1}{2M_2}\left(R_2^2+\frac{\Theta_2^2}{r_2^2}\right) \\
&- \frac{m_1m_2}{r_1} - \frac{m_1m_3}{\rho_1} - \frac{m_2m_3}{\rho_2}\ , \qquad (4.31)
\end{aligned}
$$

where

$$
\begin{aligned}
\rho_1 &\equiv \sqrt{r_2^2 + M_3^2r_1^2 + 2M_3r_1r_2\cos(\vartheta_2-\vartheta_1)}\ , \\
\rho_2 &\equiv \sqrt{r_2^2 + M_4^2r_1^2 + 2M_4r_1r_2\cos(\vartheta_2-\vartheta_1)}\ .
\end{aligned}
$$

Since (4.31) depends on ϑ_1, ϑ_2 through the difference $\vartheta_2-\vartheta_1$, we can perform the canonical change of variables

$$
\begin{aligned}
\xi &= \vartheta_1 & \varXi &= \Theta_1+\Theta_2 \\
\lambda &= \vartheta_2-\vartheta_1 & \varLambda &= \Theta_2\ ,
\end{aligned}
$$

which makes the Hamiltonian (4.31) independent of ξ. Therefore, setting $\Xi = h$, where h denotes the constant angular momentum, the transformed Hamiltonian becomes:

$$\begin{aligned}\mathcal{H}_4(R_1,R_2,\Lambda,r_1,r_2,\lambda) &= \frac{1}{2M_1}\left(R_1^2+\frac{(h-\Lambda)^2}{r_1^2}\right)+\frac{1}{2M_2}\left(R_2^2+\frac{\Lambda^2}{r_2^2}\right)\\ &- \frac{m_1m_2}{r_1}-\frac{m_1m_3}{\delta_1}-\frac{m_2m_3}{\delta_2}\ , \end{aligned} \qquad (4.32)$$

where

$$\delta_1 \equiv \sqrt{r_2^2+M_3^2r_1^2+2M_3r_1r_2\cos\lambda}\ , \qquad \delta_2 \equiv \sqrt{r_2^2+M_4^2r_1^2+2M_4r_1r_2\cos\lambda}\ .$$

Remark. In the planetary case one assumes that one mass is much larger than the others, say $m_1 = \mu_1$, $m_2 = \varepsilon\mu_2$, $m_3 = \varepsilon\mu_3$, where ε is a small quantity and μ_i $(i = 1, 2, 3)$ is of the order of unity. Then, applying the change of variables

$$\tilde{R}_i = \frac{R_i}{\varepsilon}\ , \qquad \tilde{r}_i = r_i\ , \qquad \tilde{\Lambda} = \frac{\Lambda}{\varepsilon}\ , \qquad \tilde{\lambda} = \lambda\ , \qquad \tilde{h} = \frac{h}{\varepsilon}\ ,$$

one obtains the Hamiltonian

$$\begin{aligned}\mathcal{H}_5(\tilde{R}_1,\tilde{R}_2,\tilde{\Lambda},\tilde{r}_1,\tilde{r}_2,\tilde{\lambda}) &= \frac{\varepsilon}{2M_1}\left(\tilde{R}_1^2+\frac{(\tilde{h}-\tilde{\Lambda})^2}{\tilde{r}_1^2}\right)+\frac{\varepsilon}{2M_2}\left(\tilde{R}_2^2+\frac{\tilde{\Lambda}^2}{\tilde{r}_2^2}\right)\\ &- \frac{\mu_1\mu_2}{\tilde{r}_1}-\frac{\mu_1\mu_3}{\tilde{\delta}_1}-\varepsilon\frac{\mu_2\mu_3}{\tilde{\delta}_2}\ , \end{aligned} \qquad (4.33)$$

where $\tilde{\delta}_i$ are the quantities δ_i with r_i, λ replaced by $\tilde{r}_i$, $\tilde{\lambda}$. Observing that

$$\frac{\varepsilon}{M_1} = \frac{1}{\mu_2} + O(\varepsilon)\ , \qquad \frac{\varepsilon}{M_2} = \frac{1}{\mu_3} + O(\varepsilon)\ ,$$

one finds that the Hamiltonian (4.33) can be written as

$$\begin{aligned}\mathcal{H}_6(\tilde{R}_1,\tilde{R}_2,\tilde{\Lambda},\tilde{r}_1,\tilde{r}_2,\tilde{\lambda}) &= \frac{1}{2\mu_2}\left(\tilde{R}_1^2+\frac{(\tilde{h}-\tilde{\Lambda})^2}{\tilde{r}_1^2}\right)+\frac{1}{2\mu_3}\left(\tilde{R}_2^2+\frac{\tilde{\Lambda}^2}{\tilde{r}_2^2}\right)\\ &- \frac{\mu_1\mu_2}{\tilde{r}_1}-\frac{\mu_1\mu_3}{\tilde{\delta}_1}+\varepsilon F(\tilde{R}_1,\tilde{R}_2,\tilde{\Lambda},\tilde{r}_1,\tilde{r}_2,\tilde{\lambda})\ , \end{aligned} \qquad (4.34)$$

for a suitable function $F = F(\tilde{R}_1,\tilde{R}_2,\tilde{\Lambda},\tilde{r}_1,\tilde{r}_2,\tilde{\lambda})$. The Hamiltonian (4.34) is equal to the sum of two decoupled Kepler's motions, perturbed by a function of order ε, which can be considered as a small parameter. This model fits the planetary case where one mass (corresponding to the Sun) is much larger than those of the other bodies (the planets), which can be assumed to be of the same order of magnitude.

Coming back to the Lagrangian positions, let us denote by $\gamma = \gamma(t)$ the periodic orbit corresponding to the triangular configuration with sides of length d as in (4.27). Following [154] the stability of such configurations is investigated by linearizing the equations of motion associated to the Hamiltonian function (4.32) around the periodic solution. One obtains a time–dependent, periodic, linear system in the overall set of variables $\underline{z} \in \mathbf{R}^6$ of the form:

$$\dot{\underline{z}} = JD^2\mathcal{H}_4(\gamma(t))\underline{z} \ , \tag{4.35}$$

where J is the 6×6–dimensional matrix $J = \begin{pmatrix} 0 & I_3 \\ -I_3 & 0 \end{pmatrix}$ (being I_n with $n \in \mathbf{Z}_+$ the $n \times n$ identity matrix) and $D^2\mathcal{H}_4(\gamma(t))$ denotes the Hessian of $\mathcal{H}_4$ computed along the periodic orbit. Let T be its period; the linear stability analysis involves the determination of the monodromy matrix $C = Z(T)$, where Z is the 6×6–dimensional matrix, solution of (4.35) with initial data $Z(0) = I_6$. The eigenvalues of C are the so–called *characteristic multipliers,* which are symmetric about the unit circle, due to the Hamiltonian character of the dynamics. The system is linearly stable if all multipliers have modulus one. In particular, two multipliers are unity: one of them is associated to the periodic orbit and the other to the Hamiltonian. Therefore, the linear stability is determined by the remaining four eigenvalues. Indeed, a suitable change of variables allows us to decouple the system: one part is associated to the unitary eigenvalues and a second part is a 4×4–dimensional system associated to the other eigenvalues (see [154]). In the latter case, the secular equation of order 4 depends on two parameters: the eccentricity and the mass parameter β defined as

$$\beta \equiv 27 \ \frac{m_1 m_2 + m_1 m_3 + m_2 m_3}{(m_1 + m_2 + m_3)^2} \ .$$

In the circular case the characteristic multipliers can be analytically computed and it is shown that they are purely imaginary if $0 \leq \beta < 1$; this is a classical result, already obtained by E.J. Gascheau in the $19th$ century ([74], see also [156]).

In the elliptic case the characteristic multipliers are obtained through a numerical integration; the results show that the triangular configuration becomes unstable as the eccentricity increases (see [154]). In particular, the stability is lost through a *period–doubling bifurcation* (namely two multipliers collide at -1 and move off along the real axis). For $\beta = \frac{3}{4}$ the system becomes unstable for any value of the eccentricity; afterwards there is an interval where the stability is maintained locally for non–zero values of the eccentricity, even though the circular solution is unstable. Finally the stability is lost through a *Krein bifurcation,* according to which two multipliers collide on the unitary circle and move off in the complex plane (see Figure 4.3).

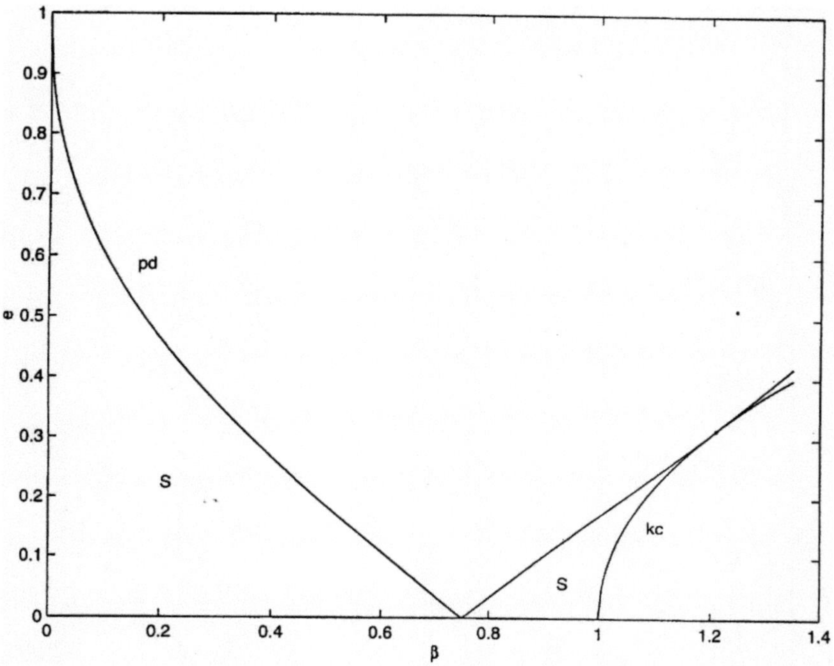

Fig. 4.3. The linear stability of the elliptic, unrestricted, triangular solutions within the plane (β,e). The meaning of the labels is the following: S denotes a linear stability region, pd is the period doubling curve ending at $\beta = \frac{3}{4}$, kc is the Krein collision curve starting at $\beta = 1$ (reprinted from [154] with permission from Elsevier).

5 Rotational dynamics

The dynamics of celestial bodies is essentially ruled by two motions: the revolution around a primary body and the rotation about an internal spin–axis. To a first approximation one can deal with a rigid body motion, where no internal deformations are considered. A suitable model for the rotational dynamics is obtained through the introduction of the Euler angles (Section 5.1), while the Hamiltonian formulation can be derived in terms of the Andoyer–Deprit variables (Section 5.2). We consider first the free rigid body motion (Section 5.3), and then the perturbed one under the effect of the gravitational influence of a primary body (Section 5.4).

A simplified model is provided by the *spin–orbit problem* (Section 5.5), where the rigid body moves on a Keplerian orbit around the primary; moreover, one assumes that the spin–axis is perpendicular to the orbital plane and that it coincides with the shortest physical axis. The corresponding Hamiltonian function is one–dimensional with an explicit time dependence. The dynamics around the resonances can be conveniently described by averaging such a Hamiltonian. Relaxing the assumption that the body is rigid, one needs to consider the effect of tidal forces which provide the so–called *dissipative spin–orbit problem.* We also discuss the motion of a point–mass around an oblate primary (Section 5.6), and the interaction between two bodies of finite dimensions (Section 5.7).

Another model of rotational dynamics is given by the dumbbell satellite composed by two point–masses connected by a rigid rod, whose barycenter moves around a primary body (Section 5.9); relaxing the assumption that the rod is rigid, one obtains the tether satellite (Section 5.8).

5.1 Euler angles

Let us consider a rigid body $\mathcal{S}$ with ellipsoidal shape, rotating about a fixed point $\mathcal{O}$. We introduce a reference frame $(\mathcal{O}, \underline{i}, \underline{j}, \underline{k})$, whose axes coincide with the directions of the principal axes of inertia of the rigid body; let $(\mathcal{O}, \underline{I}, \underline{J}, \underline{K})$ be a fixed reference frame with origin coinciding with that of the body frame. We denote by $\underline{n}$ the line of nodes, defined as the intersection between the planes $(\underline{i}, \underline{j})$ and $(\underline{I}, \underline{J})$. The Euler angles $(\varphi, \psi, \vartheta)$ (see Figure 5.1) are introduced as follows:

$0 \le \varphi \le 2\pi$ is the *precession* angle formed by the directions of $\underline{I}$ and $\underline{n}$;
$0 \le \psi \le 2\pi$ is the *proper rotation* angle formed by the directions of $\underline{n}$ and $\underline{i}$;
$0 \le \vartheta \le \pi$ is the *nutation* angle formed by the directions of $\underline{K}$ and $\underline{k}$.

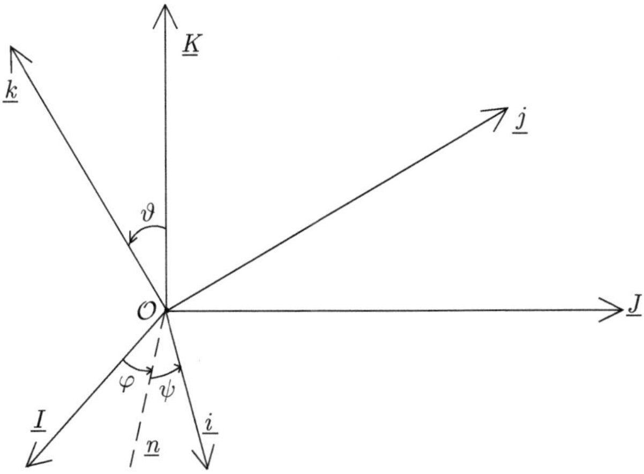

Fig. 5.1. Euler angles.

In the body reference frame the components $(\omega_1, \omega_2, \omega_3)$ of the angular velocity of rotation are derived as follows. Since $\dot{\vartheta}$ has the same direction as $\underline{n}$, while $\dot{\varphi}$ is aligned with $\underline{K}$ and $\dot{\psi}$ with $\underline{k}$, their projections on the body reference frame are given by:

$$\begin{array}{lll} \dot{\vartheta}_1 = \dot{\vartheta}\cos\psi & \dot{\varphi}_1 = \dot{\varphi}\sin\vartheta\sin\psi & \dot{\psi}_1 = 0 \\ \dot{\vartheta}_2 = -\dot{\vartheta}\sin\psi & \dot{\varphi}_2 = \dot{\varphi}\sin\vartheta\cos\psi & \dot{\psi}_2 = 0 \\ \dot{\vartheta}_3 = 0 & \dot{\varphi}_3 = \dot{\varphi}\cos\vartheta & \dot{\psi}_3 = \dot{\psi}\,. \end{array}$$

Therefore we get:

$$\begin{aligned} \omega_1 &= \dot{\varphi}\sin\vartheta\sin\psi + \dot{\vartheta}\cos\psi \\ \omega_2 &= \dot{\varphi}\sin\vartheta\cos\psi - \dot{\vartheta}\sin\psi \\ \omega_3 &= \dot{\varphi}\cos\vartheta + \dot{\psi}\,. \end{aligned} \tag{5.1}$$

Let I_1, I_2, I_3 be the principal moments of inertia of the rigid body $\mathcal{S}$, which is assumed to be subject to a force with potential energy $V(\varphi, \psi, \vartheta)$ [5]. The Lagrangian function describing the motion is given by

$$\begin{aligned} \mathcal{L}(\dot{\vartheta}, \dot{\varphi}, \dot{\psi}, \vartheta, \varphi, \psi) &= \frac{1}{2}(I_1\omega_1^2 + I_2\omega_2^2 + I_3\omega_3^2) - V(\varphi, \psi, \vartheta) \\ &= \frac{1}{2}I_1(\dot{\varphi}\sin\vartheta\sin\psi + \dot{\vartheta}\cos\psi)^2 \\ &\quad + \frac{1}{2}I_2(\dot{\varphi}\sin\vartheta\cos\psi - \dot{\vartheta}\sin\psi)^2 \\ &\quad + \frac{1}{2}I_3(\dot{\varphi}\cos\vartheta + \dot{\psi})^2 - V(\varphi, \psi, \vartheta)\,. \end{aligned}$$

It is readily seen that the energy is preserved. The rigid body is said to have a gyroscopic structure whenever two moments of inertia coincide, say $I_1 = I_2$ with

$I_1 \neq I_3$; in this case the Lagrangian function takes the form

$$\mathcal{L}(\dot{\vartheta}, \dot{\varphi}, \dot{\psi}, \varphi, \psi, \vartheta) = \frac{1}{2} I_1 (\dot{\vartheta}^2 + \dot{\varphi}^2 \sin^2 \vartheta) + \frac{1}{2} I_3 (\dot{\varphi} \cos \vartheta + \dot{\psi})^2 - V(\varphi, \psi, \vartheta) .$$

For the free rigid body motion the momenta conjugated to φ and ψ are constants, since the kinetic part depends only on the angle ϑ.

5.2 Andoyer–Deprit variables

The rigid body dynamics can be conveniently described through a set of action–angle coordinates, known as *Andoyer–Deprit* variables [55]. In the general case in which the rotation axis does not necessarily coincide with a principal direction, we introduce three reference frames with the same origin $\mathcal{O}$ located at the center of mass:

- $(\mathcal{O}, \underline{I}, \underline{J}, \underline{K})$ is an *inertial* reference frame, for example with the plane $(\underline{I}, \underline{J})$ coinciding with the ecliptic plane (or the Laplace plane[1]) at some epoch;
- $(\mathcal{O}, \underline{i}, \underline{j}, \underline{k})$ is a *body* frame, with the axes coinciding with the principal axes;
- $(\mathcal{O}, \underline{i}_1, \underline{i}_2, \underline{i}_3)$ is a *spin* frame, with the vertical axis aligned with the angular momentum.

We introduce the following lines of nodes:

- $\underline{m}$ is the intersection of $(\underline{I}, \underline{J})$ with the plane $(\underline{i}_1, \underline{i}_2)$;
- $\overline{\underline{n}}$ is the intersection of $(\underline{i}, \underline{j})$ with the plane $(\underline{i}_1, \underline{i}_2)$;
- $\underline{n}$ is the intersection of $(\underline{I}, \underline{J})$ with the plane $(\underline{i}, \underline{j})$.

The angles φ, ψ, ϑ, g, ℓ, h, J, K are defined as follows (see also Figure 5.2):

- $(\vartheta, \varphi, \psi)$ are the Euler angles of the body frame with respect to the inertial frame;
- (J, g, ℓ) are the Euler angles of the body frame with respect to the spin frame;
- $(K, h, 0)$ are the Euler angles of the inertial frame with respect to the spin frame, where we assume that $\underline{i}_1$ coincides with $\underline{m}$.

We also remark that $\underline{m} \equiv \underline{i}_3 \wedge \underline{K}$ denotes the line of nodes between the inertial plane and the plane orthogonal to the angular momentum; $\underline{n} \equiv \underline{K} \wedge \underline{k}$ is the line of nodes between the inertial and equatorial planes; $\overline{\underline{n}} \equiv \underline{k} \wedge \underline{i}_3$ is the line of nodes between the equatorial plane and the plane orthogonal to the angular momentum; the angle g provides the motion of the equatorial axis with respect to the inertial frame; the angle ℓ gives the motion of the angular momentum with respect to the body frame; the angle h provides the motion of the angular momentum with respect to the inertial frame; the angle ϑ is formed by the direction of the vertical principal axis of the ellipsoid with the vertical axis of the inertial frame.

[1] The Laplace plane can be defined as the plane with respect to which the orbital inclination has a constant value.

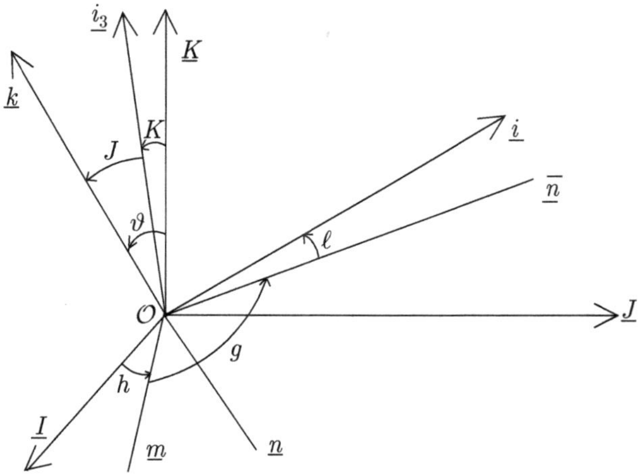

Fig. 5.2. Andoyer–Deprit angles.

If $\underline{M}_0$ denotes the angular momentum, we introduce the following variables defined as the projections on the axes $\underline{i}_3$, $\underline{k}$, $\underline{K}$:

$$\begin{aligned} G &\equiv \underline{M}_0 \cdot \underline{i}_3 = M_0 \\ L &\equiv \underline{M}_0 \cdot \underline{k} = G\cos J \\ H &\equiv \underline{M}_0 \cdot \underline{K} = G\cos K \ , \end{aligned} \tag{5.2}$$

where J is the so–called *non–principal rotation angle* formed by the vertical axes of the spin and body frames, while K is the so–called *obliquity* angle formed by the vertical axes of the spin and inertial frames. We denote by

$$(G, g), \qquad (L, \ell), \qquad (H, h)$$

the Andoyer–Deprit variables. Let p_φ, p_ψ, p_ϑ be the momenta conjugated to the Euler angles. Their relations with the Andoyer–Deprit variables are the following:

$$\begin{aligned} p_\varphi &= H \\ p_\psi &= L \\ p_\vartheta &= G\sin J \sin(\ell - \psi) \ . \end{aligned} \tag{5.3}$$

Proposition. *The transformation from* $(p_\varphi, p_\psi, p_\vartheta, \varphi, \psi, \vartheta)$ *to* (G, L, H, g, ℓ, h) *is canonical.*

Proof. To prove the canonicity of the transformation we need to show that (see, e.g. [5])

$$G dg + L d\ell + H dh = p_\varphi d\varphi + p_\psi d\psi + p_\vartheta d\vartheta \ . \tag{5.4}$$

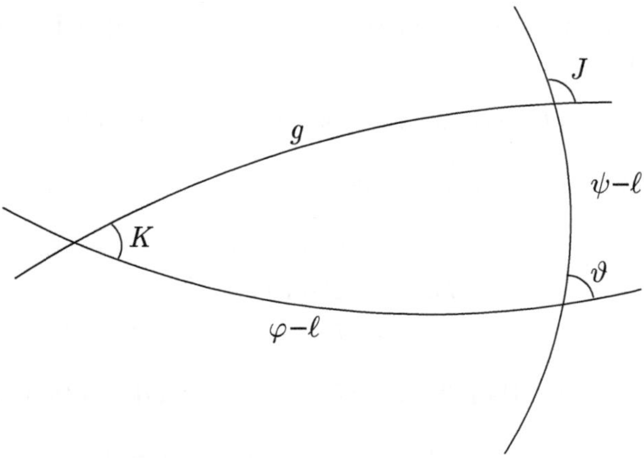

Fig. 5.3. A spherical triangle.

From properties of spherical trigonometry applied to the spherical triangle of Figure 5.3, one obtains:

$$dg = -\sin J \sin(\psi - \ell) d\vartheta + \cos K d(\varphi - h) + \cos J d(\psi - \ell) \ .$$

Using (5.3) it is readily proved that (5.4) holds. □

5.3 Free rigid body motion

Taking into account the relations (5.1) we can express the momenta conjugated to the Euler angles as

$$\begin{aligned} p_\varphi &= (I_1\omega_1 \sin\psi + I_2\omega_2 \cos\psi)\sin\vartheta + I_3\omega_3\cos\vartheta \\ p_\vartheta &= I_1\omega_1\cos\psi - I_2\omega_2\sin\psi \\ p_\psi &= I_3\omega_3 \ . \end{aligned} \tag{5.5}$$

Inverting (5.5) one obtains

$$\begin{aligned} I_1\omega_1 &= \frac{p_\varphi - p_\psi\cos\vartheta}{\sin\vartheta}\sin\psi + p_\vartheta\cos\psi \\ I_2\omega_2 &= \frac{p_\varphi - p_\psi\cos\vartheta}{\sin\vartheta}\cos\psi - p_\vartheta\sin\psi \\ I_3\omega_3 &= p_\psi \ . \end{aligned} \tag{5.6}$$

Using the relations (5.2), (5.3) and (5.6) one finds:

$$\begin{aligned} I_1\omega_1 &= G\left[\frac{\cos K - \cos J\cos\vartheta}{\sin\vartheta}\sin\psi + \sin J\sin(\ell-\psi)\cos\psi\right] \\ I_2\omega_2 &= G\left[\frac{\cos K - \cos J\cos\vartheta}{\sin\vartheta}\cos\psi - \sin J\sin(\ell-\psi)\sin\psi\right] \\ I_3\omega_3 &= L \ . \end{aligned} \tag{5.7}$$

From elementary properties of spherical trigonometry applied to the triangle of Figure 5.3, one finds[2]

$$\cos K = \cos J \cos \vartheta + \sin J \sin \vartheta \cos(\psi - \ell) \ ;$$

together with (5.7) this relation provides

$$\begin{aligned} I_1\omega_1 &= G \sin J \sin \ell \\ I_2\omega_2 &= G \sin J \cos \ell \\ I_3\omega_3 &= L \ . \end{aligned}$$

Finally, we can write the Hamiltonian of a free rigid body motion as

$$\mathcal{H}(G, L, H, g, \ell, h) = \frac{1}{2}(G^2 - L^2)\Big(\frac{\sin^2 \ell}{I_1} + \frac{\cos^2 \ell}{I_2}\Big) + \frac{1}{2}\frac{L^2}{I_3} \ ,$$

which does not depend on H as well as on the angles g and h. Notice that in the gyroscopic case $I_1 = I_2$ the Hamiltonian function becomes:

$$\mathcal{H}(G, L, H, g, \ell, h) = \frac{G^2 - L^2}{2I_1} + \frac{1}{2}\frac{L^2}{I_3} \ , \tag{5.8}$$

which does not depend on H as well as on the angles g, ℓ and h. Therefore, the associated Hamilton's equations provide that the actions (G, L, H) are constants. As for the angles, one has

$$\dot g = \frac{G}{I_1} \ , \qquad \dot\ell = \Big(\frac{1}{I_3} - \frac{1}{I_1}\Big)L \ , \qquad \dot h = 0 \ ,$$

which means that the rigid body rotates around the symmetry axis with constant velocity and that it precedes with uniform velocity.

Remark. The Andoyer–Deprit variables are not well–defined if $K = 0$ or $J = 0$; in these situations one can introduce a transformation to non–singular variables defined as

$$\begin{aligned} \lambda_1 &= \ell + g + h & \Lambda_1 &= G \\ \lambda_2 &= -\ell & \Lambda_2 &= G - L = G(1 - \cos J) \\ \lambda_3 &= -h & \Lambda_3 &= G - H = G(1 - \cos K) \ . \end{aligned} \tag{5.9}$$

The corresponding Hamiltonian is given by

$$\mathcal{H}(\Lambda_1, \Lambda_2, \Lambda_3, \lambda_1, \lambda_2, \lambda_3) = \frac{(\Lambda_1 - \Lambda_2)^2}{2I_3} + \frac{1}{2}\Big(\Lambda_1^2 - (\Lambda_1 - \Lambda_2)^2\Big)\Big(\frac{\sin^2 \lambda_2}{I_1} + \frac{\cos^2 \lambda_2}{I_2}\Big) \ .$$

[2] Consider a spherical triangle with angles A, B, C and opposite sides a, b, c; then: $\cos A = -\cos B \cos C + \sin B \sin C \cos a$.

5.4 Perturbed rigid body motion

Let the rigid body $\mathcal{S}$ of mass m be subject to the gravitational attraction of a perturber point–mass body $\mathcal{P}$ with mass M. We assume that the rigid body moves on a Keplerian orbit with instantaneous orbital radius vector $\underline{r}$, joining $\mathcal{P}$ with the barycenter of $\mathcal{S}$; denoting by $|\mathcal{S}|$ the volume of $\mathcal{S}$ and by $\underline{x}$ the position vector with respect to the barycenter of a generic point of the rigid body, the potential takes the form

$$\tilde{V} \equiv -\int_{\mathcal{S}} \frac{\mathcal{G}Mm}{|\underline{r}+\underline{x}|}\frac{d\underline{x}}{|\mathcal{S}|} \ . \tag{5.10}$$

The development of $\tilde{V}$ in spherical harmonics is given by (see [57,103,112])

$$\tilde{V} = -\mathcal{G}M\sum_{i=0}^{\infty}\sum_{j=0}^{i}\frac{1}{r^{i+1}}\,P_{ij}(\sin\theta)(C_{ij}\cos j\phi + S_{ij}\sin j\phi) \ ,$$

where (r,ϕ,θ) are, respectively, the modulus of $\underline{r}$, the longitude ϕ and the latitude θ measured eastward in a reference frame with origin in the barycenter of the rigid body, C_{ij} and S_{ij} are the *potential coefficients* which depend on the density distribution of the rigid body, P_{ij} are the Legendre associated functions defined by the set of equations

$$\begin{aligned} P_{ij}(\sin\theta) &\equiv \cos^j\theta\sum_{k=0}^{n} T_{ijk}\sin^{i-j-2k}\theta \ , \qquad n=\left[\frac{i-j}{2}\right] \\ T_{ijk} &\equiv \frac{(-1)^k(2i-2k)!}{2^i k!(i-k)!(i-j-2k)!} \ . \end{aligned}$$

The lowest significative order of the development of the potential, say $\tilde{V}_2$, is given by

$$\tilde{V}_2 = -\frac{\mathcal{G}Mm}{r}\left(\frac{R_e}{r}\right)^2\Big[C_{20}P_2(\sin\theta) + C_{22}P_{22}(\sin\theta)\cos 2\phi\Big] \ , \tag{5.11}$$

where R_e is the body's equatorial radius and $P_2(\sin\theta)$ is the Legendre polynomial of second order; the spherical harmonic terms C_{20}, C_{22} are given by

$$C_{20} \equiv \frac{1}{2}\frac{1}{mR_e^2}(I_1+I_2-2I_3) \ , \qquad C_{22} \equiv \frac{1}{4}\frac{1}{mR_e^2}(I_2-I_1) \ .$$

Let us denote by $(\tilde{x},\tilde{y},\tilde{z})$ the coordinates of the unitary vector oriented toward $\mathcal{P}$ in the body frame, namely

$$\begin{aligned} \tilde{x} &= \cos\phi\cos\theta \\ \tilde{y} &= \sin\phi\cos\theta \\ \tilde{z} &= \sin\theta \ . \end{aligned}$$

Then, we can write an expression for $P_2(\sin\theta)$, $P_{22}(\sin\theta)\cos 2\phi$ in terms of $(\tilde{x}, \tilde{y}, \tilde{z})$ as

$$P_2(\sin\theta) = \frac{2\tilde{z}^2 - \tilde{x}^2 - \tilde{y}^2}{2}\ , \qquad P_{22}(\sin\theta)\cos 2\phi = 3(\tilde{x}^2 - \tilde{y}^2)\ .$$

In order to express $(\tilde{x}, \tilde{y}, \tilde{z})$ with respect to the inertial frame we need to define the matrices $R_1(\alpha)$, $R_2(\alpha)$, $R_3(\alpha)$ as the rotations of angle α with the subscript denoting the rotation axis. Denoting by f the true anomaly of $\mathcal{S}$ along its Keplerian orbit, we obtain:

$$\begin{pmatrix} \tilde{x} \\ \tilde{y} \\ \tilde{z} \end{pmatrix} = R_3(\ell)R_1(J)R_3(g)R_1(K)R_3(h) \begin{pmatrix} \cos f \\ \sin f \\ 0 \end{pmatrix}\ . \tag{5.12}$$

In the general case, encompassing also J and K close to zero, one introduces non–singular angles as in (5.9) by defining $\lambda_1 = \ell + g + h$, in terms of which we provide the following definition of spin–orbit resonance.

Definition. A spin–orbit resonance of order $p : q$, for $p, q \in \mathbf{Z}$ with $q \neq 0$, occurs whenever the ratio of the rates of variation of λ_1 and of the mean anomaly ℓ_0 is equal to p/q, namely

$$\frac{\dot{\lambda}_1}{\dot{\ell}_0} = \frac{p}{q}\ . \tag{5.13}$$

The relation (5.13) means that during q revolutions around $\mathcal{P}$, the body $\mathcal{S}$ makes p rotations about its spin–axis. The associated *resonant angle* is equal to

$$\sigma \equiv \frac{q\lambda_1 - p\ell_0}{q}\ .$$

The synchronous resonance is characterized by $p = q = 1$, which implies that the body always points the same face toward the perturber, since the period of rotation around the spin–axis coincides with the period of revolution about $\mathcal{P}$. This is the case of the Moon–Earth system, where astronomical data provide that the period of revolution of the Moon around the Earth and the period of rotation about the spin–axis are equal to 27.322 days.

At the end of the 17th century G.D. Cassini formulated three laws about the motion of the Moon around the Earth. Observing that an equilibrium state is provided by $J = K = 0$ and by $\sigma = \ell = h = 0$, the Cassini laws can be expressed as follows:

(1) the Moon rotates around a principal axis in a synchronous resonance;
(2) the spin–axis and the orbit normal form a constant angle (which means that K is constant);
(3) the spin–axis, the orbit normal and the ecliptic normal lie in the same plane.

5.5 The spin–orbit problem

A simple paradigmatic model of rotational dynamics is provided by the spin–orbit problem (see, e.g., [23, 172]) obtained by setting to zero the non–principal rotation angle as well as the obliquity: $J = K = 0$. In this case $G = L = H$, namely $\Lambda_2 = \Lambda_3 = 0$, while the kinetic part reduces to $\frac{\Lambda_1^2}{2I_3}$. Also the potential energy simplifies, since for $J = K = 0$, taking into account (5.12), we obtain

$$\tilde{x} = \cos(f - \lambda_1) \ , \qquad \tilde{y} = \sin(f - \lambda_1) \ , \qquad \tilde{z} = 0 \ .$$

The above relations imply that the non–trivial term in $\tilde{V}_2$ is provided by

$$\begin{aligned} \tilde{V}_2 &= -\frac{\mathcal{G}Mm}{a^3} R_e^2 \left(\frac{a}{r}\right)^3 \left[3C_{22}(\tilde{x}^2 - \tilde{y}^2)\right] \\ &= -\frac{1}{2}\frac{\mathcal{G}M}{a^3}\left(\frac{a}{r}\right)^3 \frac{3}{2}\frac{I_2 - I_1}{I_3} I_3 \cos(2\lambda_1 - 2f) \ . \end{aligned}$$

We adopt the units of measure such that the mean motion, which by Kepler's law coincides with $\frac{\mathcal{G}M}{a^3}$, is equal to one. We define the *equatorial ellipticity* as

$$\varepsilon \equiv \frac{3}{2}\frac{I_2 - I_1}{I_3} \ ;$$

as a consequence, the potential can be written as:

$$\tilde{V}_2 = -\frac{\varepsilon}{2}\left(\frac{a}{r}\right)^3 I_3 \cos(2\lambda_1 - 2f) \ .$$

We will specify two different spin–orbit models: the first one is subject only to the gravitational attraction of the perturber, while the second model takes into account a dissipative contribution due to the internal non–rigidity of the body. The two models will be referred to as the *conservative* and *dissipative* spin–orbit problems.

5.5.1 The conservative spin–orbit problem

Consider the motion of a rigid body $\mathcal{S}$ with triaxial structure, under the gravitational influence of a point–mass perturber $\mathcal{P}$. We assume that

(*i*) $\mathcal{S}$ moves on an elliptic Keplerian orbit with semimajor axis a and eccentricity e (namely, the perturbations due to other bodies are neglected);
(*ii*) the spin–axis coincides with the smallest physical axis of the ellipsoid (namely, with the largest principal axis of inertia so that $J = 0$);
(*iii*) the spin–axis is perpendicular to the orbital plane (namely, the obliquity K is zero);
(*iv*) dissipative effects are neglected.

Under such assumptions the Hamiltonian governing the dynamics is given by

$$\mathcal{H}_0(\Lambda_1, \lambda_1, t) = \frac{\Lambda_1^2}{2I_3} - \frac{\varepsilon}{2}\left(\frac{a}{r}\right)^3 I_3 \cos(2\lambda_1 - 2f) , \tag{5.14}$$

where it is intended that the Hamiltonian depends upon the time through the quantities r and f. Making the change of variables $y = \frac{\Lambda_1}{I_3}$, $x = \lambda_1$, then (5.14) reduces to

$$\mathcal{H}(y, x, t) = \frac{y^2}{2} - \frac{\varepsilon}{2}\left(\frac{a}{r}\right)^3 \cos(2x - 2f) .$$

The corresponding Hamilton's equations are

$$\begin{aligned} \dot{y} &= -\varepsilon\left(\frac{a}{r}\right)^3 \sin(2x - 2f) \\ \dot{x} &= y , \end{aligned} \tag{5.15}$$

which are equivalent to

$$\ddot{x} + \varepsilon\left(\frac{a}{r}\right)^3 \sin(2x - 2f) = 0 . \tag{5.16}$$

In the spin–orbit problem the quantity x denotes the angle formed by the direction of the largest physical axis (which by assumptions (*ii*) and (*iii*) belongs to the orbital plane) with a reference axis, say the perihelion line (see Figure 5.4).

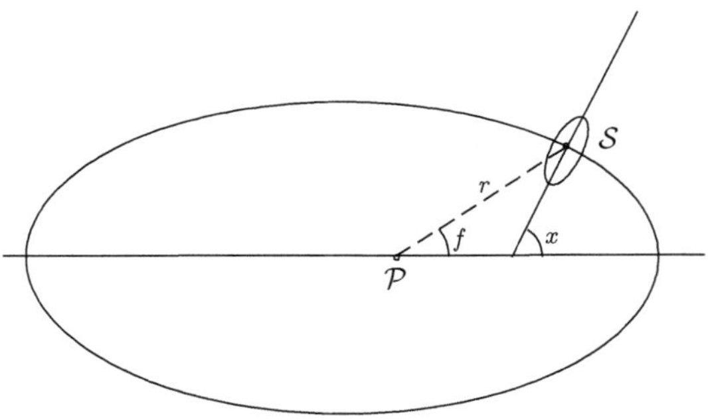

Fig. 5.4. The geometry of the spin–orbit problem.

Remarks.

(*a*) The Moon, as well as most of the evolved satellites of the solar system, are observed to move in a synchronous resonance; the only exception is provided by Mercury [48], which moves in a 3:2 spin–orbit resonance, since the orbital period is equal to the rotational period within an error of the order of 10^{-4}. It means that, almost exactly, during two orbital revolutions around the Sun, Mercury makes three rotations about its spin–axis.

(*b*) The parameter $\varepsilon = \frac{3}{2}\frac{I_2 - I_1}{I_3}$ is zero in the case of equatorial symmetry $I_1 = I_2$. In this case the equation of motion is trivially integrable. For regular bodies like the

Moon or Mercury, the parameter ε is of the order of 10^{-4} and the corresponding dynamical system is nearly–integrable.

(*c*) The orbital radius and the true anomaly are known functions of time, being determined through the Keplerian relations:

$$\begin{aligned} r &= a(1 - \mathrm{e}\cos u) \\ f &= 2\arctan\left(\sqrt{\frac{1+\mathrm{e}}{1-\mathrm{e}}}\tan\frac{u}{2}\right) , \end{aligned} \tag{5.17}$$

where u is related to the mean anomaly ℓ_0 by means of Kepler's equation $\ell_0 = u - \mathrm{e}\sin u$.

(*d*) The orbital radius and the true anomaly depend on the eccentricity e; for $\mathrm{e} = 0$ one has $r = a$, $f = t + t_0$ for a suitable constant t_0; henceforth, for circular orbits (5.16) reduces to the integrable equation $\ddot{x} + \varepsilon\sin(2x - 2t - 2t_0) = 0$.

(*e*) Considering the lift of the angle x on $\mathbf{R}$, we interpret a $p : q$ spin–orbit resonance for $p, q \in \mathbf{Z}$ with $q > 0$ as a periodic solution for (5.16), say $t \in \mathbf{R} \to x = x(t) \in \mathbf{R}$, such that

$$x(t + 2\pi q) = x(t) + 2\pi p \qquad \text{for any } \ t \in \mathbf{R} \, .$$

Expanding (5.17) in power series of the orbital eccentricity, equation (5.16) can be developed in Fourier series as

$$\ddot{x} \ + \ \varepsilon \sum_{m\neq 0, m=-\infty}^{+\infty} W\left(\frac{m}{2}, \mathrm{e}\right) \ \sin(2x - mt) \ = \ 0 \, , \tag{5.18}$$

where the expressions of the first few coefficients $W(\frac{m}{2}, \mathrm{e})$ are reported in Table 5.1; the coefficients have been expanded in power series of the orbital eccentricity as $W(\frac{m}{2}, \mathrm{e}) \equiv W_0^m(\mathrm{e}) + W_1^m(\mathrm{e}) + W_2^m(\mathrm{e}) + \ldots$, being $W_j^m(\mathrm{e}) = O(\mathrm{e}^j)$.

Table 5.1. Expansion in powers of the orbital eccentricity of the coefficients $W(\frac{m}{2}, \mathrm{e})$ appearing in (5.18).

$\frac{m}{2}$	$W_0^m(\mathrm{e})$	$W_1^m(\mathrm{e})$	$W_2^m(\mathrm{e})$	$W_3^m(\mathrm{e})$	$W_4^m(\mathrm{e})$	$W_5^m(\mathrm{e})$	$W_6^m(\mathrm{e})$	$W_7^m(\mathrm{e})$
-1					$\frac{\mathrm{e}^4}{24}$		$\frac{7\mathrm{e}^6}{240}$	
$-\frac{1}{2}$				$\frac{\mathrm{e}^3}{48}$		$\frac{11\mathrm{e}^5}{768}$		$\frac{313\mathrm{e}^7}{30720}$
$\frac{1}{2}$		$-\frac{\mathrm{e}}{2}$		$\frac{\mathrm{e}^3}{16}$		$-\frac{5\mathrm{e}^5}{384}$		$-\frac{143\mathrm{e}^7}{18432}$
1	1		$-\frac{5\mathrm{e}^2}{2}$		$\frac{13\mathrm{e}^4}{16}$		$-\frac{35\mathrm{e}^6}{288}$	
$\frac{3}{2}$		$\frac{7\mathrm{e}}{2}$		$-\frac{123\mathrm{e}^3}{16}$		$\frac{489\mathrm{e}^5}{128}$		$-\frac{1763\mathrm{e}^7}{2048}$
2			$\frac{17\mathrm{e}^2}{2}$		$-\frac{115\mathrm{e}^4}{6}$		$\frac{601\mathrm{e}^6}{48}$	
$\frac{5}{2}$				$\frac{845\mathrm{e}^3}{48}$		$-\frac{32525\mathrm{e}^5}{768}$		$\frac{208225\mathrm{e}^7}{6144}$
3					$\frac{533\mathrm{e}^4}{16}$		$-\frac{13827\mathrm{e}^6}{160}$	

Considering the series expansion (5.18) up to the order 4 in the eccentricity, one obtains

$$\begin{aligned}\ddot{x} \ + \ \varepsilon \Bigg[& \frac{\mathrm{e}^4}{24} \sin(2x+2t) + \frac{\mathrm{e}^3}{48} \sin(2x+t) + \Big(-\frac{\mathrm{e}}{2} + \frac{\mathrm{e}^3}{16} \Big) \ \sin(2x-t) \ + \\ + & \Big(1 - \frac{5}{2}\mathrm{e}^2 + \frac{13}{16}\mathrm{e}^4 \Big) \ \sin(2x-2t) \ + \ \Big(\frac{7}{2}\mathrm{e} - \frac{123}{16}\mathrm{e}^3 \Big) \ \sin(2x-3t) \ + \\ + & \Big(\frac{17}{2}\mathrm{e}^2 - \frac{115}{6}\mathrm{e}^4 \Big) \ \sin(2x-4t) \ + \ \frac{845}{48}\mathrm{e}^3 \ \sin(2x-5t) \ + \\ + & \frac{533}{16}\mathrm{e}^4 \ \sin(2x-6t) \Bigg] \ = \ 0 \ , \end{aligned} \tag{5.19}$$

which can be written in the form

$$\ddot{x} + \varepsilon V(x,t) = 0 \ ,$$

for a suitable periodic function $V = V(x,t)$. Such an equation corresponds to that of a pendulum subject to a forcing term, depending periodically upon the time.

5.5.2 The averaged equation

For a given resonance of order $p:2$, let us define the resonant angle $\gamma \equiv x - \frac{p}{2}t$; in terms of γ equation (5.18) becomes

$$\ddot{\gamma} + \varepsilon W\Big(\frac{p}{2},\mathrm{e}\Big) \sin 2\gamma + \varepsilon \sum_{m \neq 0,p, \ m=-\infty}^{+\infty} W\Big(\frac{m}{2},\mathrm{e}\Big) \ \sin(2\gamma + (p-m)t) \ = \ 0 \ .$$

Averaging over the time one obtains the pendulum equation

$$\ddot{\gamma} + \varepsilon W\Big(\frac{p}{2},\mathrm{e}\Big) \sin 2\gamma = 0 \ ,$$

which admits the energy integral

$$\frac{1}{2}\dot{\gamma}^2 - \frac{\varepsilon}{2} W\Big(\frac{p}{2},\mathrm{e}\Big) \ \cos 2\gamma = E \ ,$$

being E the total mechanical energy. The equilibrium points correspond to $\gamma = 0$ and $\gamma = \frac{\pi}{2}$ (modulus π). If $W(\frac{p}{2},\mathrm{e}) > 0$ then $\gamma = 0$ is stable, while $\gamma = \frac{\pi}{2}$ is unstable. The maximum width $\eta_p(\mathrm{e})$ of the resonant region around the stable point amounts to $\eta_p(\mathrm{e}) \equiv 2\sqrt{\varepsilon W(\frac{p}{2},\mathrm{e})}$. For example, taking into account the expressions provided in Table 5.1, one has

$$\begin{aligned} \eta_2(0.0549) \ &= \ 0.037 \ , \qquad & \eta_3(0.0549) = 0.016 \ , \\ \eta_2(0.2056) \ &= \ 0.023 \ , \qquad & \eta_3(0.2056) = 0.020 \ , \end{aligned}$$

where $p = 2$ corresponds to the 1:1 resonance, $p = 3$ to the 3:2 resonance, $\mathrm{e} = 0.0549$ is the eccentricity of the Moon and $\mathrm{e} = 0.2056$ is that of Mercury; we have assumed $\varepsilon = 3.45 \cdot 10^{-4}$ for the Moon and $\varepsilon = 1.5 \cdot 10^{-4}$ for Mercury (see also [33]).

5.5.3 The dissipative spin–orbit problem

In writing equation (5.16) we have explicitly neglected the dissipative effects; among these terms, the most important contribution is typically due to the internal non–rigidity of the body and it can be described by a model of tidal torque [78, 79, 123, 147]). Within the different expressions existing in the literature, we quote a tidal torque depending linearly on the relative angular velocity as [147]

$$\mathcal{T}(\dot{x};t) = -K_d\Big[L(\mathrm{e},t)\dot{x} - N(\mathrm{e},t)\Big] ,$$

where

$$L(\mathrm{e},t) = \frac{a^6}{r^6} , \qquad N(\mathrm{e},t) = \frac{a^6}{r^6}\dot{f} .$$

Moreover, the coefficient K_d is named the *dissipative constant;* it depends on the physical and orbital features of the body. More precisely, K_d takes the form

$$K_d \equiv 3n \,\frac{k_2}{\xi Q}\left(\frac{R_e}{a}\right)^3 \frac{M}{m} ,$$

where n is the mean motion, k_2 is the *Love number* (depending on the structure of the body), Q is the so–called *quality factor* (which compares the frequency of oscillation of the system to the rate of dissipation of energy), ξ is a structure constant such that $I_3 = \xi m R_e^2$ and, as before, R_e is the equatorial radius, M is the mass of the central body $\mathcal{P}$, m is the mass of $\mathcal{S}$. Astronomical observations suggest that for bodies like the Moon or Mercury the dissipative constant K_d is of the order of 10^{-8}.

The expression for the tidal torque can be simplified by assuming (as in [50]) that the dynamics is essentially ruled by the averages of $L(\mathrm{e},t)$ and $N(\mathrm{e},t)$ over one orbital period:

$$\bar{\mathcal{T}} \equiv \bar{\mathcal{T}}(\dot{x}) = -K_d\Big[\bar{L}(\mathrm{e})\dot{x} - \bar{N}(\mathrm{e})\Big] \tag{5.20}$$

where (compare with [147])

$$\begin{aligned}
\bar{L}(\mathrm{e}) &\equiv \frac{1}{(1-\mathrm{e}^2)^{9/2}}\left(1 + 3\mathrm{e}^2 + \frac{3}{8}\mathrm{e}^4\right) \\
\bar{N}(\mathrm{e}) &\equiv \frac{1}{(1-\mathrm{e}^2)^6}\left(1 + \frac{15}{2}\mathrm{e}^2 + \frac{45}{8}\mathrm{e}^4 + \frac{5}{16}\mathrm{e}^6\right) .
\end{aligned}$$

Thus, we are led to consider the following equation of motion for the dissipative spin–orbit problem:

$$\ddot{x} + \varepsilon\left(\frac{a}{r}\right)^3 \sin(2x - 2f) = -K_d\Big[\bar{L}(\mathrm{e})\dot{x} - \bar{N}(\mathrm{e})\Big] . \tag{5.21}$$

Due to (5.20) the tidal torque vanishes provided

$$\dot{x} \equiv \frac{\bar{N}(\mathrm{e})}{\bar{L}(\mathrm{e})} = \frac{1 + \frac{15}{2}\mathrm{e}^2 + \frac{45}{8}\mathrm{e}^4 + \frac{5}{16}\mathrm{e}^6}{(1-\mathrm{e}^2)^{\frac{3}{2}}(1 + 3\mathrm{e}^2 + \frac{3}{8}\mathrm{e}^4)} .$$

It is readily shown that for circular orbits the angular velocity of rotation corresponds to the synchronous resonance, being $\dot{x} = 1$. For Mercury's eccentricity $e = 0.2056$, it turns out that $\dot{x} = 1.256$. We refer the reader to [9, 97, 144] for a discussion on the problem of the capture into resonance under the effect of the dissipation.

5.5.4 The discrete spin–orbit problem

Let us introduce the resonant angle of order $p:2$ defined by the relation $\gamma \equiv x - \frac{p}{2}t$; averaging also the conservative contribution over one orbital period, up to constants one obtains

$$\ddot{\gamma} + \varepsilon W\left(\frac{p}{2}, e\right) \sin 2\gamma = -K_d\left[\bar{L}(e)\dot{\gamma} - \bar{N}(e)\right],$$

which can be written as

$$\begin{aligned} \dot{\gamma} &= \Gamma \\ \dot{\Gamma} &= -\varepsilon W\left(\frac{p}{2}, e\right) \sin 2\gamma - K_d\left[\bar{L}(e)\Gamma - \bar{N}(e)\right] . \end{aligned} \tag{5.22}$$

Integrating (5.22) through Euler's symplectic method, one obtains

$$\begin{aligned} \Gamma' &= \Gamma + h\left[-\varepsilon W\left(\frac{p}{2}, e\right) \sin 2\gamma - K_d\left(\bar{L}(e)\Gamma - \bar{N}(e)\right)\right] \\ \gamma' &= \gamma + h\Gamma' \\ t' &= t + h , \end{aligned} \tag{5.23}$$

where h denotes the integration step and (Γ', γ') is the solution at time $t+h$. Let us write (5.23) as

$$\begin{aligned} \Gamma' &= (1 - hK_d\bar{L}(e))\Gamma + hK_d\bar{N}(e) - h\varepsilon W\left(\frac{p}{2}, e\right) \sin 2\gamma \\ \gamma' &= \gamma + h\Gamma' \\ t' &= t + h ; \end{aligned} \tag{5.24}$$

notice that the determinant of the Jacobian of (5.24) is equal to $1 - hK_d\bar{L}(e)$. The quantity $1 - hK_d\bar{L}(e)$ is positive for typical values of the parameters; for example, if $h = 2\pi$ and $K_d = 10^{-3}$, then the Jacobian is positive for any value of the eccentricity less than 0.767.

In (5.24) take the integration step as $h = 2\pi$ and define $b \equiv 1 - 2\pi K_d\bar{L}(e)$, $c_0 \equiv 2\pi K_d\bar{N}(e)$, $g(\gamma) \equiv -2\pi W(\frac{p}{2}, e) \sin 2\gamma$. Then, considering the first two equations of (5.24) one obtains the following map at times 2π:

$$\begin{aligned} \Gamma' &= b\Gamma + c_0 + \varepsilon g(\gamma) \\ \gamma' &= \gamma + 2\pi\Gamma' . \end{aligned}$$

Performing the change of variables $\xi = \gamma$, $\eta = 2\pi\Gamma$, we obtain the dissipative standard map (see Section 1.7):

$$\begin{aligned} \eta' &= b\eta + c + \varepsilon f(\xi) \\ \xi' &= \xi + \eta' , \end{aligned}$$

where $c \equiv 2\pi c_0$ and $f(\xi) = 2\pi g(\xi)$.

Remarks.
(a) Notice that the parameter c depends on $\bar{N}(\mathrm{e})$, which represents the average over an orbital period of $\frac{a^6}{r^6}\dot{f}$. Therefore $\bar{N}(\mathrm{e})$ is zero whenever $\dot{f} = 0$, namely when the true anomaly is constant.

(b) Let us focus on the case when the dissipation is zero, namely $\bar{L}(\mathrm{e})\Gamma - \bar{N}(\mathrm{e}) = 0$. As far as $\Gamma = 1$ (corresponding to the synchronous resonance) one has that $\bar{L}(\mathrm{e}) - \bar{N}(\mathrm{e}) = 0$, which is satisfied only if $\mathrm{e} = 0$; for $\Gamma = 1.5$ (corresponding to the 3:2 resonance) one has $1.5\bar{L}(\mathrm{e}) - \bar{N}(\mathrm{e}) = 0$ which admits the solution $\mathrm{e} = 0.285$.

5.6 Motion around an oblate primary

Let $\mathcal{P}$ be a body of finite dimensions with principal moments of inertia I_1, I_2, I_3 and let $\mathcal{S}$ be a point–mass satellite orbiting under the gravitational influence of $\mathcal{P}$. The potential of this model is the same as in (5.10), (5.11) and therefore we just limit ourselves to quoting the main formulae, which allow us to write the equations of motion. Let M and m be the masses of $\mathcal{P}$ and $\mathcal{S}$, respectively. Denote by $(\mathcal{O}, x, y, z)$ a reference frame with origin in the center of mass $\mathcal{O}$ of $\mathcal{P}$; let Q be a generic point of $\mathcal{P}$ with coordinates (x', y', z') and let $\mathcal{S}$ have coordinates (x, y, z). If $\delta(x', y', z')$ denotes the density of $\mathcal{P}$, then the potential energy acting on $\mathcal{S}$ takes the form

$$U(x, y, z) = \mathcal{G} \int_{|\mathcal{P}|} \frac{\delta(x', y', z')}{\Delta}\, dx'dy'dz' ,$$

where $|\mathcal{P}|$ represents the volume of $\mathcal{P}$ and Δ is the distance between Q and $\mathcal{S}$:

$$\Delta \equiv \sqrt{(x - x')^2 + (y - y')^2 + (z - z')^2} .$$

The equations of motion are given by

$$\ddot{x} = \frac{\partial U}{\partial x} , \qquad \ddot{y} = \frac{\partial U}{\partial y} , \qquad \ddot{z} = \frac{\partial U}{\partial z} .$$

If r denotes the distance between $\mathcal{S}$ and $\mathcal{O}$, ρ is the distance $\mathcal{O}Q$ and α is the angle formed by $\mathcal{O}\mathcal{S}$ and $\mathcal{O}Q$, then $\Delta^2 = r^2 + \rho^2 - 2r\rho\cos\alpha$. Therefore, we can expand $\frac{1}{\Delta}$ using the Legendre polynomials P_n and we can write the potential function as

$$U \equiv \sum_{n=0}^{\infty} U_n = \frac{\mathcal{G}}{r} \int_{|\mathcal{P}|} \delta(x', y', z') \sum_{n=0}^{\infty} \left(\frac{\rho}{r}\right)^n P_n(\cos\alpha)\, dx'dy'dz' . \qquad (5.25)$$

Explicit computations for the first few terms provide

$$\begin{aligned} U_0 &= \frac{\mathcal{G}M}{r} \\ U_1 &= 0 \\ U_2 &= \frac{\mathcal{G}}{r^3}\left[\frac{1}{2}(I_1+I_2+I_3) - \frac{3}{2r^2}(I_1x^2+I_2y^2+I_3z^2)\right] . \end{aligned}$$

Introducing the polar coordinates (r, ϕ, θ), where (ϕ, θ) are the satellite's longitude and latitude, defined through the relations

$$\begin{aligned} x &= r\cos\phi\cos\theta \\ y &= r\sin\phi\cos\theta \\ z &= r\sin\theta\ , \end{aligned}$$

one finds

$$U_2 = \frac{\mathcal{G}}{r^3}\left[\left(I_3 - \frac{I_1+I_2}{2}\right)\left(\frac{1}{2}-\frac{3}{2}\sin^2\theta\right) - \frac{3}{4}(I_1-I_2)\cos^2\theta\cos 2\phi\right] . \tag{5.26}$$

In the gyroscopic case $I_1 = I_2$, the potential (5.26) reduces to

$$U_2 = \frac{\mathcal{G}MR_e^2}{r^3}\ J_2\left(\frac{1}{2}-\frac{3}{2}\sin^2\theta\right) , \tag{5.27}$$

where

$$J_2 \equiv \frac{I_3 - I_1}{MR_e^2}$$

and R_e is the equatorial's radius of $\mathcal{P}$.

For a body with spherical symmetry $I_1 = I_2 = I_3$, one finds

$$U = \frac{\mathcal{G}M}{r} .$$

In general, for a body symmetric with respect to the equatorial plane, the odd terms of the development (5.25) are zero and one can write the potential energy as

$$U = \frac{\mathcal{G}M}{r}\left[1 - \sum_{n=1}^{\infty}\left(\frac{R_e}{r}\right)^{2n} J_{2n}P_{2n}(\sin\theta)\right]$$

for suitable coefficients J_{2n}, called the *zonal coefficients* [103].

5.7 Interaction between two bodies of finite dimensions

Let $\mathcal{P}$ and $\mathcal{P}'$ be two bodies of finite dimensions with masses M, M' and centers of mass $\mathcal{O}$, $\mathcal{O}'$. Let $(\mathcal{O}, x, y, z)$ be a reference frame centered in $\mathcal{O}$ with the x–axis along the direction $\mathcal{O}\mathcal{O}'$. Let r be the distance between $\mathcal{O}$ and $\mathcal{O}'$. If dM represents an element of mass of $\mathcal{P}$ with coordinates (x_M, y_M, z_M) with respect to $(\mathcal{O}, x, y, z)$

and dM' is an element of mass of $\mathcal{P}'$ with coordinates (x'_M, y'_M, z'_M) with respect to $(\mathcal{O}', x, y, z)$, then the potential energy can be written as

$$U = \mathcal{G} \int_{\mathcal{P}} \int_{\mathcal{P}'} \frac{dMdM'}{\Delta} ,$$

where Δ denotes the distance between dM and dM', namely

$$\Delta = \sqrt{(r + x'_M - x_M)^2 + (y'_M - y_M)^2 + (z'_M - z_M)^2} .$$

Introducing the quantities

$$\rho = \frac{1}{r} \sqrt{(x'_M - x_M)^2 + (y'_M - y_M)^2 + (z'_M - z_M)^2} , \qquad \alpha = \frac{x_M - x'_M}{r\rho} ,$$

we obtain

$$\Delta = r\sqrt{1 - 2\alpha\rho + \rho^2} .$$

Finally, using Legendre polynomials one can expand the potential energy as

$$U = \mathcal{G} \int_{\mathcal{P}} \int_{\mathcal{P}'} \frac{dMdM'}{r} \sum_{n=0}^{\infty} P_n(\alpha)\rho^n . \qquad (5.28)$$

If (a, b, c) and (a', b', c') denote the direction cosines of $\mathcal{O}\mathcal{O}'$ relative to the principal axes of, respectively, $\mathcal{P}$ and $\mathcal{P}'$, then the first few terms of (5.28) can be written as

$$\begin{aligned} U_0 &= \frac{\mathcal{G}MM'}{r} \\ U_1 &= 0 \\ U_2 &= \frac{\mathcal{G}M'}{r^3} \left[\frac{1}{2}(I_1 + I_2 + I_3) - \frac{3}{2}(I_1 a^2 + I_2 b^2 + I_3 c^2) \right] \\ &+ \frac{\mathcal{G}M}{r^3} \left[\frac{1}{2}(I'_1 + I'_2 + I'_3) - \frac{3}{2}(I'_1 a'^2 + I'_2 b'^2 + I'_3 c'^2) \right] , \end{aligned}$$

where (I_1, I_2, I_3) are the principal moments of inertia of $\mathcal{P}$, while (I'_1, I'_2, I'_3) refer to $\mathcal{P}'$. Higher–order terms can be constructed in a similar way.

5.8 The tether satellite

The tether satellite is formed by two points $\mathcal{P}_1$ and $\mathcal{P}_2$ with masses m_1 and m_2 connected by an extensible, massless tether (see [161] and references by the same author); the center of mass $\mathcal{O}$ of $\mathcal{P}_1$ and $\mathcal{P}_2$ is supposed to move under the gravitational attraction of a body $\mathcal{P}$ with mass M. The average distance R between $\mathcal{O}$ and $\mathcal{P}$ is much larger then the maximal distance d between m_1 and m_2. We assume that $\mathcal{O}$ moves on a Keplerian orbit with eccentricity e and semimajor axis a; its position on the ellipse is characterized by the true anomaly f. Let $(\mathcal{O}, x, y, z)$ be the orbital reference frame oriented as in Figure 5.5 and let $\underline{e}_x, \underline{e}_y, \underline{e}_z$ be the

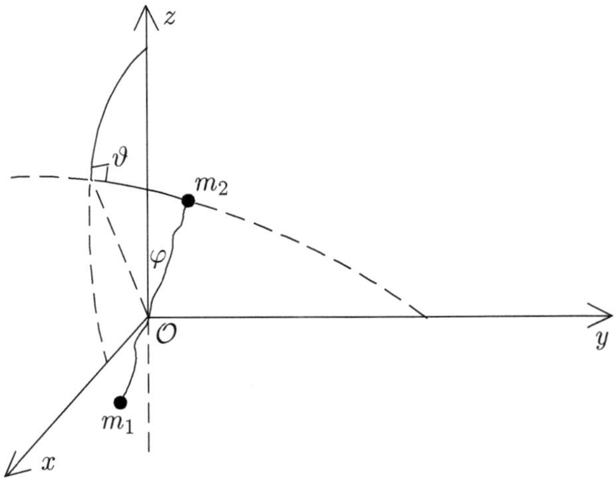

Fig. 5.5. The orbital reference frame $(\mathcal{O}, x, y, z)$ of the tether satellite.

corresponding unit vectors. If d_0 is the maximal length of the tether, we denote by ξ the relative extension defined as

$$\xi = \frac{d}{d_0} - 1 ,$$

where we assume that $\xi > 0$. Moreover, let ϑ be the angle formed by the z–axis with the projection of the tether on the orbital plane xz and let φ be the angle formed by the direction of the tether with this plane.

The expressions for the kinetic and potential energy of the attitude dynamics can be obtained as follows (see, e.g., [41]). Let us write the potential energy as

$$\Pi = \Pi^g + \Pi^e ,$$

where Π^g is the gravitational contribution, while Π^e is due to the elastic energy caused by the tether extension. Let $\underline{R} = R\underline{e}_z$ be the radius vector $\mathcal{OP}$ and $\underline{r}_i$, $i = 1, 2$, be the radius vector joining the mass m_i with $\mathcal{O}$:

$$\underline{r}_i = \frac{(-1)^i m_{3-i} d_0}{m_1 + m_2}(1 + \xi)\underline{e}_d , \qquad \underline{e}_d = \begin{pmatrix} \cos\varphi \sin\vartheta \\ -\sin\varphi \\ \cos\varphi \cos\vartheta \end{pmatrix} . \tag{5.29}$$

We can write Π^g as the sum of the contributions due to m_1 and m_2:

$$\Pi^g = \Pi_1^g + \Pi_2^g ,$$

where

$$\Pi_i^g = -\frac{\mathcal{G}Mm_i}{|\underline{R} + \underline{r}_i|} \qquad i = 1, 2 ,$$

which can be expanded as

$$\Pi_i^g = -\frac{\mathcal{G}Mm_i}{R}\left\{1 - \frac{\underline{e}_z \cdot \underline{r}_i}{R} - \frac{1}{2R^2}[r_i^2 - 3(\underline{e}_z \cdot \underline{r}_i)^2] + \dots\right\}\ , \qquad i = 1, 2\ .$$

Let us write the gravitational energy Π^g as the sum

$$\Pi^g = \Pi_0^g + \Pi_r^g\ ,$$

where Π_0^g is independent of the tether orientation and extension, namely

$$\Pi_0^g = -\frac{\mathcal{G}M(m_1 + m_2)}{R}\ ,$$

while Π_r^g is given by

$$\Pi_r^g = \frac{\mathcal{G}Mm_0 d_0^2}{2R^3}(1+\xi)^2\left[1 - (\underline{e}_z \cdot \underline{e}_d)^2\right] + \dots\ , \tag{5.30}$$

where m_0 is the reduced mass:

$$m_0 = \frac{m_1 m_2}{m_1 + m_2}.$$

Notice that the following Keplerian relation holds:

$$\frac{\mathcal{G}M}{R^3} = n^2\left(\frac{a}{R}\right)^3 = \frac{n^2(1 + \mathrm{e}\cos f)^3}{(1 - \mathrm{e}^2)^3}\ ,$$

where n is the mean motion and f is the true anomaly of the Keplerian orbit of $\mathcal{O}$ with respect to $\mathcal{P}$. According to Hooke's law, the contribution due to the elastic deformation is given by

$$\Pi^e = \frac{c d_0^2 \xi^2}{2}\ ,$$

where c denotes the stiffness of the tether.

The kinetic energy T is given by the sum of the contributions due to m_1 and m_2, namely

$$T = T_1 + T_2\ , \qquad \text{where} \quad T_i = \frac{m_i V_i^2}{2}\ , \quad i = 1, 2\ .$$

We denote by $V_i = |\underline{V}_i|$ the modulus of the velocity vector

$$\underline{V}_i = \underline{V}_0 + \underline{v}_i\ ,$$

where $\underline{V}_0$ represents the velocity of the barycenter and $\underline{v}_i$ is the velocity of the mass m_i with respect to $\mathcal{O}$. Let $\underline{\Omega} = \underline{\Omega}(f)$ denote the angular velocity of the Keplerian motion:

$$\underline{\Omega} = \Omega(f)\underline{e}_y\ , \qquad \text{where} \quad \Omega(f) = \frac{n(1 + \mathrm{e}\cos f)^2}{(1 - \mathrm{e}^2)^{3/2}}\ .$$

Then $\underline{v}_i$ can be written as

$$\underline{v}_i = \frac{d^* \underline{r}_i}{dt} + \underline{\Omega} \wedge \underline{r}_i \ ,$$

where the operator d^*/dt denotes differentiation in the non–inertial orbital reference frame $(\mathcal{O}, x, y, z)$. Using (5.29) we obtain

$$\frac{d^* \underline{r}_i}{dt} = \frac{(-1)^i m_{3-i} d_0}{m_1 + m_2} \left[\dot{\xi} \underline{e}_d + (1+\xi) \frac{d^* \underline{e}_d}{dt} \right] ,$$

where

$$\frac{d^* \underline{e}_d}{dt} = \begin{pmatrix} -\dot{\varphi} \sin\varphi \sin\vartheta \ + \dot{\vartheta} \cos\varphi \cos\vartheta \\ -\dot{\varphi} \cos\varphi \\ -\dot{\varphi} \sin\varphi \cos\vartheta - \dot{\vartheta} \cos\varphi \sin\vartheta \end{pmatrix} .$$

Finally, the kinetic energy can be expressed as the sum (see [41])

$$T = T_0 + T_r \ ,$$

where

$$T_0 = \frac{(m_1 + m_2) V_0^2}{2}$$

and

$$T_r = \frac{m_0 d_0^2}{2} \left\{ \dot{\xi}^2 + (1+\xi)^2 \left[\dot{\varphi}^2 + (\dot{\vartheta} + \Omega(f))^2 \cos^2\varphi \right] \right\} .$$

The Lagrangian function describing the tether's attitude dynamics[3] is given by

$$\mathcal{L} = T_r - (\Pi_r^g + \Pi^e) \ , \tag{5.31}$$

where the terms T_0 and Π_0^g have been omitted, since they do not depend on the attitude variables. It is convenient to adopt the true anomaly f as the independent variable. Let us denote by p_ϑ, p_φ, p_ξ the canonical momenta conjugated to ϑ, φ, ξ. Retaining only the lowest–order terms in (5.30), the Hamiltonian function associated to (5.31) takes the form

$$\begin{aligned} \mathcal{H}(p_\vartheta, p_\varphi, p_\xi, \vartheta, \varphi, \xi, f) &= \frac{1}{2(1+\mathrm{e}\cos f)^2 (1+\xi)^2} \left[\frac{(p_\vartheta + 1)^2}{\cos^2\varphi} + p_\varphi^2 \right. \\ &+ \left. (1+\xi)^2 p_\xi^2 \right] - p_\vartheta \\ &+ \frac{1}{2} (1+\xi)^2 (1+\mathrm{e}\cos f)(1 - 3\cos^2\vartheta \cos^2\varphi) \\ &+ \frac{(1-\mathrm{e}^2)^3}{2\beta (1+\mathrm{e}\cos f)^2} \xi^2 \ , \end{aligned} \tag{5.32}$$

where the dimensionless parameter β in (5.32) denotes the relative stiffness of the tether, namely

$$\beta = \frac{m_0 n^2}{c} \ .$$

[3] By *attitude dynamics* we intend the study of the orientation of the satellite in a given reference frame.

5.9 The dumbbell satellite

A particular case of tether dynamics is obtained whenever β goes to zero, thus yielding a dumbbell satellite composed by two points $\mathcal{P}_1$ and $\mathcal{P}_2$ with masses m_1 and m_2 connected by a non–extensible, massless rod of length d (see, e.g., [41] for full details). The center of mass $\mathcal{O}$ of $\mathcal{P}_1$ and $\mathcal{P}_2$ is assumed to move under the influence of the gravitational attraction of a body $\mathcal{P}$ with mass M. Let R be the average distance between $\mathcal{O}$ and $\mathcal{P}$, which is taken to be much larger than d. We assume that $\mathcal{O}$ moves on a Keplerian orbit with eccentricity e and semimajor axis a. Let $(\mathcal{O}, x, y, z)$ be a reference frame with the z–axis oriented along the local vertical, the y–axis perpendicular to the orbital plane and the x–axis forming an oriented frame (see Figure 5.6). Denote by $\underline{e}_x, \underline{e}_y, \underline{e}_z$ the corresponding unit vectors.

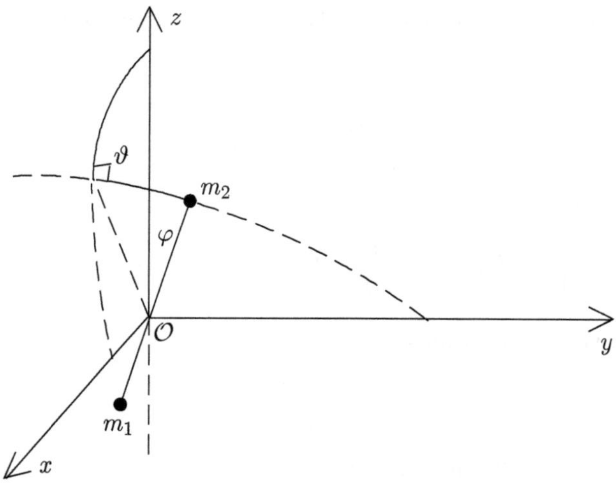

Fig. 5.6. The definition of the orbital reference frame $(\mathcal{O}, x, y, z)$.

In order to study the attitude dynamics of the dumbbell satellite, we introduce the angle φ formed by the direction of the rod with the xz–plane and the angle ϑ formed by the z–axis with the projection of the rod on the orbital plane (Figure 5.7). As usual, let f be the true anomaly of the Keplerian orbit.

The Lagrangian function describing the motion of the dumbbell satellite is the difference between the kinetic energy and the potential energy, which are defined as follows [41]. Let us write the potential energy as the sum

$$\Pi = \Pi_1 + \Pi_2 \ , \tag{5.33}$$

where Π_1 and Π_2 are the contributions due to m_1 and m_2. Set $\underline{R} = R\underline{e}_z$ equal to the radius vector $\mathcal{OP}$, $\underline{r}_i$ as the radius vector of the mass m_i with respect to $\mathcal{O}$:

$$\underline{r}_i = \frac{(-1)^i m_{3-i} d}{m_1 + m_2} \underline{e}_d \ , \qquad \underline{e}_d = \begin{pmatrix} \cos\varphi \sin\vartheta \\ -\sin\varphi \\ \cos\varphi\cos\vartheta \end{pmatrix} \ ;$$

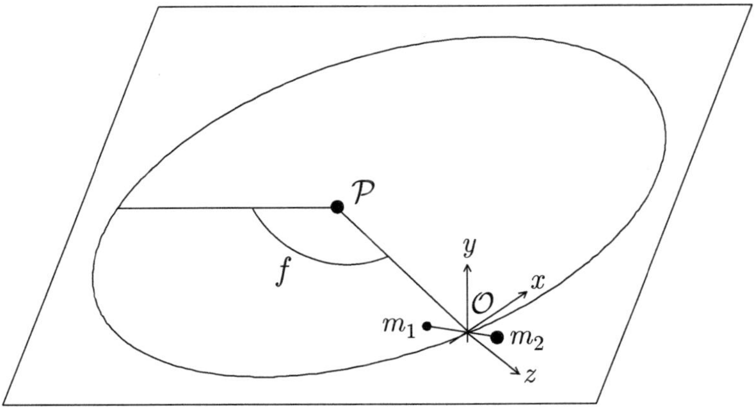

Fig. 5.7. The angles used to define the dumbbell satellite attitude position.

then, one has

$$\Pi_i = -\frac{\mathcal{G}Mm_i}{|\underline{R}+\underline{r}_i|} , \qquad i = 1, 2 ,$$

which can be expanded as

$$\Pi = -\frac{\mathcal{G}Mm_i}{R}\left\{1 - \frac{\underline{r}_i \cdot \underline{e}_z}{R} - \frac{1}{2R^2}\left[r_i^2 - 3(\underline{r}_i \cdot \underline{e}_z)^2\right] + \ldots\right\} , \qquad i = 1, 2 .$$

Denoting by m_0 the reduced mass

$$m_0 = \frac{m_1 m_2}{m_1 + m_2} ,$$

the potential energy (5.33) can also be written as

$$\Pi = \Pi_0 + \Pi_r , \tag{5.34}$$

where Π_0 is independent of the satellite's orientation, being

$$\Pi_0 = -\frac{\mathcal{G}M}{R}\left[(m_1 + m_2) - \frac{m_0 d^2}{2R^3}\right] ,$$

while Π_r is given by

$$\Pi_r = -\frac{3\mathcal{G}Mm_0d^2}{2R^3}(\underline{e}_z \cdot \underline{e}_d)^2 + \ldots , \tag{5.35}$$

where the dots in (5.35) denote higher–order terms. Concerning the kinetic energy, we can again decompose its expression as the sum of the contributions due to m_1 and m_2 (see [41]), namely

$$T = T_1 + T_2 ,$$

with

$$T_i = \frac{m_i V_i^2}{2} , \quad i = 1, 2 .$$

Here the velocity of $\mathcal{P}_i$ is expressed as

$$\underline{V}_i = \underline{V}_0 + \underline{v}_i \ ,$$

where $\underline{V}_0$ is the velocity of the dumbbell barycenter and $\underline{v}_i$ is the velocity with respect to $\mathcal{O}$. Let $\underline{\Omega}$ be the angular velocity associated to the orbital motion:

$$\underline{\Omega} = \Omega(f)\underline{e}_y \ , \qquad \Omega(f) = \frac{n(1+\mathrm{e}\cos f)^2}{(1-\mathrm{e}^2)^{3/2}} \ .$$

Then, $\underline{v}_i$ can be written as

$$\underline{v}_i = \frac{d^*\underline{r}_i}{dt} + \underline{\Omega} \wedge \underline{r}_i \ ,$$

where the operator d^*/dt denotes differentiation in the non–inertial orbital reference frame $(\mathcal{O}, x, y, z)$, namely

$$\frac{d^*\underline{r}_i}{dt} = \frac{(-1)^i m_{3-i} d}{m_1+m_2}\frac{d^*\underline{e}_d}{dt} \ , \qquad \text{where} \qquad \frac{d^*\underline{e}_d}{dt} = \begin{pmatrix} -\dot\varphi\sin\varphi\sin\vartheta \ + \dot\vartheta\cos\varphi\cos\vartheta \\ -\dot\varphi\cos\varphi \\ -\dot\varphi\sin\varphi\cos\vartheta - \dot\vartheta\cos\varphi\sin\vartheta \end{pmatrix} .$$

By analogy with (5.34) we split the kinetic energy as

$$T = T_0 + T_r \ ,$$

where T_0 and T_r are defined by

$$T_0 = \frac{(m_1+m_2)V_0^2}{2} \ , \qquad T_r = \frac{m_0 d^2}{2}\left[\dot\varphi^2 + (\dot\vartheta + \Omega(f))^2\cos^2\varphi\right] \ .$$

Finally, the attitude dynamics of the dumbbell satellite is described by the Lagrangian function

$$\mathcal{L} = T_r - \Pi_r \ , \tag{5.36}$$

where the terms T_0 and Π_0 have been omitted, since they do not depend on the attitude variables. The equations of motion associated to (5.36) are given by

$$\begin{aligned} \ddot\vartheta + \dot\Omega(f) - 2\dot\varphi(\dot\vartheta + \Omega(f))\,\mathrm{tg}\,\varphi + 3n^2\left(\frac{a}{R}\right)^3\sin\vartheta\cos\vartheta &= 0 \\ \ddot\varphi + \left[(\dot\vartheta+\Omega(f))^2 + 3n^2\left(\frac{a}{R}\right)^3\cos^2\vartheta\right]\cos\varphi\sin\varphi &= 0 \ . \end{aligned} \tag{5.37}$$

Referring to [41], an alternative formulation can be obtained by introducing the true anomaly f as the independent variable. We denote by a prime the derivative with respect to f, so that one has

$$\frac{d}{dt} = \Omega(f)\frac{d}{df} \ , \qquad \frac{d^2}{dt^2} = \Omega^2\frac{d^2}{df^2} + \Omega\Omega'\frac{d}{df} \ ;$$

the equations of motion (5.37) are transformed to

$$\begin{aligned}
\vartheta'' - 2(\vartheta'+1)\left[\varphi' \,\mathrm{tg}\,\varphi + \frac{\mathrm{e}\sin f}{1+\mathrm{e}\cos f}\right] + \frac{3\sin\vartheta\cos\vartheta}{1+\mathrm{e}\cos f} &= 0 \\
\varphi'' - \frac{2\mathrm{e}\sin f}{1+\mathrm{e}\cos f}\varphi' + \left[(\vartheta'+1)^2 + \frac{3\cos^2\vartheta}{1+\mathrm{e}\cos f}\right]\cos\varphi\sin\varphi &= 0\ , \qquad (5.38)
\end{aligned}$$

which are associated to the Hamiltonian function

$$\begin{aligned}
\mathcal{H}(p_\vartheta, p_\varphi, \vartheta, \varphi, f) &= \frac{1}{2(1+\mathrm{e}\cos f)^2}\left[\frac{(p_\vartheta+1)^2}{\cos^2\varphi} + p_\varphi^2\right] \\
&- p_\vartheta - \frac{3}{2}(1+\mathrm{e}\cos f)\cos^2\vartheta\cos^2\varphi\ .
\end{aligned}$$

If the dynamics of the dumbbell satellite takes place on the xz–plane, then $\varphi = 0$ and $p_\varphi = 0$. Therefore the equations of motion (5.38) reduce to the set of equations in the variables ϑ and p_ϑ:

$$\begin{aligned}
\vartheta' &= \frac{p_\vartheta + 1}{(1+\mathrm{e}\cos f)^2} - 1 \\
p'_\vartheta &= -3(1+\mathrm{e}\cos f)\cos\vartheta\sin\vartheta\ . \qquad (5.39)
\end{aligned}$$

We remark that the equations (5.39) can be written as the second–order differential equation

$$(1+\mathrm{e}\cos f)\vartheta'' - 2\vartheta'\mathrm{e}\sin f + 3\sin\vartheta\cos\vartheta = 2\mathrm{e}\sin f\ , \qquad (5.40)$$

which is known as the *Beletsky equation* [10]. For a non–zero eccentricity, the dynamics associated to (5.40) is non–integrable [22], eventually giving place to chaotic motions. We refer to [41] for a discussion of the stability properties of dumbbell satellite dynamics.

6 Perturbation theory

Perturbation theory is an efficient tool for investigating the dynamics of nearly–integrable Hamiltonian systems. The restricted three–body problem is the prototype of a nearly–integrable mechanical system (Section 6.1); the integrable part is given by the two–body approximation, while the perturbation is due to the gravitational influence of the other primary. A typical example is represented by the motion of an asteroid under the gravitational attraction of the Sun and Jupiter. The mass of the asteroid is so small, that one can assume that the primaries move on Keplerian orbits. The dynamics of the asteroid is essentially driven by the Sun and it is perturbed by Jupiter, where the Jupiter–Sun mass–ratio is observed to be about 10^{-3}. The solution of the restricted three–body problem can be investigated through perturbation theories, which were developed in the 18th and 19th centuries; they are used nowadays in many contexts of Celestial Mechanics, from ephemeris computations to astrodynamics.
Perturbation theory in Celestial Mechanics is based on the implementation of a canonical transformation, which allows us to find the solution of a nearly–integrable system within a better degree of approximation [66]. We review classical perturbation theory (Section 6.2), as well as in the presence of a resonance relation (Section 6.3) and in the context of degenerate systems (Section 6.4). We discuss also the Birkhoff normal form (Section 6.5) around equilibrium positions and around closed trajectories; we conclude with some results concerning the averaging theorem (Section 6.6).

6.1 Nearly–integrable Hamiltonian systems

Let us consider an n–dimensional Hamiltonian system described in terms of a set of conjugated action–angle variables $(\underline{I},\underline{\varphi})$ with $\underline{I}\in V$, V being an open set of $\mathbf{R}^n$, and $\underline{\varphi}\in\mathbf{T}^n$. A nearly–integrable Hamiltonian function $\mathcal{H}(\underline{I},\underline{\varphi})$ can be written in the form

$$\mathcal{H}(\underline{I},\underline{\varphi}) = h(\underline{I}) + \varepsilon f(\underline{I},\underline{\varphi}) \ , \tag{6.1}$$

where h and f are analytic functions called, respectively, the unperturbed (or integrable) Hamiltonian and the perturbing function, while ε is a small parameter measuring the strength of the perturbation. Indeed, for $\varepsilon = 0$ the Hamiltonian function reduces to

$$\mathcal{H}(\underline{I},\underline{\varphi}) = h(\underline{I}) \ .$$

The associated Hamilton's equations are simply

$$\begin{aligned}\dot{\underline{I}} &= \underline{0}\\ \dot{\underline{\varphi}} &= \underline{\omega}(\underline{I}) \ ,\end{aligned} \tag{6.2}$$

where we have introduced the *frequency vector*

$$\underline{\omega}(\underline{I}) \equiv \frac{\partial h(\underline{I})}{\partial \underline{I}} \ .$$

Equations (6.2) can be trivially integrated as

$$\begin{aligned}\underline{I}(t) &= \underline{I}(0)\\ \underline{\varphi}(t) &= \underline{\omega}(\underline{I}(0))t + \underline{\varphi}(0) \ ,\end{aligned}$$

thus showing that the actions are constants, while the angle variables vary linearly with the time. For $\varepsilon \neq 0$ the equations of motion

$$\begin{aligned}\dot{\underline{I}} &= -\varepsilon \frac{\partial f}{\partial \underline{\varphi}}(\underline{I}, \underline{\varphi})\\ \dot{\underline{\varphi}} &= \underline{\omega}(\underline{I}) + \varepsilon \frac{\partial f}{\partial \underline{I}}(\underline{I}, \underline{\varphi})\end{aligned}$$

might no longer be integrable and chaotic motions could appear.

6.2 Classical perturbation theory

The aim of *classical perturbation theory* is to construct a canonical transformation, which allows us to push the perturbation to higher orders in the perturbing parameter. With reference to the Hamiltonian (6.1), we introduce a canonical change of variables $\mathcal{C} : (\underline{I}, \underline{\varphi}) \to (\underline{I}', \underline{\varphi}')$, such that (6.1) in the transformed variables takes the form

$$\mathcal{H}'(\underline{I}', \underline{\varphi}') = \mathcal{H} \circ \mathcal{C}(\underline{I}, \underline{\varphi}) \equiv h'(\underline{I}') + \varepsilon^2 f'(\underline{I}', \underline{\varphi}') \ , \tag{6.3}$$

where h' and f' denote, respectively, the new unperturbed Hamiltonian and the new perturbing function. The result is obtained through the following steps: define a suitable canonical transformation close to the identity, perform a Taylor series expansion in the perturbing parameter, require that the change of variables removes the dependence on the angles up to second–order terms; finally an expansion in Fourier series allows us to construct the explicit form of the canonical transformation. Let us describe in detail this procedure.

Define a change of variables through a close–to–identity generating function of the form $\underline{I}' \cdot \underline{\varphi} + \varepsilon \Phi(\underline{I}', \underline{\varphi})$ providing

$$\begin{aligned}\underline{I} &= \underline{I}' + \varepsilon \frac{\partial \Phi(\underline{I}', \underline{\varphi})}{\partial \underline{\varphi}}\\ \underline{\varphi}' &= \underline{\varphi} + \varepsilon \frac{\partial \Phi(\underline{I}', \underline{\varphi})}{\partial \underline{I}'} \ ,\end{aligned} \tag{6.4}$$

where $\Phi = \Phi(\underline{I}', \underline{\varphi})$ is an unknown function, which is determined in order that (6.1) be transformed to (6.3). Let us split the perturbing function as

$$f(\underline{I}, \underline{\varphi}) = \overline{f}(\underline{I}) + \tilde{f}(\underline{I}, \underline{\varphi}) ,$$

where $\overline{f}(\underline{I})$ is the average over the angle variables and $\tilde{f}(\underline{I}, \underline{\varphi})$ is the remainder function defined as $\tilde{f}(\underline{I}, \underline{\varphi}) \equiv f(\underline{I}, \underline{\varphi}) - \overline{f}(\underline{I})$. Inserting (6.4) into (6.1) and expanding in Taylor series around $\varepsilon = 0$ up to the second order, one gets

$$\begin{aligned} h\left(\underline{I}' + \varepsilon \frac{\partial \Phi(\underline{I}', \underline{\varphi})}{\partial \underline{\varphi}}\right) + \varepsilon f\left(\underline{I}' + \varepsilon \frac{\partial \Phi(\underline{I}', \underline{\varphi})}{\partial \underline{\varphi}}, \underline{\varphi}\right) \\ = h(\underline{I}') + \underline{\omega}(\underline{I}') \cdot \varepsilon \frac{\partial \Phi(\underline{I}', \underline{\varphi})}{\partial \underline{\varphi}} + \varepsilon \overline{f}(\underline{I}') + \varepsilon \tilde{f}(\underline{I}', \underline{\varphi}) + O(\varepsilon^2) . \end{aligned}$$

The transformed Hamiltonian is integrable up to the second order in ε provided that the function Φ satisfies:

$$\underline{\omega}(\underline{I}') \cdot \frac{\partial \Phi(\underline{I}', \underline{\varphi})}{\partial \underline{\varphi}} + \tilde{f}(\underline{I}', \underline{\varphi}) = \underline{0} . \tag{6.5}$$

The new unperturbed Hamiltonian becomes

$$h'(\underline{I}') = h(\underline{I}') + \varepsilon \overline{f}(\underline{I}') ,$$

which provides a better integrable approximation with respect to that associated to (6.1). An explicit expression of the generating function is obtained solving (6.5). To this end, let us expand Φ and $\tilde{f}$ in Fourier series as

$$\begin{aligned} \Phi(\underline{I}', \underline{\varphi}) &= \sum_{\underline{m} \in \mathbf{Z}^n \setminus \{\underline{0}\}} \hat{\Phi}_{\underline{m}}(\underline{I}') \, e^{i\underline{m} \cdot \underline{\varphi}} , \\ \tilde{f}(\underline{I}', \underline{\varphi}) &= \sum_{\underline{m} \in \mathcal{I}} \hat{f}_{\underline{m}}(\underline{I}') \, e^{i\underline{m} \cdot \underline{\varphi}} , \end{aligned} \tag{6.6}$$

where $\mathcal{I}$ denotes a suitable set of integer vectors defining the Fourier indexes of $\tilde{f}$. Inserting (6.6) in (6.5) one obtains

$$i \sum_{\underline{m} \in \mathbf{Z}^n \setminus \{\underline{0}\}} \underline{\omega}(\underline{I}') \cdot \underline{m} \, \hat{\Phi}_{\underline{m}}(\underline{I}') \, e^{i\underline{m} \cdot \underline{\varphi}} = - \sum_{\underline{m} \in \mathcal{I}} \hat{f}_{\underline{m}}(\underline{I}') \, e^{i\underline{m} \cdot \underline{\varphi}} ,$$

which provides

$$\hat{\Phi}_{\underline{m}}(\underline{I}') = - \frac{\hat{f}_{\underline{m}}(\underline{I}')}{i \, \underline{\omega}(\underline{I}') \cdot \underline{m}} . \tag{6.7}$$

Using (6.6) and (6.7), the generating function is given by

$$\Phi(\underline{I}', \underline{\varphi}) = i \sum_{\underline{m} \in \mathcal{I}} \frac{\hat{f}_{\underline{m}}(\underline{I}')}{\underline{\omega}(\underline{I}') \cdot \underline{m}} \, e^{i\underline{m} \cdot \underline{\varphi}} . \tag{6.8}$$

The algorithm described above is constructive in the sense that it provides an explicit expression for the generating function and for the transformed Hamiltonian. We stress that (6.8) is well defined unless there exists an integer vector $\underline{m} \in \mathcal{I}$ such that

$$\underline{\omega}(\underline{I}') \cdot \underline{m} = 0 \ . \tag{6.9}$$

On the other hand if, for a given value of the actions, $\underline{\omega} = \underline{\omega}(\underline{I})$ is rationally independent (which means that (6.9) is satisfied only for $\underline{m} = \underline{0}$), then there do not appear zero divisors in (6.8), though the divisors can become arbitrarily small with a proper choice of the vector $\underline{m}$. For this reason, terms of the form $\underline{\omega}(\underline{I}') \cdot \underline{m}$ are called *small divisors* and they can prevent the implementation of perturbation theory.

6.2.1 An example

We apply classical perturbation theory to the two–dimensional Hamiltonian function

$$\mathcal{H}(I_1, I_2, \varphi_1, \varphi_2) = \frac{I_1^2}{2} + \frac{I_2^2}{2} + \varepsilon \Big[\cos(\varphi_1 + \varphi_2) + 2\cos(\varphi_1 - \varphi_2) \Big] \ ,$$

which can be shortly written as

$$\mathcal{H}(I_1, I_2, \varphi_1, \varphi_2) = h(I_1, I_2) + \varepsilon f(\varphi_1, \varphi_2), \tag{6.10}$$

where

$$h(I_1, I_2) = \frac{I_1^2}{2} + \frac{I_2^2}{2}$$

and

$$f(\varphi_1, \varphi_2) = \cos(\varphi_1 + \varphi_2) + 2\cos(\varphi_1 - \varphi_2).$$

Let us perform the change of coordinates

$$\begin{aligned}
I_1 &= I_1' + \varepsilon \frac{\partial \Phi}{\partial \varphi_1}(I_1', I_2', \varphi_1, \varphi_2) \\
I_2 &= I_2' + \varepsilon \frac{\partial \Phi}{\partial \varphi_2}(I_1', I_2', \varphi_1, \varphi_2) \\
\varphi_1' &= \varphi_1 + \varepsilon \frac{\partial \Phi}{\partial I_1'}(I_1', I_2', \varphi_1, \varphi_2) \\
\varphi_2' &= \varphi_2 + \varepsilon \frac{\partial \Phi}{\partial I_2'}(I_1', I_2', \varphi_1, \varphi_2) \ .
\end{aligned}$$

Expanding the Hamiltonian (6.10) in Taylor series up to the second order, one obtains:

$$\begin{aligned}
& h\left(I_1' + \varepsilon \frac{\partial \Phi}{\partial \varphi_1}, I_2' + \varepsilon \frac{\partial \Phi}{\partial \varphi_2}\right) + \varepsilon f(\varphi_1, \varphi_2) \\
&= h(I_1', I_2') + \varepsilon \frac{\partial h}{\partial I_1}(I_1', I_2') \frac{\partial \Phi}{\partial \varphi_1} + \varepsilon \frac{\partial h}{\partial I_2}(I_1', I_2') \frac{\partial \Phi}{\partial \varphi_2} + \varepsilon f(\varphi_1, \varphi_2) + O(\varepsilon^2) \ ,
\end{aligned}$$

where

$$\frac{\partial h}{\partial I_1}(I_1',I_2') = I_1' \equiv \omega_1 \ , \qquad \frac{\partial h}{\partial I_2}(I_1',I_2') = I_2' \equiv \omega_2 .$$

The first–order terms in ε must be zero; this yields the generating function as the solution of the equation

$$\omega_1 \frac{\partial \Phi}{\partial \varphi_1} + \omega_2 \frac{\partial \Phi}{\partial \varphi_2} = -f(\varphi_1,\varphi_2) \ .$$

Expanding in Fourier series and taking into account the explicit form of the perturbation, one obtains

$$\sum_{m,n} i(\omega_1 m + \omega_2 n)\Phi_{m,n}(I_1',I_2')e^{i(m\varphi_1+n\varphi_2)} = -\Big[\cos(\varphi_1+\varphi_2) + 2\cos(\varphi_1-\varphi_2)\Big] \ .$$

Using the relations $\cos(k_1\varphi_1+k_2\varphi_2) = \frac{1}{2}(e^{i(k_1\varphi_1+k_2\varphi_2)} + e^{-i(k_1\varphi_1+k_2\varphi_2)})$ for some integers k_1, k_2, and equating the coefficients with the same Fourier indexes, one gets:

$$\Phi_{1,1} = -\frac{1}{2i(\omega_1+\omega_2)} \ , \qquad \Phi_{-1,-1} = \frac{1}{2i(\omega_1+\omega_2)} \ ,$$
$$\Phi_{1,-1} = -\frac{1}{i(\omega_1-\omega_2)} \ , \qquad \Phi_{-1,1} = -\frac{1}{i(-\omega_1+\omega_2)} \ .$$

Casting together the above terms, the generating function is given by

$$\begin{aligned}\Phi(I_1',I_2',\varphi_1,\varphi_2) &= -\frac{1}{\omega_1+\omega_2}\left(\frac{e^{i(\varphi_1+\varphi_2)} - e^{-i(\varphi_1+\varphi_2)}}{2i}\right) \\ &- \frac{2}{\omega_1-\omega_2}\left(\frac{e^{i(\varphi_1-\varphi_2)} - e^{-i(\varphi_1-\varphi_2)}}{2i}\right) \ ,\end{aligned}$$

namely

$$\Phi(I_1',I_2',\varphi_1,\varphi_2) = -\frac{1}{\omega_1+\omega_2}\sin(\varphi_1+\varphi_2) - \frac{2}{\omega_1-\omega_2}\sin(\varphi_1-\varphi_2) \ .$$

Notice that the generating function is not defined when there appear the following zero divisors:

$$\omega_1 \pm \omega_2 = 0 \ , \qquad \text{namely} \qquad I_1' = \pm I_2' \ .$$

The new unperturbed Hamiltonian coincides with the old unperturbed Hamiltonian (expressed in the new set of variables), since the average of the perturbing function is zero:

$$h'(I_1',I_2') = \frac{I_1'^2}{2} + \frac{I_2'^2}{2} \ .$$

6.2.2 Computation of the precession of the perihelion

A straightforward application of classical perturbation theory allows us to compute the amount of the precession of the perihelion. A first–order computation is obtained starting with the restricted, planar, circular three–body model. In particular, we identify the three bodies $\mathcal{P}_0$, $\mathcal{P}_1$ and $\mathcal{P}_2$ with the Sun, Mercury and Jupiter. In terms of the Delaunay action–angle variables, the perturbing function can be expanded as in (4.7). The perturbing parameter ε represents the Jupiter–Sun mass ratio. We implement a first–order perturbation theory, which provides a new integrable Hamiltonian function of the form

$$h'(L',G') = -\frac{1}{2L'^2} - G' + \varepsilon\, R_{00}(L',G')\ ,$$

where $R_{00}(L,G) = -\frac{L^4}{4}(1+\frac{9}{16}L^4+\frac{3}{2}\mathrm{e}^2)+O(\mathrm{e}^3)$. Hamilton's equations yield

$$\dot{g}' = \frac{\partial h'(L',G')}{\partial G'} = -1+\varepsilon\frac{\partial R_{00}(L',G')}{\partial G'}\ .$$

Recall that $g = g_0 - t$, being g_0 the argument of the perihelion. Neglecting terms of order $O(\mathrm{e}^3)$ in R_{00}, one gets that to the lowest order the argument of perihelion g_0 varies as

$$\dot{g}_0 = \varepsilon\ \frac{\partial R_{00}(L',G')}{\partial G'} = \frac{3}{4}\varepsilon L'^2 G'\ .$$

Notice that up to the first order in ε one has $L' = L$, $G' = G$. Taking $\varepsilon = 9.54\cdot 10^{-4}$ (the actual value of the Jupiter–Sun mass ratio), $a = 0.0744$ (setting to one the Jupiter–Sun distance) and $\mathrm{e} = 0.2056$, one obtains

$$\dot{g}_0 = 155.25\ \frac{\text{arcsecond}}{\text{century}}\ ,$$

which represents the contribution due to Jupiter to the precession of the perihelion of Mercury. A more refined value is obtained taking into account higher–order terms in the eccentricity.

6.3 Resonant perturbation theory

Consider the following Hamiltonian system with n degrees of freedom

$$\mathcal{H}(\underline{I},\underline{\varphi}) = h(\underline{I}) + \varepsilon f(\underline{I},\underline{\varphi})\ , \qquad \underline{I}\in\mathbf{R}^n\ ,\ \ \underline{\varphi}\in\mathbf{T}^n$$

and let $\underline{\omega}(\underline{I}) = \frac{\partial h(\underline{I})}{\partial \underline{I}}$ be the frequency vector of the motion. We assume that the frequencies satisfy ℓ resonance relations, with $\ell < n$, of the form

$$\underline{\omega}\cdot\underline{m}_k = 0 \qquad \text{for } k = 1,\dots,\ell\ ,$$

for some vectors $\underline{m}_1$, ..., $\underline{m}_\ell \in \mathbf{Z}^n$. A *resonant perturbation theory* can be implemented to eliminate the non–resonant terms. More precisely, the aim is to construct a change of variables $\mathcal{C} : (\underline{I}, \underline{\varphi}) \to (\underline{I}', \underline{\varphi}')$ such that the new Hamiltonian takes the form

$$\mathcal{H}'(\underline{I}', \underline{\varphi}') = h'(\underline{I}', \underline{m}_1 \cdot \underline{\varphi}', \ldots, \underline{m}_\ell \cdot \underline{\varphi}') + \varepsilon^2 f'(\underline{I}', \underline{\varphi}') , \tag{6.11}$$

where h' depends on $\underline{\varphi}'$ only through the combinations $\underline{m}_k \cdot \underline{\varphi}'$ with $k = 1, \ldots, \ell$. To this end, let us first define the angles

$$\begin{aligned} \vartheta_j &= \underline{m}_j \cdot \underline{\varphi} , \qquad & j = 1, \ldots, \ell \\ \vartheta_{j'} &= \underline{m}_{j'} \cdot \underline{\varphi} , \qquad & j' = \ell + 1, \ldots, n , \end{aligned}$$

where the first ℓ angle variables are the resonant angles, while the latter $n - \ell$ angles are defined as arbitrary linear combinations with integer coefficients $\underline{m}_{j'}$. The corresponding actions are defined as

$$\begin{aligned} I_j &= \underline{m}_j \cdot \underline{J} , \qquad & j = 1, \ldots, \ell \\ I_{j'} &= \underline{m}_{j'} \cdot \underline{J} , \qquad & j' = \ell + 1, \ldots, n . \end{aligned}$$

Next we construct a canonical transformation which removes (to higher orders) the dependence on the short–period angles $(\vartheta_{\ell+1}, \ldots, \vartheta_n)$, while the lowest–order Hamiltonian will necessarily depend upon the resonant angles. To this end, let us first decompose the perturbation, expressed in terms of the variables $(\underline{J}, \underline{\vartheta})$, as

$$f(\underline{J}, \underline{\vartheta}) = \overline{f}(\underline{J}) + f_r(\underline{J}, \vartheta_1, \ldots, \vartheta_\ell) + f_n(\underline{J}, \underline{\vartheta}) , \tag{6.12}$$

where $\overline{f}(\underline{J})$ is the average of the perturbation over the angles, $f_r(\underline{J}, \vartheta_1, \ldots, \vartheta_\ell)$ is the part depending on the resonant angles and $f_n(\underline{J}, \underline{\vartheta})$ is the non–resonant part. By analogy with classical perturbation theory, we implement a canonical transformation of the form (6.4), such that the new Hamiltonian takes the form (6.11). Using (6.12) and expanding up to the second order in the perturbing parameter, one obtains:

$$\begin{aligned} h\left(\underline{J}' + \varepsilon \frac{\partial \Phi}{\partial \underline{\vartheta}}\right) + \varepsilon f(\underline{J}', \underline{\vartheta}) + O(\varepsilon^2) &= h(\underline{J}') + \varepsilon \sum_{k=1}^{n} \frac{\partial h}{\partial J'_k} \frac{\partial \Phi}{\partial \vartheta_k} \\ &+ \varepsilon \overline{f}(\underline{J}') + \varepsilon f_r(\underline{J}', \vartheta_1, \ldots, \vartheta_\ell) + \varepsilon f_n(\underline{J}', \underline{\vartheta}) + O(\varepsilon^2) . \end{aligned}$$

Recalling (6.11) and equating terms of the same orders is ε, one gets that

$$h'(\underline{J}', \vartheta_1, \ldots, \vartheta_\ell) = h(\underline{J}') + \varepsilon \overline{f}(\underline{J}') + \varepsilon f_r(\underline{J}', \vartheta_1, \ldots, \vartheta_\ell) , \tag{6.13}$$

provided

$$\sum_{k=1}^{n} \omega'_k \frac{\partial \Phi}{\partial \vartheta_k} = -f_n(\underline{J}', \underline{\vartheta}) , \tag{6.14}$$

where $\omega'_k = \omega'_k(\underline{J}') \equiv \frac{\partial h(\underline{J}')}{\partial J_k}$. The solution of (6.14) provides the generating function allowing us to reduce the Hamiltonian to the required form (6.11); moreover, the conjugated action variables, say $J'_{\ell+1}$, ..., J'_n, are constants of the motion up to the second order in ε. We remark that using the new frequencies ω'_k, the resonant relations take the form $\omega'_k = 0$ for $k = 1, \ldots, \ell$.

6.3.1 Three–body resonance

As an example of the application of resonant perturbation theory we consider the three–body Hamiltonian (4.2) with the perturbing function expanded as in (4.7). Let $\underline{\omega} \equiv (\omega_\ell, \omega_g)$ be the frequency of motion; we assume that the following resonance relation is satisfied:

$$\omega_\ell + 2\omega_g = 0 \ .$$

Next, we perform the canonical change of variables

$$\begin{aligned} \vartheta_1 &= \ell + 2g \ , \qquad & J_1 &= \frac{1}{2}G \ , \\ \vartheta_2 &= 2\ell \ , \qquad & J_2 &= \frac{1}{2}L - \frac{1}{4}G \ . \end{aligned}$$

In the new coordinates the unperturbed Hamiltonian takes the form

$$h(\underline{J}) \equiv -\frac{\mu^2}{2(J_1+2J_2)^2} - 2J_1 \ ,$$

while the perturbing function is given by

$$\begin{aligned} R(J_1,J_2,\vartheta_1,\vartheta_2) \equiv\ & R_{00}(\underline{J}) + R_{10}(\underline{J})\cos\left(\frac{1}{2}\vartheta_2\right) + R_{11}(\underline{J})\cos\left(\frac{1}{2}\vartheta_1 + \frac{1}{4}\vartheta_2\right) \\ &+ R_{12}(\underline{J})\cos(\vartheta_1) + R_{22}(\underline{J})\cos\left(\vartheta_1 + \frac{1}{2}\vartheta_2\right) \\ &+ R_{32}(\underline{J})\cos(\vartheta_1+\vartheta_2) + R_{33}(\underline{J})\cos\left(\frac{3}{2}\vartheta_1 + \frac{3}{4}\vartheta_2\right) \\ &+ R_{44}(\underline{J})\cos(2\vartheta_1+\vartheta_2) + R_{55}(\underline{J})\cos\left(\frac{5}{2}\vartheta_1 + \frac{5}{4}\vartheta_2\right) + \ldots \end{aligned}$$

with the coefficients R_{ij} as in (4.8). Let us split the perturbation as $R = \overline{R}(\underline{J}) + R_r(\underline{J},\vartheta_1) + R_n(\underline{J},\underline{\vartheta})$, where $\overline{R}(\underline{J})$ is the average over the angles, $R_r(\underline{J},\vartheta_1) = R_{12}(\underline{J})\cos(\vartheta_1)$ is the resonant part, while R_n contains all the remaining non–resonant terms. We look for a change of coordinates close to the identity with generating function $\Phi = \Phi(\underline{J}',\underline{\vartheta})$ such that

$$\underline{\omega}'(\underline{J}') \cdot \frac{\partial \Phi(\underline{J}',\underline{\vartheta})}{\partial \underline{\vartheta}} = -R_n(\underline{J}',\underline{\vartheta}) \ ,$$

being $\underline{\omega}'(\underline{J}') \equiv \frac{\partial h(\underline{J}')}{\partial \underline{J}}$. The above expression is well defined since $\underline{\omega}'$ is non–resonant for the Fourier components appearing in R_n. Finally, according to (6.13) the new unperturbed Hamiltonian is given by

$$h'(\underline{J}',\vartheta_1) \equiv h(\underline{J}') + \varepsilon R_{00}(\underline{J}') + \varepsilon R_{12}(\underline{J}')\cos(\vartheta_1) \ .$$

6.4 Degenerate perturbation theory

Consider the Hamiltonian function with n degrees of freedom

$$\mathcal{H}(\underline{I},\underline{\varphi}) = h(I_1,\dots,I_d) + \varepsilon f(\underline{I},\underline{\varphi}) \ , \qquad d<n \ , \tag{6.15}$$

where the unperturbed Hamiltonian depends on a subset of the action variables, being degenerate in $I_{d+1},\dots,I_n$. As in the resonant perturbation theory, we look for a canonical transformation $\mathcal{C} : (\underline{I},\underline{\varphi}) \to (\underline{I}',\underline{\varphi}')$ such that the new Hamiltonian becomes

$$\mathcal{H}'(\underline{I}',\underline{\varphi}') = h'(\underline{I}') + \varepsilon h_1'(\underline{I}',\varphi_{d+1}',\dots,\varphi_n') + \varepsilon^2 f'(\underline{I}',\underline{\varphi}') \ , \tag{6.16}$$

where the term $h' + \varepsilon h_1'$ admits d integrals of motion. Let us split the perturbing function in (6.15) as

$$f(\underline{I},\underline{\varphi}) = \overline{f}(\underline{I}) + f_d(\underline{I},\varphi_{d+1},..,\varphi_n) + f_n(\underline{I},\underline{\varphi}) \ , \tag{6.17}$$

where $\overline{f}$ is the average over the angle variables, f_d is independent of φ_1, ..., φ_d and f_n is the remainder, namely $f_n = f - \overline{f} - f_d$. We want to determine a near–to–identity change of variables of the form (6.4), such that in view of (6.17) the Hamiltonian (6.15) is transformed into (6.16), namely

$$\begin{aligned} & h\Big(I_1' + \varepsilon\frac{\partial\Phi}{\partial\varphi_1},\dots,I_d' + \varepsilon\frac{\partial\Phi}{\partial\varphi_d}\Big) + \varepsilon f\Big(\underline{I}' + \varepsilon\frac{\partial\Phi}{\partial\underline{\varphi}},\underline{\varphi}\Big) \\ &= h(I_1',\dots,I_d') + \varepsilon\sum_{k=1}^d \frac{\partial h}{\partial I_k}\frac{\partial\Phi}{\partial\varphi_k} + \varepsilon\overline{f}(\underline{I}') + \varepsilon f_d(\underline{I}',\varphi_{d+1},\dots,\varphi_n) \\ &\quad + \varepsilon f_n(\underline{I}',\underline{\varphi}) + O(\varepsilon^2) \\ &\equiv h'(\underline{I}') + \varepsilon h_1'(\underline{I}',\varphi_{d+1},\dots,\varphi_n) + O(\varepsilon^2) \ , \end{aligned}$$

where

$$\begin{aligned} h'(\underline{I}') &\equiv h(I_1',\dots,I_d') + \varepsilon\overline{f}(\underline{I}') \\ h_1'(\underline{I}',\varphi_{d+1},\dots,\varphi_n) &\equiv f_d(\underline{I}',\varphi_{d+1},\dots,\varphi_n) \ , \end{aligned}$$

provided Φ is determined so that

$$\sum_{k=1}^d \frac{\partial h}{\partial I_k}\frac{\partial\Phi}{\partial\varphi_k} + f_n(\underline{I}',\underline{\varphi}) = 0 \ . \tag{6.18}$$

As in the previous sections, let us expand Φ and f_n in Fourier series as

$$\begin{aligned} \Phi(\underline{I}',\underline{\varphi}) &= \sum_{\underline{m}\in\mathbf{Z}^n\backslash\{\underline{0}\}} \hat{\Phi}_{\underline{m}}(\underline{I}')\, e^{i\underline{m}\cdot\underline{\varphi}} \\ f_n(\underline{I}',\underline{\varphi}) &= \sum_{\underline{m}\in\mathcal{I}_n} \hat{f}_{n,\underline{m}}(\underline{I}')\, e^{i\underline{m}\cdot\underline{\varphi}} \ , \end{aligned}$$

where $\mathcal{I}_n$ denotes a suitable set of integer vectors defining the Fourier indexes of f_n. From (6.18) and setting $\omega_k \equiv \frac{\partial h}{\partial I_k}$, one obtains

$$i \sum_{\underline{m} \in \mathbf{Z}^n \setminus \{\underline{0}\}} \hat{\Phi}_{\underline{m}}(\underline{I}') \sum_{k=1}^{d} m_k \omega_k \, e^{i\underline{m} \cdot \underline{\varphi}} = - \sum_{\underline{m} \in \mathcal{I}_n} \hat{f}_{n,\underline{m}}(\underline{I}') \, e^{i\underline{m} \cdot \underline{\varphi}} . \tag{6.19}$$

Due to the fact that $\omega_k = 0$ for $k = d+1, \ldots, n$, we obtain that

$$\underline{\omega} \cdot \underline{m} = \sum_{k=1}^{d} m_k \omega_k . \tag{6.20}$$

Equation (6.19) yields that the generating function takes the form

$$\Phi(\underline{I}', \underline{\varphi}) = i \sum_{\underline{m} \in \mathcal{I}_n} \frac{\hat{f}_{n,\underline{m}}(\underline{I}')}{\underline{\omega} \cdot \underline{m}} \, e^{i\underline{m} \cdot \underline{\varphi}} .$$

The generating function is well defined provided that $\underline{\omega} \cdot \underline{m} \neq 0$ for any $\underline{m} \in \mathcal{I}_n$, which in view of (6.20) is equivalent to requiring that

$$\sum_{k=1}^{d} m_k \omega_k \neq 0 \qquad \text{for } \underline{m} \in \mathcal{I}_n .$$

6.4.1 The precession of the equinoxes

An application of the degenerate perturbation theory to Celestial Mechanics is offered by the computation of the precession of the equinoxes, namely the constant retrograde precession of the spin–axis provoked by gravitational interactions. In particular, we compute the Earth's equinox precession due to the influence of the Sun and of the Moon. Assume that the Earth $\mathcal{E}$ is an oblate rigid body moving around the (point–mass) Sun $\mathcal{S}$ on a Keplerian orbit with semimajor axis a and eccentricity e; recalling (5.8) and (5.10), in the gyroscopic case $I_1 = I_2$ the Hamiltonian describing the motion of $\mathcal{E}$ around $\mathcal{S}$ is given by

$$\mathcal{H}(L, G, H, \ell, g, h, t) = \frac{G^2}{2I_1} + \frac{I_1 - I_3}{2I_1 I_3} L^2 + \tilde{V}(L, G, H, \ell, g, h, t) ,$$

where I_1, I_2, I_3 are the principal moments of inertia and where the perturbation is implicitly defined by

$$\tilde{V} \equiv - \int_{\mathcal{E}} \frac{\mathcal{G} m_{\mathcal{S}} m_{\mathcal{E}}}{|\underline{r}_{\mathcal{E}} + \underline{x}|} \frac{d\underline{x}}{|\mathcal{E}|} ,$$

being $\underline{r}_{\mathcal{E}}$ the orbital radius of the Earth and $|\mathcal{E}|$ the volume of $\mathcal{E}$. Setting $r_{\mathcal{E}} = |\underline{r}_{\mathcal{E}}|$ and $x = |\underline{x}|$, we can expand $\tilde{V}$ using the Legendre polynomials as

$$\tilde{V} = - \frac{\mathcal{G} m_{\mathcal{S}} m_{\mathcal{E}}}{r_{\mathcal{E}}} \int_{\mathcal{E}} \frac{d\underline{x}}{|\mathcal{E}|} \left[1 - \frac{\underline{x} \cdot \underline{r}_{\mathcal{E}}}{r_{\mathcal{E}}^2} + \frac{1}{2 r_{\mathcal{E}}^2} \left(3 \frac{(\underline{x} \cdot \underline{r}_{\mathcal{E}})^2}{r_{\mathcal{E}}^2} - x^2 \right) \right] + O\left[\left(\frac{\underline{x}}{\underline{r}_{\mathcal{E}}} \right)^3 \right] .$$

Assume that the Earth rotates around a principal axis, namely that the non–principal rotation angle J is zero or, equivalently, that the actions G and L are equal. Let $\bar{G}$ and $\bar{H}$ be the initial values of G and H at $t=0$ and let α denotes the angle between $\underline{r}_{\mathcal{E}}$ and $\underline{k}$ ($\underline{k}$ being the vertical axis of the body frame). Retaining only the second order of the development of the perturbing function in terms of the Legendre polynomials, one obtains

$$\tilde{V} = \varepsilon\bar{\omega}\frac{\bar{G}^2}{\bar{H}}\ \frac{(1-\mathrm{e}\cos\lambda_{\mathcal{E}})^3}{(1-\mathrm{e}^2)^3}\ \cos^2\alpha$$

with $\varepsilon = \frac{3}{2}\frac{I_3-I_1}{I_3}$, $\bar{\omega} = \frac{\mathcal{G}m_{\mathcal{S}}}{a^3}\ I_3\ \frac{\bar{H}}{\bar{G}^2}$ and where $\lambda_{\mathcal{E}}$ is the longitude of the Earth. Elementary computations show that

$$\cos\alpha = \sin(\lambda_{\mathcal{E}} - h)\ \sqrt{1-\frac{H^2}{G^2}}\ .$$

Neglecting first–order terms in the orbital eccentricity, we have that $\frac{(1-\mathrm{e}\cos\lambda_{\mathcal{E}})^3}{(1-\mathrm{e}^2)^3} \simeq 1$. A first–order degenerate perturbation theory provides the new unperturbed Hamiltonian in the form (we omit the primes to denote new variables):

$$\mathcal{H}_1(G,H) = \frac{G^2}{2I_3} + \varepsilon\bar{\omega}\ \frac{\bar{G}^2}{\bar{H}}\frac{G^2-H^2}{2G^2}\ .$$

Finally, the average angular velocity of precession is given by

$$\dot{h} = \frac{\partial\mathcal{H}_1(G,H)}{\partial H} = -\varepsilon\bar{\omega}\ \frac{\bar{G}^2}{\bar{H}}\frac{H}{G^2}\ .$$

At $t=0$ it is

$$\dot{h} = -\varepsilon\bar{\omega} = -\varepsilon\omega_y^2\omega_d^{-1}\cos K\ , \tag{6.21}$$

where we used $\bar{\omega} = \omega_y^2\omega_d^{-1}\cos K$ with ω_y being the frequency of revolution and ω_d the frequency of rotation, while K denotes the obliquity.

Astronomical measurements show that $\frac{I_3-I_1}{I_3} \simeq \frac{1}{298.25}$, $K \simeq 23.45^o$. The contribution $\dot{h}^{(S)}$ due to the Sun is thus obtained inserting $\omega_y = 1$ year, $\omega_d = 1$ day in (6.21), yielding $\dot{h}^{(S)} = -2.51857\cdot 10^{-12}$ rad/sec, which corresponds to a retrograde precessional period of 79 107.9 years. A similar computation shows that the contribution $\dot{h}^{(M)}$ of the Moon amounts to $\dot{h}^{(M)} = -5.49028\cdot 10^{-12}$ rad/sec, corresponding to a precessional period of 36 289.3 years. The total precessional period is obtained as the sum of $\dot{h}^{(S)}$ and $\dot{h}^{(M)}$, providing an overall retrograde precessional period of 24 877.3 years, in good agreement with the value corresponding to astronomical observations and amounting to 25 700 years for the precession of the equinoxes.

6.5 Birkhoff's normal form

6.5.1 Normal form around an equilibrium position

Assume that the Hamiltonian $\mathcal{H} = \mathcal{H}(\underline{p},\underline{q})$, $(\underline{p},\underline{q}) \in \mathbf{R}^{2n}$, admits the origin as a stable equilibrium position; as a consequence, the eigenvalues of the quadratic part are all distinct and purely imaginary. In a neighborhood of the equilibrium position, after a series expansion and eventual diagonalization of the quadratic terms, we can write the Hamiltonian in the form

$$\mathcal{H}(\underline{p},\underline{q}) = \frac{1}{2}\sum_{j=1}^{n} \omega_j(p_j^2+q_j^2) + \mathcal{H}_3(\underline{p},\underline{q}) + \mathcal{H}_4(\underline{p},\underline{q}) + \dots \ , \tag{6.22}$$

where $\omega_j \in \mathbf{R}$ for $j=1,\dots,n$ are called the *frequencies* of the motion and the terms $\mathcal{H}_k$ are polynomials of degree k in $\underline{p}$ and $\underline{q}$. The following definitions introduce the resonant relations and the Birkhoff normal form associated to the Hamiltonian (6.22).

Definition. The frequencies $\omega_1, \dots, \omega_n$ are said to satisfy a *resonance relation* of order $K > 0$, if there exists a non–zero integer vector $(k_1,\dots,k_n)$ such that $k_1\omega_1 + \dots + k_n\omega_n = 0$ and $|k_1| + \dots + |k_n| = K$.

Definition. Let K be a positive number; a *Birkhoff normal form* for the Hamiltonian (6.22) is a polynomial of degree K in a set of variables $\underline{P}$, $\underline{Q}$, such that it is a polynomial of degree $[\frac{K}{2}]$ in the quantity $I'_j = \frac{1}{2}(P_j^2+Q_j^2)$ for $j=1,\dots,n$.

The construction of the Birkhoff normal form is the content of the following theorem.

Theorem. *Let K be a positive integer; assume that the frequencies $\omega_1, \dots, \omega_n$ do not satisfy any resonance relation of order less than or equal to K. Then, there exists a canonical transformation from $(\underline{p},\underline{q})$ to $(\underline{P},\underline{Q})$ such that the Hamiltonian (6.22) reduces to a Birkhoff normal form of degree K.*

Proof. Let us introduce action–angle variables $\underline{I} = (I_1,\dots,I_n) \in \mathbf{R}^n$, $\underline{\varphi} = (\varphi_1,\dots,\varphi_n) \in \mathbf{T}^n$, such that

$$\begin{aligned} p_j &= \sqrt{2I_j}\cos\varphi_j \\ q_j &= \sqrt{2I_j}\sin\varphi_j \ , \qquad j=1,\dots,n \ . \end{aligned} \tag{6.23}$$

Then, the Hamiltonian (6.22) can be written as

$$\mathcal{H}_1(\underline{I},\underline{\varphi}) = \sum_{j=1}^{n} \omega_j I_j + \mathcal{H}_{1,3}(\underline{I},\underline{\varphi}) + \mathcal{H}_{1,4}(\underline{I},\underline{\varphi}) + \dots \ , \tag{6.24}$$

where the terms $\mathcal{H}_{1,k}$ are polynomials of degree $[k/2]$ in $I_1,\ldots,I_n$. Let $\underline{I}'\cdot\underline{\varphi}+\Phi(\underline{I}',\underline{\varphi})$ be the generating function of a canonical transformation close to the identity from $(\underline{I},\underline{\varphi})$ to $(\underline{I}',\underline{\varphi}')$:

$$\begin{aligned}\underline{I} &= \underline{I}' + \frac{\partial\Phi}{\partial\underline{\varphi}}\\ \underline{\varphi}' &= \underline{\varphi} + \frac{\partial\Phi}{\partial\underline{I}'}\ .\end{aligned} \tag{6.25}$$

Let us decompose Φ as the sum of polynomials

$$\Phi = \Phi_3 + \Phi_4 + \cdots + \Phi_K\ ,$$

where Φ_k, $k = 3,\ldots,K$, is a polynomial of order $[\frac{k}{2}]$ in $I_1,\ldots,I_n$. Inserting the first of (6.25) in the Hamiltonian (6.24), one obtains the transformed Hamiltonian

$$\mathcal{H}_2(\underline{I}',\underline{\varphi}) = \underline{\omega}\cdot\underline{I}' + \underline{\omega}\cdot\frac{\partial\Phi}{\partial\underline{\varphi}} + \mathcal{H}_{1,3}\Big(\underline{I}' + \frac{\partial\Phi}{\partial\underline{\varphi}},\varphi\Big) + \mathcal{H}_{1,4}\Big(\underline{I}' + \frac{\partial\Phi}{\partial\underline{\varphi}},\varphi\Big) + \ldots \tag{6.26}$$

Let us determine Φ_3 such that the Hamiltonian (6.26) reduces to the Birkhoff normal form up to degree 3. To this end, split $\mathcal{H}_3$ as

$$\mathcal{H}_{1,3}(\underline{I}',\underline{\varphi}) = \bar{\mathcal{H}}_{1,3}(\underline{I}') + \tilde{\mathcal{H}}_{1,3}(\underline{I}',\underline{\varphi})\ ,$$

where $\bar{\mathcal{H}}_{1,3}(\underline{I}')$ is the average of $\mathcal{H}_{1,3}$ over the angles and $\tilde{\mathcal{H}}_{1,3}(\underline{I}',\underline{\varphi})$ is the remainder. Using (6.26) we obtain

$$\mathcal{H}_2(\underline{I}',\underline{\varphi}) = \underline{\omega}\cdot\underline{I}' + \bar{\mathcal{H}}_{1,3}(\underline{I}') + \Big[\underline{\omega}\cdot\frac{\partial\Phi_3}{\partial\underline{\varphi}} + \tilde{\mathcal{H}}_{1,3}(\underline{I}',\underline{\varphi})\Big] + \underline{\omega}\cdot\frac{\partial\Phi_4}{\partial\underline{\varphi}} + \mathcal{H}_{1,4}\Big(\underline{I}' + \frac{\partial\Phi}{\partial\underline{\varphi}},\varphi\Big) + \ldots$$

Expanding Φ_3 and $\tilde{\mathcal{H}}_{1,3}$ in Fourier series as

$$\begin{aligned}\Phi_3(\underline{I}',\underline{\varphi}) &= \sum_{\underline{m}\in\mathbf{Z}^n} \hat{\Phi}_{3,\underline{m}}(\underline{I}')\ e^{i\underline{m}\cdot\underline{\varphi}}\\ \tilde{\mathcal{H}}_{1,3}(\underline{I}',\underline{\varphi}) &= \sum_{\underline{m}\in\mathbf{Z}^n\setminus\{\underline{0}\}} \hat{\mathcal{H}}_{1,3,\underline{m}}(\underline{I}')\ e^{i\underline{m}\cdot\underline{\varphi}}\ ,\end{aligned} \tag{6.27}$$

one obtains

$$i\sum_{\underline{m}\in\mathbf{Z}^n} \underline{\omega}\cdot\underline{m}\ \hat{\Phi}_{3,\underline{m}}(\underline{I}')e^{i\underline{m}\cdot\underline{\varphi}} + \sum_{\underline{m}\in\mathbf{Z}^n\setminus\{\underline{0}\}} \hat{\mathcal{H}}_{1,3,\underline{m}}(\underline{I}')\ e^{i\underline{m}\cdot\underline{\varphi}} = 0\ .$$

Therefore the unknown Fourier coefficients of Φ_3 are given by

$$\hat{\Phi}_{3,\underline{m}}(\underline{I}') = -\frac{\hat{\mathcal{H}}_{1,3,\underline{m}}(\underline{I}')}{i\ \underline{\omega}\cdot\underline{m}}\ . \tag{6.28}$$

Casting together (6.27) and (6.28), one obtains

$$\Phi_3(\underline{I}',\underline{\varphi}) = -\sum_{\underline{m}\in\mathbf{Z}^n\setminus\{\underline{0}\}} \frac{\hat{\mathcal{H}}_{1,3,\underline{m}}(\underline{I}')}{i\,\underline{\omega}\cdot\underline{m}}\, e^{i\underline{m}\cdot\underline{\varphi}}\ .$$

Therefore the transformed Hamiltonian depends only on the actions $\underline{I}'$ up to terms of the fourth order, thus yielding a Birkhoff normal form of degree 3; the normalized terms define an integrable system in the set of action–angle variables $(\underline{I}',\underline{\varphi}')$ which provide the set of variables $(\underline{P},\underline{Q})$ through the transformation $P_j = \sqrt{2I'_j}\cos\varphi'_j$, $Q_j = \sqrt{2I'_j}\sin\varphi'_j$, $j = 1,\ldots,n$. The same procedure applied to higher orders leads to the determination of the generating function associated to the Birkhoff normal form of degree K. □

The Birkhoff normal form can be applied to the resonant case (see [6]) as the classical perturbation theory extends to the resonant perturbation theory. More precisely, recalling the action–angle variables introduced in (6.23), one has the following definition.

Definition. Let $\mathcal{K}$ be a sublattice of $\mathbf{Z}^n$; a *resonant Birkhoff normal form* of degree K for resonances in $\mathcal{K}$ is a polynomial of degree $[\frac{K}{2}]$ in $I_1,\ldots,I_n$, depending on the angles only through combinations of the form $\underline{k}\cdot\underline{\varphi}$ for $\underline{k}\in\mathcal{K}$.

The extension of the Birkhoff normal form to the resonant case is the content of the following theorem.

Theorem. *Let K be a positive integer and let $\mathcal{K}$ be a sublattice of $\mathbf{Z}^n$; assume that the frequencies $\omega_1, \ldots, \omega_n$ do not satisfy any resonance relation of order less than or equal to K, except for combinations of the form $\underline{k}\cdot\underline{\varphi}$ for $\underline{k}\in\mathcal{K}$. Then, there exists a canonical transformation such that the Hamiltonian* (6.22) *reduces to a resonant Birkhoff normal form of degree K for resonances in $\mathcal{K}$.*

Remark. The above results extend straightforwardly to mapping systems having the origin as an elliptic stable fixed point, so that all eigenvalues lie on the unitary circle of the complex plane. We briefly quote here the main result, referring to [162] for further details. Let $(p',q') = \mathcal{M}(p,q)$ be a two–dimensional area preserving map with $(p,q)\in\mathbf{R}^2$.

Definition. Let K be a positive number; close to an elliptic fixed point, a Birkhoff normal form of degree K for $\mathcal{M}$ is a polynomial in a set of variables P, Q, which is a polynomial of degree $[\frac{K}{2}]-1$ in the quantity $I' = \frac{1}{2}(P^2+Q^2)$.

The Birkhoff normal form for mappings is the content of the following theorem.

Theorem. *If the eigenvalue of the linear part of $\mathcal{M}$ at the elliptic fixed point is not a root of unity of degree less than or equal to K, then there exists a canonical change of variables which reduces the map to a Birkhoff normal form of degree K.*

6.5.2 Normal form around closed trajectories

Let us consider a non–autonomous Hamiltonian system of the form

$$\mathcal{H} = \mathcal{H}(\underline{p}, \underline{q}, t) ,$$

where $(\underline{p}, \underline{q}) \in \mathbf{R}^{2n}$ and $\mathcal{H}$ is a 2π–periodic function of the time. Closed trajectories for $\mathcal{H}$ are generally not isolated, but they rather form families. In a neighborhood of a closed trajectory one can reduce the Hamiltonian to the form

$$\mathcal{H}(\underline{p}, \underline{q}, t) = \frac{1}{2}\sum_{j=1}^{n} \omega_j (p_j^2 + q_j^2) + \mathcal{H}_3(\underline{p}, \underline{q}, t) + \mathcal{H}_4(\underline{p}, \underline{q}, t) + \dots , \tag{6.29}$$

where $\underline{\omega} = (\omega_1, \dots, \omega_n)$ is the so–called frequency vector. Referring the reader to [6], we introduce the notion of resonance relation and a result on the construction of the Birkhoff normal form for the Hamiltonian (6.29).

Definition. The frequencies $\omega_1, \dots, \omega_n$ are said to satisfy a *resonance relation* of order K, with $K > 0$, if there exists a non–zero integer vector $(k_0, k_1, \dots, k_n)$ such that $k_0 + k_1\omega_1 + \dots + k_n\omega_n = 0$ and $|k_1| + \dots + |k_n| = K$.

Theorem. *Assume that K is a positive integer and that the frequencies $\omega_1, \dots, \omega_n$ do not satisfy any resonance relation of order less than or equal to K. Then, there exists a canonical transformation 2π–periodic in time, such that (6.29) is reduced to an autonomous Birkhoff normal form of degree K with a time–dependent remainder of order $K + 1$.*

The extension of such result to the resonant case is formulated as follows (see [6]).

Definition. Let $\mathcal{K}$ be a sublattice of $\mathbf{Z}^{n+1}$; a *resonant Birkhoff normal form* of degree K for resonances in $\mathcal{K}$ is a polynomial of degree $[\frac{K}{2}]$ in the actions $I_1, \dots, I_n$, depending on the angles and on the time only through combinations of the form $k_0 t + \underline{k} \cdot \underline{\varphi}$ for $(k_0, \underline{k}) \in \mathcal{K}$.

Theorem. *Let K be a positive integer and let $\mathcal{K}$ be a sublattice of $\mathbf{Z}^{n+1}$; assume that the frequencies $\omega_1, \dots, \omega_n$ do not satisfy any resonance relation of order less than or equal to K, except for combinations of the form $k_0 + \underline{k} \cdot \underline{\varphi}$ for $(k_0, \underline{k}) \in \mathcal{K}$. Then, there exists a canonical transformation reducing the Hamiltonian to a resonant Birkhoff normal form of degree K in $\mathcal{K}$ up to terms of order $K + 1$.*

6.6 The averaging theorem

Consider the n–dimensional nearly–integrable Hamiltonian system

$$\mathcal{H}(\underline{I}, \underline{\varphi}) = h(\underline{I}) + \varepsilon f(\underline{I}, \underline{\varphi}) , \qquad \underline{I} \in \mathbf{R}^n , \ \underline{\varphi} \in \mathbf{T}^n ,$$

with associated Hamilton's equations

$$\begin{aligned} \dot{\underline{I}} &= \varepsilon \underline{F}(\underline{I}, \underline{\varphi}) \\ \dot{\underline{\varphi}} &= \underline{\omega}(\underline{I}) + \varepsilon \underline{G}(\underline{I}, \underline{\varphi}) , \end{aligned} \tag{6.30}$$

where $\underline{F}(\underline{I},\underline{\varphi}) \equiv -\frac{\partial f(\underline{I},\underline{\varphi})}{\partial \underline{\varphi}}$, $\underline{\omega}(\underline{I}) \equiv \frac{\partial h(\underline{I})}{\partial \underline{I}}$, $\underline{G}(\underline{I},\underline{\varphi}) = \frac{\partial f(\underline{I},\underline{\varphi})}{\partial \underline{I}}$. Let us decompose $\underline{F}$ as its average plus an oscillating part, say $\underline{F}(\underline{I},\underline{\varphi}) = \overline{\underline{F}}(\underline{I}) + \tilde{\underline{F}}(\underline{I},\underline{\varphi})$, so that we can write (6.30) as

$$\begin{aligned} \dot{\underline{I}} &= \varepsilon\overline{\underline{F}}(\underline{I}) + \varepsilon\tilde{\underline{F}}(\underline{I},\underline{\varphi}) \\ \dot{\underline{\varphi}} &= \underline{\omega}(\underline{I}) + \varepsilon G(\underline{I},\underline{\varphi}) \ . \end{aligned} \tag{6.31}$$

Averaging (6.31) with respect to the angles, we obtain the following differential equations in a new set of coordinates $\underline{J}$:

$$\dot{\underline{J}} = \varepsilon\overline{\underline{F}}(\underline{J}) \ . \tag{6.32}$$

Denoting by $\underline{I}_\varepsilon(t)$ the solution of (6.31) with initial data $\underline{I}_\varepsilon(0)$ and by $\underline{J}_\varepsilon(t)$ the solution of (6.32) with initial data $\underline{J}_\varepsilon(0) = \underline{I}_\varepsilon(0)$, we want to investigate the conditions for which the averaged system is a good approximation of the full system (see for example [83] for applications to Celestial Mechanics). More precisely, we aim to study the conditions for which

$$\lim_{\varepsilon\to 0} |\underline{I}_\varepsilon(t) - \underline{J}_\varepsilon(t)| = 0 \qquad \text{for } t \in \left[0, \frac{1}{\varepsilon}\right] \ . \tag{6.33}$$

We prove such statement in some particular cases. Let us consider first the one–dimensional case described by the Hamiltonian function

$$\mathcal{H}(I,\varphi) = \omega I + \varepsilon f(\varphi) \ ,$$

where ω is a non–zero real number and where the perturbation does not depend on the action. Setting $F(\varphi) = -\frac{df(\varphi)}{d\varphi}$, the equations of motion are given by

$$\begin{aligned} \dot{I} &= \varepsilon F(\varphi) \\ \dot{\varphi} &= \omega \ . \end{aligned} \tag{6.34}$$

In this case (6.33) is guaranteed by the following result.

Proposition. *Let $I_\varepsilon(t)$ and $J_\varepsilon(t)$ denote, respectively, the solutions at time t of (6.34) and of the averaged equation with initial conditions, respectively, $I_\varepsilon(0)$ and $J_\varepsilon(0) = I_\varepsilon(0)$. Then, for any $0 \le t \le \frac{1}{\varepsilon}$, one has*

$$\lim_{\varepsilon\to 0} |I_\varepsilon(t) - J_\varepsilon(t)| = 0 \ . \tag{6.35}$$

Proof. Let c be the average of $F(\varphi)$; then $J_\varepsilon(t) = I_\varepsilon(0) + \varepsilon ct$. Defining $\tilde{F}(\varphi) \equiv F(\varphi) - c$, we have

$$\begin{aligned} I_\varepsilon(t) - J_\varepsilon(t) &= I_\varepsilon(0) + \int_0^t \dot{I}_\varepsilon(\tau)d\tau - (I_\varepsilon(0) + \varepsilon ct) \\ &= \varepsilon\int_0^t \tilde{F}(\varphi(0) + \omega\tau)d\tau = \frac{\varepsilon}{\omega}\int_{\varphi(0)}^{\varphi(0)+\omega t} \tilde{F}(\psi)d\psi \ . \end{aligned}$$

If M denotes an upper bound on $\int_{\varphi(0)}^{\varphi(0)+\omega t} \tilde{F}(\psi)d\psi$ for $0 \leq t \leq \frac{1}{\varepsilon}$, then

$$|I_\varepsilon(t) - J_\varepsilon(t)| \leq \frac{\varepsilon}{\omega} M \ ,$$

which yields (6.35). □

For higher–dimensional systems, let us consider the Hamiltonian function with $n > 1$ degrees of freedom:

$$\mathcal{H}(\underline{I}, \underline{\varphi}) = \underline{\omega} \cdot \underline{I} + \varepsilon f(\underline{\varphi}) \ ,$$

where $\underline{I} \in \mathbf{R}^n$, $\underline{\varphi} \in \mathbf{T}^n$, $\underline{\omega} \in \mathbf{R}^n \backslash \{\underline{0}\}$. The equations of motion are

$$\begin{aligned} \dot{\underline{I}} &= \varepsilon \underline{F}(\underline{\varphi}) \\ \dot{\underline{\varphi}} &= \underline{\omega} \ , \end{aligned} \tag{6.36}$$

with $\underline{F}(\underline{\varphi}) \equiv -\frac{\partial f(\underline{\varphi})}{\partial \underline{\varphi}}$. Let $\underline{c}$ be the average of $\underline{F}(\underline{\varphi})$; more precisely, for a suitable sublattice $\mathcal{K}$ of $\mathbf{Z}^n \backslash \{\underline{0}\}$, let

$$\underline{F}(\underline{\varphi}) = \underline{c} + \sum_{\underline{k} \in \mathcal{K}} \hat{\underline{F}}_{\underline{k}} e^{i\underline{k}\cdot\underline{\varphi}} \ .$$

Proposition. *Let $\underline{I}_\varepsilon(t)$ and $\underline{J}_\varepsilon(t)$ denote, respectively, the solutions at time t of (6.36) and of the averaged equations with initial conditions, respectively, $\underline{I}_\varepsilon(0)$ and $\underline{J}_\varepsilon(0) = \underline{I}_\varepsilon(0)$. If the set $\mathcal{K}_0 \equiv \{\underline{k} \in \mathcal{K} : \ \underline{k} \cdot \underline{\omega} = 0\}$ is empty, then for any $0 \leq t \leq \frac{1}{\varepsilon}$, one has*

$$\lim_{\varepsilon \to 0} |\underline{I}_\varepsilon(t) - \underline{J}_\varepsilon(t)| = 0 \ .$$

Proof. We can write

$$\begin{aligned} \dot{\underline{I}} &= \varepsilon \underline{c} + \varepsilon \sum_{\underline{k} \in \mathcal{K}} \hat{F}_{\underline{k}} e^{i\underline{k}\cdot\underline{\varphi}(0)} \ e^{i\underline{k}\cdot\underline{\omega}t} \\ &= \varepsilon \underline{c} + \varepsilon \sum_{\underline{k} \in \mathcal{K}_0} \hat{F}_{\underline{k}} e^{i\underline{k}\cdot\underline{\varphi}(0)} + \varepsilon \sum_{\underline{k} \in \mathcal{K} \backslash \mathcal{K}_0} \hat{F}_{\underline{k}} e^{i\underline{k}\cdot\underline{\varphi}(0)} e^{i\underline{k}\cdot\underline{\omega}t} \ , \end{aligned}$$

whose integration yields

$$\underline{I}_\varepsilon(t) - \underline{I}_\varepsilon(0) = \varepsilon \underline{c} t + t\varepsilon \sum_{\underline{k} \in \mathcal{K}_0} \hat{F}_{\underline{k}} e^{i\underline{k}\cdot\underline{\varphi}(0)} + \varepsilon \sum_{\underline{k} \in \mathcal{K} \backslash \mathcal{K}_0} \hat{F}_{\underline{k}} e^{i\underline{k}\cdot\underline{\varphi}(0)} \ \frac{e^{i\underline{k}\cdot\underline{\omega}t} - 1}{i\underline{k} \cdot \underline{\omega}} \ .$$

The sum over $\mathcal{K}_0$ generates secular terms; nevertheless, by assumption the set $\mathcal{K}_0$ is empty. As a consequence, the distance between the complete and averaged solutions becomes:

$$\underline{I}_\varepsilon(t) - \underline{J}_\varepsilon(t) = \varepsilon \sum_{\underline{k} \in \mathcal{K} \backslash \mathcal{K}_0} \hat{F}_{\underline{k}} e^{i\underline{k}\cdot\underline{\varphi}(0)} \ \frac{e^{i\underline{k}\cdot\underline{\omega}t} - 1}{i\underline{k} \cdot \underline{\omega}} \ ,$$

which vanishes as ε tends to zero. □

6.6.1 An example

Let us consider the Hamiltonian function with two degrees of freedom:

$$\begin{aligned}\mathcal{H}(L,G,\ell,g) &= L^2 - G + \varepsilon R(L,G,\ell,g)\\ &= L^2 - G + \varepsilon\Big(R_{00}(L,G) + R_{10}(L,G)\cos 2\ell\\ &+ R_{12}(L,G)\cos(\ell+2g)\Big)\ ,\end{aligned}$$

for some real functions $R_{00}(L,G)$, $R_{10}(L,G)$, $R_{12}(L,G)$. The frequency vector is $(\omega_\ell,\omega_g)=(2L,-1)$; assume that the following resonance condition holds:

$$\omega_\ell + 2\omega_g = 0\ .$$

We perform the symplectic change of variables from (L,G,ℓ,g) to $(I_1,I_2,\vartheta_1,\vartheta_2)$ defined as

$$\begin{aligned}\vartheta_1 &= \ell+2g\ , \qquad & I_1 &= \frac{1}{2}G\ ,\\ \vartheta_2 &= 2\ell\ , \qquad & I_2 &= \frac{1}{2}L - \frac{1}{4}G\ ;\end{aligned} \tag{6.37}$$

due to the resonance, ϑ_1 is a slow variable, while ϑ_2 is a fast variable. The new Hamiltonian becomes

$$\begin{aligned}\mathcal{H}(I_1,I_2,\vartheta_1,\vartheta_2) &= (I_1+2I_2)^2 - 2I_1 + \varepsilon R(I_1,I_2,\vartheta_1,\vartheta_2)\\ &= (I_1+2I_2)^2 - 2I_1 + \varepsilon\Big(R_{00}(I_1,I_2)\\ &+ R_{10}(I_1,I_2)\cos\vartheta_2 + R_{12}(I_1,I_2)\cos\vartheta_1\Big)\ ,\end{aligned} \tag{6.38}$$

where $R(I_1,I_2,\vartheta_1,\vartheta_2)$ (and its coefficients) is the transformed function of $R(L,G,\ell,g)$ (and of its coefficients). Hamilton's equations are

$$\begin{aligned}\dot I_1 &= \varepsilon R_{12}(I_1,I_2)\sin\vartheta_1\\ \dot I_2 &= \varepsilon R_{10}(I_1,I_2)\sin\vartheta_2\\ \dot\vartheta_1 &= 2(I_1+2I_2) - 2 + \varepsilon\frac{\partial R(I_1,I_2,\vartheta_1,\vartheta_2)}{\partial I_1}\\ \dot\vartheta_2 &= 4(I_1+2I_2) + \varepsilon\frac{\partial R(I_1,I_2,\vartheta_1,\vartheta_2)}{\partial I_2}\ .\end{aligned}$$

Averaging over the fast variable ϑ_2 and denoting by $(J_1,J_2,\varphi_1,\varphi_2)$ the averaged variables, one obtains the Hamiltonian

$$\begin{aligned}\overline{\mathcal{H}}(J_1,J_2,\varphi_1,\varphi_2) &= (J_1+2J_2)^2 - 2J_1 + \varepsilon\Big(R_{00}(J_1,J_2) + R_{12}(J_1,J_2)\cos\varphi_1\Big)\\ &= (J_1+2J_2)^2 - 2J_1 + \varepsilon\bar R(J_1,J_2,\varphi_1)\ ,\end{aligned}$$

where $\bar{R}(J_1, J_2, \varphi_1) \equiv R_{00}(J_1, J_2) + R_{12}(J_1, J_2) \cos \varphi_1$. The associated Hamilton's equations are

$$\begin{aligned}
\dot{J}_1 &= \varepsilon R_{12}(J_1, J_2) \sin \varphi_1 \\
\dot{J}_2 &= 0 \\
\dot{\varphi}_1 &= 2(J_1 + 2J_2) - 2 + \varepsilon \frac{\partial \bar{R}(J_1, J_2, \varphi_1)}{\partial J_1} \\
\dot{\varphi}_2 &= 4(J_1 + 2J_2) + \varepsilon \frac{\partial \bar{R}(J_1, J_2, \varphi_1)}{\partial J_2} .
\end{aligned}$$

As a special case we set $R_{00}(L, G) = L$, $R_{12}(L, G) = L^2 G$, $R_{10}(L, G) = LG^2$; taking $\varepsilon = 0.01$ and setting the initial conditions in the transformed variables (6.37) as $I_1(0) = 0.9$, $I_2(0) = 0.5$, $\vartheta_1(0) = 0$, $\vartheta_2(0) = 0$, one obtains that the difference between the complete and averaged solutions (see Figure 6.1) is $|I_1(t) - J_1(t)| < 0.076$ for any $0 \leq t \leq 100$ in agreement with the averaging results discussed in Section 6.6.

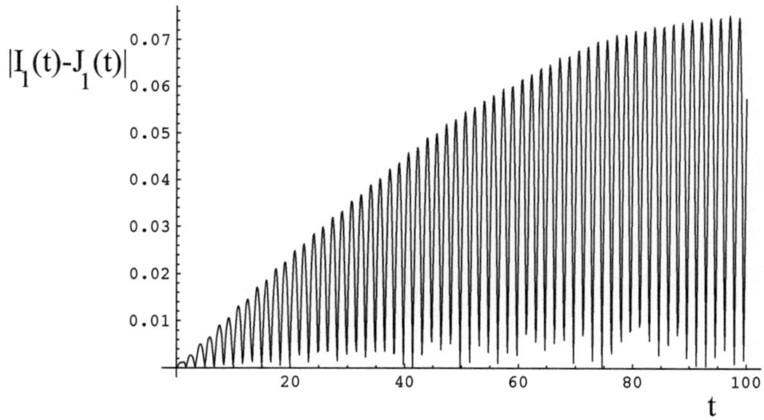

Fig. 6.1. The difference between the complete and averaged solutions associated to (6.38) for the special case $R_{00}(L, G) = L$, $R_{12}(L, G) = L^2 G$, $R_{10}(L, G) = LG^2$ with $\varepsilon = 0.01$ and with initial conditions $I_1(0) = 0.9$, $I_2(0) = 0.5$, $\vartheta_1(0) = 0$, $\vartheta_2(0) = 0$.

7 Invariant tori

Perturbation theory fails whenever a resonance condition is met; however, even if the non–resonance condition is fulfilled, there could be linear combinations with integer coefficients of the frequency vector which become arbitrarily small. These quantities, which are called the *small divisors,* appear at the denominator of the series defining the canonical transformation needed to implement perturbation theory. Small divisors might prevent the convergence of the series and therefore the application of perturbation theory. To overcome this problem, a breakthrough came with the work of Kolmogorov [105], later proved in different mathematical settings by Arnold [3] and Moser [138]. The overall theory is known with the acronym of KAM theory (Section 7.2) and it allows us to prove the persistence of invariant tori (Section 7.1) under perturbation (compare with [28,31,117,118]). KAM theory was applied to several physical models of interest in Celestial Mechanics (Section 7.3). However, the original versions of the theory gave concrete results very far from the physical measurements of the parameters involved in the proof. The implementation of computer–assisted KAM proofs allowed us to obtain realistic results in simple models of Celestial Mechanics, like the spin–orbit problem or the planar, circular, restricted three–body problem. The validity of such results is also attested by numerical methods for the determination of the breakdown threshold, like the well–known Greene's method (Section 7.4). KAM theory can also be extended to encompass the case of lower–dimensional tori (Section 7.5) as well as of nearly–integrable, dissipative systems (see Section 7.6, [19,32]), like the dissipative spin–orbit problem introduced in Chapter 5. While KAM theory provides a lower bound on the persistence of invariant tori, converse KAM theory gives an upper bound on the non–existence of invariant tori (Section 7.7). Moreover, just above the critical breakdown threshold the invariant tori transform into cantori, which are still invariant sets though being graphs of Cantor sets. Their explicit construction is discussed in a specific example, precisely the sawtooth map where constructive formulae for the cantori can be given (Section 7.8).

7.1 The existence of KAM tori

Let us start by considering the spin–orbit equations (5.15) that we write in the form

$$\begin{aligned} \dot{y} &= -\varepsilon f_x(x,t) \\ \dot{x} &= y \ , \end{aligned} \tag{7.1}$$

where $f_x(x,t) \equiv \frac{\partial f}{\partial x} = (\frac{a}{r})^3 \sin(2x - 2f)$ with $r = r(t)$, $f = f(t)$ being known periodic functions of the time. Equations (7.1) can be viewed as Hamilton's equations associated to the Hamiltonian function

$$\mathcal{H}(y, x, t) = h(y) + \varepsilon f(x, t) ,$$

where $h(y) = \frac{y^2}{2}$ is the unperturbed Hamiltonian, ε denotes the perturbing parameter, while the perturbation $f = f(x,t)$ is a continuous periodic function whose explicit expression has been given as in (5.15). The perturbing function can be expanded in Fourier series as

$$f(x,t) = -\frac{1}{2} \sum_{m \neq 0, m=N_1}^{N_2} W\left(\frac{m}{2}, \mathrm{e}\right) \cos(2x - mt) \tag{7.2}$$

for suitable coefficients $W(\frac{m}{2}, \mathrm{e})$ listed in Table 5.1, which depend on the orbital eccentricity. For $\varepsilon = 0$ equations (7.1) can be integrated as

$$\begin{aligned} y(t) &= y(0) \\ x(t) &= x(0) + y(0)t \ ; \end{aligned}$$

henceforth, the motion takes place on a plane in the phase space $\mathbf{T}^2 \times \mathbf{R}$, labeled by the initial condition $y(0)$. The value $y(0)$ coincides with the frequency (or rotation number) $\omega = \omega(y)$ of the motion, which in general is defined as the first derivative of the unperturbed Hamiltonian: $\omega(y) = \frac{dh(y)}{dy}$. Let us fix an irrational frequency $\omega_0 = \omega(y(0))$; the surface $\{y(0)\} \times \mathbf{T}^2$ is invariant for the unperturbed system and we wonder whether for $\varepsilon \neq 0$ there still exists an invariant surface for the perturbed system with the same frequency as the unperturbed case. The answer is provided by KAM theory, which allows us to prove the persistence of invariant tori provided some generic conditions are satisfied.

In a general framework, let us consider a nearly–integrable Hamiltonian function with n degrees of freedom:

$$\mathcal{H}(\underline{y}, \underline{x}) = h(\underline{y}) + \varepsilon f(\underline{y}, \underline{x}) , \qquad \underline{y} \in \mathbf{R}^n , \quad \underline{x} \in \mathbf{T}^n ; \tag{7.3}$$

let $\underline{\omega} \equiv \frac{\partial h(\underline{y})}{\partial \underline{y}} \in \mathbf{R}^n$ be the frequency vector. The first assumption required by KAM theory concerns a non–degeneracy of the unperturbed Hamiltonian. More precisely, let us introduce the following notions.

(*i*) An n–dimensional Hamiltonian function $h = h(\underline{y})$, $\underline{y} \in V$, being V an open subset of $\mathbf{R}^n$, is said to be *non–degenerate* if

$$\det \left(\frac{\partial^2 h(\underline{y})}{\partial \underline{y}^2} \right) \neq 0 \qquad \text{for any } \ \underline{y} \in V \subset \mathbf{R}^n . \tag{7.4}$$

Condition (7.4) is equivalent to require that the frequencies vary with the actions as

$$\det \left(\frac{\partial \underline{\omega}(\underline{y})}{\partial \underline{y}} \right) \neq 0 \qquad \text{for any } \underline{y} \in V .$$

The non–degeneracy condition guarantees the persistence of invariant tori with fixed frequency.

(*ii*) An n–dimensional Hamiltonian function $h = h(\underline{y})$, $\underline{y} \in V \subset \mathbf{R}^n$, is said to be *isoenergetically non–degenerate* if

$$\det \begin{pmatrix} \frac{\partial^2 h(\underline{y})}{\partial \underline{y}^2} & \frac{\partial h(\underline{y})}{\partial \underline{y}} \\ \frac{\partial h(\underline{y})}{\partial \underline{y}} & \underline{0} \end{pmatrix} \neq 0 \qquad \text{for any } \ \underline{y} \in V \subset \mathbf{R}^n \ . \tag{7.5}$$

This condition can be written as

$$\det \begin{pmatrix} \frac{\partial \underline{\omega}(\underline{y})}{\partial \underline{y}} & \underline{\omega} \\ \underline{\omega} & \underline{0} \end{pmatrix} \neq 0 \qquad \text{for any } \ \underline{y} \in V \subset \mathbf{R}^n \ .$$

The isoenergetic non–degeneracy condition, which is independent of the non–degeneracy condition (7.4), guarantees that the frequency ratio of the invariant tori varies as one crosses the tori on fixed energy surfaces (see [6]).

(*iii*) An n–dimensional Hamiltonian function $\mathcal{H}(\underline{y}, \underline{x}) = h(\underline{y}) + \varepsilon f(\underline{y}, \underline{x})$, $\underline{y} \in \mathbf{R}^n$, $\underline{x} \in \mathbf{T}^n$, is said to be *properly degenerate* if the unperturbed Hamiltonian $h(\underline{y})$ does not depend explicitly on some action variables. In this case, the perturbation $f(\underline{y}, \underline{x})$ is said to remove the degeneracy if it can be split as the sum of two functions, say $f(\underline{y}, \underline{x}) = \bar{f}(\underline{y}) + \varepsilon f_1(\underline{y}, \underline{x})$ with the property that $h(\underline{y}) + \varepsilon \bar{f}(\underline{y})$ is non–degenerate.

In order to apply KAM theory it will be assumed that the unperturbed Hamiltonian satisfies (7.4) or (7.5). Beside non–degeneracy, the second requirement for the applicability of the KAM theorem is that the frequency $\underline{\omega}$ satisfies a strong irrationality assumption, namely the so–called diophantine condition which is defined as follows.

Definition. The frequency vector $\underline{\omega}$ satisfies a diophantine condition of type (C, τ) for some $C \in \mathbf{R}_+$, $\tau \geq 1$, if for any integer vector $\underline{m} \in \mathbf{R}^n \backslash \{\underline{0}\}$:

$$|\underline{\omega} \cdot \underline{m}| \geq \frac{1}{C|\underline{m}|^\tau} \ . \tag{7.6}$$

Under the non–degeneracy condition, the KAM theorem guarantees the persistence of invariant tori with diophantine frequency, provided the perturbing parameter is sufficiently small. More precisely, Kolmogorov [105] stated the following

Theorem (Kolmogorov). *Given the Hamiltonian system* (7.3) *satisfying the non–degeneracy condition* (7.4), *having fixed a diophantine frequency* $\underline{\omega}$ *for the unperturbed system, if* ε *is sufficiently small there still exists an invariant torus on which the motion is quasi–periodic with frequency* $\underline{\omega}$.

The theorem was later proved in different settings by V.I. Arnold [2] and J. Moser [138] and it is nowadays known by the acronym: the KAM theorem. Qualitatively, we can state that for low values of the perturbing parameter there exists an invariant surface with diophantine frequency $\underline{\omega}$; as the perturbing parameter increases the invariant torus with frequency $\underline{\omega}$ is more and more distorted and displaced, until the parameter reaches a critical value at which the torus breaks down (compare with Figure 7.1). The KAM theorem provides a lower bound on the breakdown threshold; effective KAM estimates, together with a *computer–assisted* implementation,

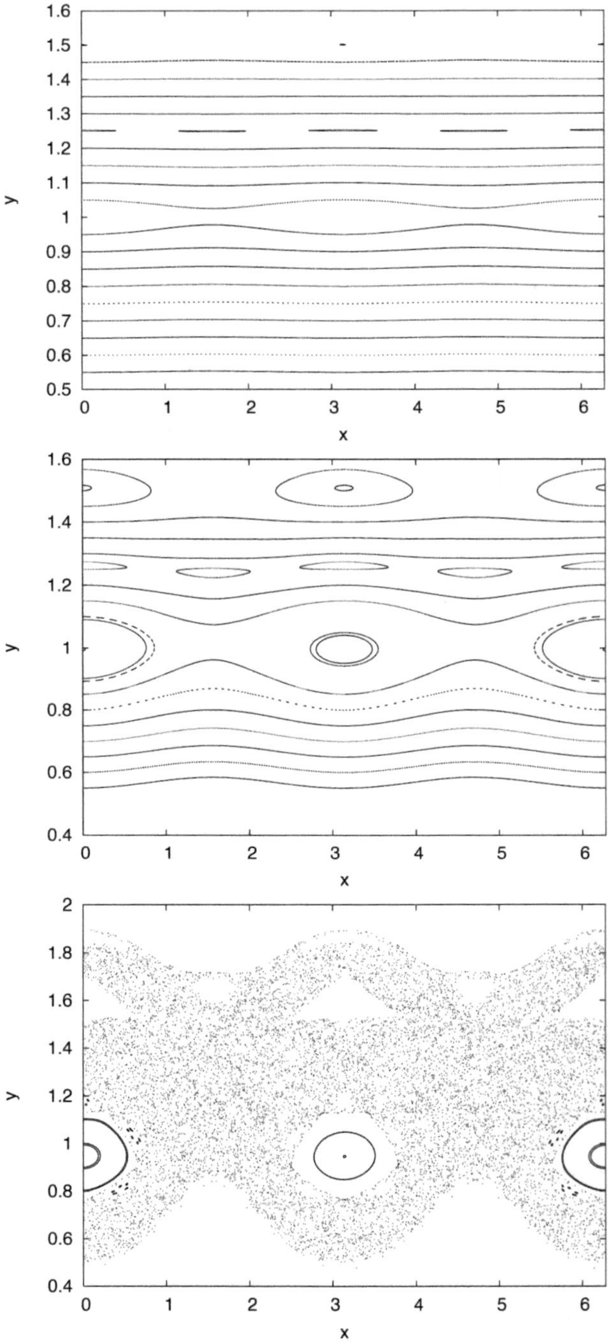

Fig. 7.1. The Poincarè section of the spin–orbit problem (5.19) for 20 different initial conditions and for e = 0.1. Top: $\varepsilon = 10^{-3}$, middle: $\varepsilon = 10^{-2}$, bottom: $\varepsilon = 10^{-1}$.

can provide, in simple examples, results on the parameters which are consistent with the physical values.

Section 7.2 will be devoted to the development of explicit estimates for the specific example of the spin–orbit model given by (7.1) with the perturbing function as in (7.2). Its unperturbed Hamiltonian satisfies the non–degeneracy condition (7.4), being $\frac{\partial^2 h(y)}{\partial y^2} = 1$. To apply the KAM theorem it is required that the frequency of the motion, say ω, satisfies a diophantine condition of type (C, τ) for some $C \in \mathbf{R}_+$, $\tau \geq 1$; therefore we assume that for any integers p and q, relatively coprime, with $q \neq 0$, the following inequality is satisfied:

$$\left| \omega - \frac{p}{q} \right| \geq \frac{1}{C|q|^{\tau+1}} \ . \tag{7.7}$$

An example of a diophantine number satisfying (7.7) with $\tau = 1$ is provided by the golden ratio $\gamma = \frac{\sqrt{5}-1}{2}$, for which (7.7) is fulfilled with the best diophantine constant given by $C = \frac{3+\sqrt{5}}{2}$. Being the system described by a one–dimensional, time–dependent Hamiltonian function, the existence of two invariant tori obtained through the KAM theorem provides a strong stability property, since the motion remains confined between such surfaces. We remark that this property is still valid for a two–dimensional system, since the phase space is four–dimensional and the two–dimensional KAM tori separate the constant energy surfaces into invariant regions. On the other hand, the confinement property is no longer valid whenever the Hamiltonian system has more than two degrees of freedom.

7.2 KAM theory

We present a version of the celebrated KAM theory by providing concrete estimates in the specific case of the spin–orbit model, following the KAM proof given in [31] to which we refer for further details (see also [28]). The goodness of the method strongly depends on the choice of the initial approximation which can be explicitly computed as a suitable truncation of the Taylor series expansion in the perturbing parameter. We also discuss how to choose the (irrational) rotation number, among those satisfying the diophantine condition. In order to obtain optimal results, it is convenient to use a computer to determine the initial approximation as well as to check the estimates provided by the theorem. The so–called interval arithmetic technique allows us to keep control of the numerical errors introduced by the machine. We also review classical and computer–assisted results of KAM applications in Celestial Mechanics.

7.2.1 The KAM theorem

The spin–orbit Hamiltonian associated to (7.1) can be written as the nearly–integrable Hamiltonian function

$$\mathcal{H}(y, x, t) = \frac{y^2}{2} + \varepsilon f(x, t) \ , \tag{7.8}$$

where $y \in \mathbf{R}$, $(x,t) \in \mathbf{T}^2$, the perturbing function $f = f(x,t)$ is assumed to be a periodic analytic function and the positive real number ε represents the perturbing parameter. Hamilton's equations associated to (7.8) can be written as the second–order differential equation

$$\ddot{x} + \varepsilon f_x(x,t) = 0 \ . \tag{7.9}$$

Definition. A KAM torus for (7.9) with rotation number ω is a two–dimensional invariant surface, described parametrically by

$$x = \vartheta + u(\vartheta,t) \ , \quad (\vartheta,t) \in \mathbf{T}^2 \ , \tag{7.10}$$

where $u = u(\vartheta,t)$ is a suitable analytic periodic function such that

$$1 + u_\vartheta(\vartheta,t) \neq 0 \quad \text{for all } (\vartheta,t) \in \mathbf{T}^2 \tag{7.11}$$

and where the flow in the parametric coordinate is linear, namely $\dot{\vartheta} = \omega$.

Notice that the requirement (7.11) ensures that (7.10) is a diffeomorphism. In this Section we want to prove the following KAM result.

Theorem. *Given the spin–orbit Hamiltonian (7.8) and having fixed for the unperturbed system a diophantine frequency ω satisfying (7.7), if ε is sufficiently small there still exists a KAM torus with frequency ω.*

Let us introduce the partial derivative operator D as

$$D \equiv \omega \frac{\partial}{\partial \vartheta} + \frac{\partial}{\partial t} \ . \tag{7.12}$$

We remark that for any function $g = g(\vartheta,t)$ the inversion of the operator D provides

$$(D^{-1}g)(\vartheta,t) = \sum_{(n,m)\in\mathbf{Z}^2\setminus\{\underline{0}\}} \frac{\hat{g}_{nm}}{i(\omega n + m)}\, e^{i(n\vartheta + mt)} \ ,$$

which provokes the appearance of the small divisors $\omega n + m$. Notice that from the second equation in (7.1) we obtain that

$$y = \omega + Du(\vartheta,t) \ .$$

Inserting the parametrization (7.10) in (7.9) and using the definition (7.12), one obtains that the function u must satisfy the differential equation

$$D^2 u(\vartheta,t) + \varepsilon f_x(\vartheta + u(\vartheta,t),t) = 0 \ . \tag{7.13}$$

To prove the existence of an invariant surface with rotation number ω is equivalent to find a solution of equation (7.13). This goal is achieved by implementing a Newton's method as follows. Let $v = v(\vartheta,t)$ be an approximate solution of (7.13) with an error term $\eta = \eta(\vartheta,t)$:

$$D^2 v(\vartheta,t) + \varepsilon f_x(\vartheta + v(\vartheta,t),t) = \eta(\vartheta,t) \ . \tag{7.14}$$

We assume that $\mathcal{M} \equiv 1 + v_\vartheta(\vartheta, t) \neq 0$ for all $(\vartheta, t) \in \mathbf{T}^2$. We want to determine a new approximate solution $v' = v'(\vartheta, t)$ which satisfies (7.13) with an error $\eta' = \eta'(\vartheta, t)$ quadratically smaller, namely

$$D^2 v'(\vartheta, t) + \varepsilon f_x(\vartheta + v'(\vartheta, t), t) = \eta'(\vartheta, t) \ , \tag{7.15}$$

where $|\eta'| = O(|\eta|^2)$. This task can be accomplished through the following Lemma (see [26]).

Lemma (New approximation). *Let z be a solution of the equation*

$$D(\mathcal{M}^2 Dz) = -\mathcal{M}\eta \ . \tag{7.16}$$

Let

$$w \equiv \mathcal{M}z \ , \qquad v' \equiv v + w \ ;$$

then v' satisfies (7.15) with

$$\eta' = \eta_\vartheta z + q_1 \tag{7.17}$$

and

$$q_1 = \varepsilon f_x(\vartheta + v + w, t) - \varepsilon f_x(\vartheta + v, t) - \varepsilon f_{xx}(\vartheta + v, t) w \ . \tag{7.18}$$

Proof. We first remark that taking the derivative of (7.14) with respect to ϑ one has

$$D^2\mathcal{M} + \varepsilon f_{xx}(\vartheta + v, t)\mathcal{M} = \eta_\vartheta \ . \tag{7.19}$$

By (7.15) and (7.17) one has

$$D^2 v + D^2(\mathcal{M}z) + \varepsilon f_x(\vartheta + v, t) + \varepsilon f_{xx}(\vartheta + v, t)\mathcal{M}z = \eta_\vartheta z \ ;$$

using (7.19) and (7.14), one obtains

$$D^2(\mathcal{M}z) - (D^2\mathcal{M})\ z = -\eta \ . \tag{7.20}$$

Multiplying (7.20) by $\mathcal{M}$ one can easily recognize that the function z must solve (7.16). □

The solution z is obtained from (7.16) in the form

$$z \equiv D^{-1}\Big(\mathcal{M}^{-2}[c_0 - D^{-1}(\mathcal{M}\eta)]\Big) + c_1 \ , \tag{7.21}$$

where c_0 and c_1 are suitable constants which take the following expressions:

$$\begin{aligned} c_0 &\equiv \langle \mathcal{M}^{-2}\rangle^{-1} \ \langle \mathcal{M}^{-2} D^{-1}(\mathcal{M}\eta)\rangle \\ c_1 &\equiv -\langle \mathcal{M}^{-1}\rangle\langle \mathcal{M} D^{-1}\Big(\mathcal{M}^{-2}[c_0 - D^{-1}(\mathcal{M}\eta)]\Big)\rangle \ , \end{aligned} \tag{7.22}$$

so that w has zero average. Let us introduce the complex domain

$$\Delta_{\xi,\rho} \equiv \{(\vartheta, t, \varepsilon) \in \mathbf{C}^3 \ : \ |\mathrm{Im}(\vartheta)| \leq \xi \ , \ \ |\mathrm{Im}(t)| \leq \xi \ , \ \ |\varepsilon| \leq \rho\} \ ;$$

then, for a function $g = g(\vartheta, t; \varepsilon)$ we define the norm

$$\|g\|_{\xi,\rho} \equiv \sup_{\Delta_{\xi,\rho}} |g(\vartheta, t; \varepsilon)| \ .$$

Now we need a technical lemma which provides bounds on the derivatives of a function $g = g(\vartheta, t; \varepsilon)$, whose Fourier series expansion is given by $g(\vartheta, t; \varepsilon) = \sum_{(n,m)\in\mathbf{Z}^2} \hat{g}_{nm} e^{i(n\vartheta + mt)}$.

Lemma (Bounds on derivatives). *Let $g = g(\vartheta, t; \varepsilon)$ be an analytic function on the domain $\Delta_{\xi,\rho}$. Then, for any $0 < \delta \le \xi$, one has*

$$\|g_\vartheta\|_{\xi-\delta,\rho} \le \|g\|_{\xi,\rho}\ \delta^{-1}\ . \tag{7.23}$$

Moreover, if $\langle g\rangle = 0$ and ∂_ϑ^ℓ denotes the derivative of order ℓ with respect to ϑ, then for $\ell = 0, 1$,

$$\|\partial_\vartheta^\ell D^{-1} g\|_{\xi-\delta,\rho} \le \sigma_\ell(2\delta)\|g\|_{\xi,\rho}\ ,$$

where

$$\sigma_\ell(\delta) \equiv 2\left[\sum_{(n,m)\in\mathbf{Z}^2\setminus\{\underline{0}\}} \left(\frac{|n|^\ell}{\omega n + m}\right)^2 e^{-\delta(|n|+|m|)}\right]^{1/2}\ . \tag{7.24}$$

Proof. Given a holomorphic function $g = g(\vartheta, t; \varepsilon)$ defined on $\Delta_{\xi,\rho}$, the estimate (7.23) is obtained through Cauchy's integral formula, i.e.

$$\|g_\vartheta\|_{\xi-\delta,\rho} = \|\frac{1}{2\pi i}\oint_{|\vartheta-\gamma|=\delta}\frac{g(\gamma,t;\varepsilon)}{(\vartheta-\gamma)^2}\,d\gamma\|_{\xi-\delta,\rho} \le \|g\|_{\xi,\rho}\ \delta^{-1}\ .$$

Under the condition $\langle g\rangle = 0$, from the maximum principle and Schwarz inequality one obtains

$$\begin{aligned}
\|\partial_\vartheta^\ell D^{-1} g\|_{\xi-\delta,\rho} &= \left\|\sum_{(n,m)\in\mathbf{Z}^2\setminus\{\underline{0}\}} \hat g_{nm}\frac{n^\ell}{\omega n + m}\ e^{i(n\vartheta+mt)}\right\|_{\xi-\delta,\rho}\\
&\le \sup_{|\varepsilon|\le\rho}\sum_{k_1,k_2\in\{-1,1\}}\left|\sum_{(n,m)\in\mathbf{Z}^2\setminus\{\underline{0}\}} \hat g_{nm}\frac{n^\ell}{\omega n + m}e^{(k_1 n + k_2 m)(\xi-\delta)}\right|\\
&\le \sup_{|\varepsilon|\le\rho}\sum_{(n,m)\in\mathbf{Z}^2\setminus\{\underline{0}\}}|\hat g_{nm}|\left(\sum_{k_1,k_2\in\{-1,1\}} e^{2(k_1 n+k_2 m)\xi}\right)^{\frac12} e^{-\delta(|n|+|m|)}\frac{|n|^\ell}{|\omega n+m|}\\
&\le \sigma_\ell(2\delta)\|g\|_{\xi,\rho}\ ,
\end{aligned}$$

with $\sigma_\ell(2\delta)$ defined according to (7.24). □

We introduce the quantities V, V_1, M, $\tilde M$, E, $s_\ell(\delta)$ as the following upper bounds:

$$\begin{aligned}
&\|v\|_{\xi,\rho} \le V\ , &&\|v_\vartheta\|_{\xi,\rho}\le V_1\ , &&\|\mathcal{M}\|_{\xi,\rho}\le M\ ,\\
&\|\mathcal{M}^{-1}\|_{\xi,\rho}\le \tilde M\ , &&\|\eta\|_{\xi,\rho}\le E\ , &&\|\sigma_\ell(\delta)\|_{\xi,\rho}\le s_\ell(\delta)\ .
\end{aligned}$$

One obtains that

$$\tilde M^{-2} \le \|\mathcal{M}^2\|_{\xi,\rho}\le M^2\ ,\quad M^{-2}\le\|\mathcal{M}^{-2}\|_{\xi,\rho}\le\tilde M^2\ ,\quad \|\langle\mathcal{M}^{-2}\rangle^{-1}\|_{\xi,\rho}\le M^2\ .$$

From (7.22), one finds that c_0, c_1 can be bounded as

$$\begin{aligned}
\|c_0\|_{\xi,\rho} &\le M^3\tilde M^2 s_0(2\xi)E\\
\|c_1\|_{\xi,\rho} &\le M\tilde M^3 s_0(\xi)\Big[M^3\tilde M^2 s_0(2\xi)E + M s_0(\xi)E\Big]\ .
\end{aligned}$$

Having introduced the quantities

$$\begin{aligned} a &\equiv (M\tilde{M}s_0(\delta))^2 \Bigg[1+(M\tilde{M})^2\frac{s_0(2\xi)}{s_0(\delta)} \\ &+ M\tilde{M}\left(\frac{s_0(\xi)}{s_0(\delta)}\right)^2\left(1+(M\tilde{M})^2\frac{s_0(2\xi)}{s_0(\xi)}\right)\Bigg] \\ b &\equiv \frac{aV_1}{M}\delta^{-1}+a\frac{s_1(\delta)}{s_0(\delta)} , \end{aligned}$$

from the definition of z in (7.21) one finds the following bounds W on w and W_1 on the derivative of w with respect to ϑ:

$$\begin{aligned} \|w\|_{\xi-\delta,\rho} &\leq Ea \equiv W \\ \|w_\vartheta\|_{\xi-\delta,\rho} &\leq Eb \equiv W_1 . \end{aligned}$$

The first inequality follows from $\|w\|_{\xi-\delta,\rho} \leq M\|z\|_{\xi-\delta,\rho}$ and from the estimate

$$\|z\|_{\xi-\delta,\rho} \leq \|c_1\|_{\xi,\rho} + s_0(\delta)\tilde{M}^2(\|c_0\|_{\xi,\rho} + s_0(\delta)ME) .$$

Similar computations hold for $\|w_\vartheta\|_{\xi-\delta,\rho}$. Finally, from (7.17) and (7.18) one obtains a bound E_1 on the new error term as

$$\|\eta'\|_{\xi-\delta,\rho} \leq E^2\left(\frac{a\delta^{-1}}{M}+\frac{a^2F}{2}\right) \equiv E_1 ,$$

where $F \equiv \|\varepsilon f_{xxx}\|_{\xi-\delta+V+W,\rho}$.

Let us assume that we start from a given initial approximation $v^{(0)}$ satisfying (7.14) with an error term $\eta^{(0)}$; we construct the solution at the jth step, say $v^{(j)}$, by an iterative application of the *New approximation Lemma* starting from the initial solution $v^{(0)}$. Let $M^{(j)}$, $\tilde{M}^{(j)}$, $E^{(j)}$, $W^{(j)}$, $W_1^{(j)}$ be the bounds corresponding to the solution $v^{(j)}$. From the previous estimates and definitions, the bounds for the solution $v^{(j+1)}$ are obtained through the following Lemma which provides the KAM algorithm needed to construct bounds on the new approximate solution.

Lemma (KAM algorithm). *Let $\xi_0 > 0$, $\xi_j \equiv \frac{\xi_0}{2^j}$ and let $\delta_j \equiv \frac{\xi_0}{2^{j+1}}$. Given the following quantities referring to the solution $v^{(j)}$ on the domain with parameters ξ_j, δ_j: $M^{(j)}$, $\tilde{M}^{(j)}$, $E^{(j)}$, $W^{(j)}$, $W_1^{(j)}$, we define the bounds corresponding to the solution $v^{(j+1)}$ as follows:*

$$
\begin{aligned}
M^{(j+1)} &\equiv M^{(j)} + W_1^{(j)} \\
\tilde{M}^{(j+1)} &\equiv \tilde{M} \left(1 - \tilde{M} \sum_{i=0}^{j} W_1^{(i)} \right)^{-1} && \text{if } \sum_{i=0}^{j} W_1^{(i)} < 1 \\
\tilde{M}^{(j+1)} &\equiv \infty && \text{if } \sum_{i=0}^{j} W_1^{(i)} \geq 1 \\
E^{(j+1)} &\equiv (E^{(j)})^2 \left(\frac{a^{(j)} \delta_j^{-1}}{M^{(j)}} + \frac{(a^{(j)})^2 F}{2} \right) \\
W^{(j+1)} &\equiv E^{(j+1)} a^{(j+1)} \\
W_1^{(j+1)} &\equiv E^{(j+1)} b^{(j+1)} .
\end{aligned}
$$

One can iterate the above algorithm for a finite number of steps; the convergence to the true solution of equation (7.13) is obtained once a suitable *KAM condition* is satisfied. To this end, let us premise the following Lemma which provides a bound on the quantity $\sigma_\ell(\delta)$ introduced in (7.24).

Lemma (Bound on $\sigma_\ell(\delta)$). *Let $0 < \delta \leq \frac{1}{2}$; for $\ell = 0, 1$, if $k_\ell \equiv \tau + \ell + 1$, then*

$$
\sigma_\ell(\delta) < K_\ell C \delta^{-k_\ell} , \tag{7.25}
$$

where $K_0 \equiv \frac{25}{2} (\frac{\Gamma(2\tau+1)}{\pi})^{1/2}$, $K_1 \equiv K_0 \sqrt{(2\tau+2)(2\tau+1)}$, with Γ being the Euler's gamma function.

Proof. For $t \geq 1$ and $0 < \delta \leq \frac{1}{2}$, one has

$$
\sum_{n \in \mathbf{Z}} |n|^t e^{-\delta |n|} < 2 e^{\frac{1}{2}} \Gamma(t+1) \delta^{-(t+1)} .
$$

Being[1] $C > 2$ and $\tau \geq 1$, one finds

$$
\begin{aligned}
\sigma_\ell(\delta) &< 2 \left(\sum_{m \neq 0} \frac{e^{-\delta|m|}}{m^2} + C^2 \sum_{n \neq 0} |n|^{2\tau+2\ell} e^{-\delta|n|} \sum_m e^{-\delta|m|} \right) \\
&< 2 \left(\frac{2}{\delta} + 2C^2 (1 + \sqrt{e}) \sqrt{e}\, \Gamma(2(\tau+\ell)+1)\, \delta^{-2(\tau+\ell+1)} \right)^{\frac{1}{2}} \\
&< 2C (1 + 2(1+\sqrt{e})\sqrt{e})^{\frac{1}{2}} (\Gamma(2(\tau+\ell)+1))^{\frac{1}{2}} \delta^{-(\tau+\ell+1)} ,
\end{aligned}
$$

which gives (7.25). □

Finally, let v, η satisfy (7.14); for some $\xi_* > 0$, $\rho > 0$, let $E \equiv \|\eta\|_{\xi_*,\rho}$, $M \equiv \|\mathcal{M}\|_{\xi_*,\rho}$, $\tilde{M} \equiv \|\mathcal{M}^{-1}\|_{\xi_*,\rho}$, $F \equiv \|f_{xxx}\|_{\xi_*+V,\rho}$. The convergence of the sequence of approximate solutions to the solution of (7.13) is obtained through the following result, which gives the persistence of the invariant torus with diophantine frequency ω, provided ε is sufficiently small (compare with (7.28) below).

[1] The smallest value of the diophantine constant corresponds to the golden ratio $\frac{\sqrt{5}-1}{2}$ and it amounts to $C \equiv \frac{3+\sqrt{5}}{2} \simeq 2.618$.

Proposition (KAM condition). *Let $\xi_* > 0$, $\rho > 0$ and let β_0, β_1, β_2, η_0, η_1, η_2 be positive constants defined as follows:*

$$\begin{aligned}
\beta_0 &\equiv \left(M\tilde{M}\ K_0 C\left(\frac{4}{\xi_*}\right)^{k_0}\right)^2 \left[1+(M\tilde{M})^2\frac{1}{8^{k_0}}+M\tilde{M}\left(\frac{1}{4}\right)^{2k_0}\left(1+(M\tilde{M})^2\frac{1}{2^{k_0}}\right)\right] \\
\beta_1 &= (M\tilde{M}C)^2 2^{4k_0+3}\xi_*^{-2k_0-1}K_1K_0 \\
&\quad \cdot \left[1+(M\tilde{M})^2\frac{1}{8^{k_0}}+M\tilde{M}\left(\frac{1}{4}\right)^{2k_0}\left(1+(M\tilde{M})^2\frac{1}{2^{k_0}}\right)\right] \\
\beta_2 &= \frac{4\beta_0}{\xi_*}+\frac{\beta_0^2 F}{2} \\
\eta_0 &= 2^{2k_0}\ , \qquad \eta_1 = 2^{2k_0+1}\ , \qquad \eta_2 = \max(2\eta_0, \eta_0^2)\ .
\end{aligned} \tag{7.26}$$

Defining

$$\mathcal{K} \equiv 2\tilde{M}\beta_1(1+2\eta_1\beta_2\eta_2)\ , \tag{7.27}$$

if

$$\mathcal{K}\ E\ <\ 1\ , \tag{7.28}$$

then (7.13) *has a unique solution u, with $\langle u\rangle = \langle v\rangle$ and*

$$\begin{aligned}
\|u-v\|_{\frac{\xi_*}{2},\rho} &< \mathcal{K}E\frac{\xi_*}{4} \\
\|u_\vartheta - v_\vartheta\|_{\frac{\xi_*}{2},\rho} &< \frac{\mathcal{K}E}{2\tilde{M}}\ .
\end{aligned} \tag{7.29}$$

Proof. Define the sequences $\{\xi_*^{(j)}\}$, $\{\delta_j\}$, $j \in \mathbf{Z}_+$, as $\xi_*^{(j)} = \frac{\xi_*}{2}+\frac{\xi_*}{2^{j+1}}$, $\delta_j = \frac{\xi_*}{2^{j+2}}$. Under the assumption (7.28), for a suitable $\mathcal{K}_0 < \mathcal{K}$ one has the following relations, valid for any $j \geq 0$:

$$\begin{aligned}
E^{(j)} &< (\mathcal{K}_0 E)^{2^j} \\
\xi_*^{(j)} + V^{(j)} &\leq \xi_* + V \\
\tilde{M}^{(j)} &\leq 2\tilde{M}\ ,
\end{aligned} \tag{7.30}$$

where $V^{(j)}$ is an upper bound on $v^{(j)}$. The first of (7.30) implies that the sequence of the error terms $\{E^{(j)}\}_{j\in\mathbf{Z}_+}$ converges to zero. Moreover, from the second of (7.30) we get that the sequence of approximate solutions $\{v^{(j)}\}_{j\in\mathbf{Z}_+}$ tends to a unique solution u. The third equation in (7.30) is equivalent to

$$\tilde{M}\sum_{i=0}^{j-1} W_1^{(i)} \leq 1\ . \tag{7.31}$$

The proof of the validity of (7.30) and (7.31) can be done by induction on j. It is readily seen that these relations are valid for $j=0$. Assume they are true for $1,..,j$; we want to prove that (7.30) and (7.31) are valid for $j+1$. We first show that the following inequalities hold:

$$E^{(i+1)} \leq (E^{(i)})^2 \beta_2 \eta_2^i$$
$$W^{(i)} \leq E^{(i)} \beta_0 \eta_0^i$$
$$W_1^{(i)} \leq E^{(i)} \beta_1 \eta_1^i \ , \qquad (7.32)$$

where the real constants β_0, β_1, β_2, η_0, η_1, η_2 are defined as in (7.26). Let $A^{(i)} \equiv \beta_0 \eta_0^i$; we prove the first in (7.32) through the following chain of inequalities:

$$\begin{aligned} E^{(i+1)} &\leq (E^{(i)})^2 \left(A^{(i)} \frac{2^{i+2}}{\xi_*} + \frac{(A^{(i)})^2 F}{2} \right) \\ &\leq (E^{(i)})^2 \left(\frac{4\beta_0 (2\eta_0)^i}{\xi_*} + \frac{\beta_0^2 \eta_0^{2i} F}{2} \right) \\ &\leq (E^{(i)})^2 \beta_2 \eta_2^i \ . \end{aligned}$$

Concerning the second relation in (7.32) one has

$$W^{(i)} \leq E^{(i)} A^{(i)} = E^{(i)} \beta_0 \eta_0^i \ .$$

Finally, the third inequality in (7.32) is obtained as follows:

$$\begin{aligned} W_1^{(i)} &\leq E^{(i)} A^{(i)} \frac{2^{i+2}}{\xi_*} \left(1 + \frac{K_1}{K_0} \right) \\ &\leq E^{(i)} \beta_1 \eta_1^i \ . \end{aligned}$$

The first relation in (7.32) yields the first in (7.30): setting

$$\mathcal{K}_0 \equiv \beta_2 \eta_2 \ ,$$

one has

$$E^{(j+1)} \leq E^{2^{j+1}} \prod_{i=0}^{j} (\beta_2 \eta_2^{j-i})^{2^i} = E^{2^{j+1}} \left[\beta_2^{\sum_{i=1}^{j+1} \frac{1}{2^i}} \eta_2^{\sum_{i=1}^{j+1} \frac{i-1}{2^i}} \right]^{2^{j+1}} < (\mathcal{K}_0 E)^{2^{j+1}} \ .$$

Let $\mathcal{K}$ satisfy the inequality

$$\sqrt{2^5 \beta_0 \eta_0 \xi_*^{-1}} \ \mathcal{K}_0 \leq \mathcal{K} \ , \qquad (7.33)$$

from the second relation in (7.32) and from (7.28) we obtain

$$\begin{aligned} \sum_{i=0}^{j} W^{(i)} &< \beta_0 E + \beta_0 \sum_{i=1}^{\infty} \eta_0^i (\mathcal{K}_0 E)^{2^i} \\ &< \beta_0 E + \beta_0 (\mathcal{K}_0 E)^2 \eta_0 \left(1 + \frac{1}{\log \frac{1}{\mathcal{K}_0 E \sqrt{\eta_0}}} \right) \\ &< \mathcal{K} E \frac{\xi_*}{4} < \xi_* \left(\frac{1}{2} - \frac{1}{2^{j+2}} \right) \ , \end{aligned} \qquad (7.34)$$

due to the following estimates:

$$\beta_0 E < \frac{\xi_*}{2^5}\mathcal{K}_0 E\ , \qquad \beta_0(\mathcal{K}_0 E)^2\eta_0 \le \frac{(\mathcal{K}E)^2\xi_*}{2^5}\ , \qquad \mathcal{K}_0 E\sqrt{\eta_0} < \frac{\mathcal{K}E}{2^{10}}\ .$$

Since

$$V^{(j+1)} \equiv V + \sum_{i=0}^{j} W^{(i)}\ ,$$

one obtains the second of (7.30). From the third in (7.32) and from $\mathcal{K}_0 E\sqrt{\eta_1} < \frac{\mathcal{K}E}{2^9}$ we get

$$\begin{aligned} 2\tilde{M}\sum_{i=0}^{j} W_1^{(i)} &= 2\tilde{M}\sum_{i=0}^{j} E^{(i)}\beta_1\eta_1^i \\ &< 2\tilde{M}\beta_1 E + 2\tilde{M}\beta_1(\mathcal{K}_0 E)^2\eta_1\left(1 + \frac{1}{\log\frac{1}{\mathcal{K}_0 E\sqrt{\eta_1}}}\right) \\ &< 2\tilde{M}\beta_1 E + 4\tilde{M}\beta_1(\mathcal{K}_0 E)^2\eta_1\ ; \end{aligned} \tag{7.35}$$

if

$$2\tilde{M}\beta_1 + 4\tilde{M}\beta_1\eta_1\mathcal{K}_0 \le \mathcal{K}\ , \tag{7.36}$$

one obtains (7.31). Notice that $\mathcal{K}$ is determined by the inequalities (7.33) and (7.36); these inequalities are satisfied provided

$$\mathcal{K} \equiv \max\{\sqrt{2^5\beta_0\eta_0\xi_*^{-1}}\beta_2\eta_2, 2\tilde{M}\beta_1(1 + 2\eta_1\beta_2\eta_2)\}\ ,$$

which is equivalent to (7.27). Finally, (7.34) and (7.35) imply (7.29). □

Remark. Let us consider the general case of a Hamiltonian function with n degrees of freedom:

$$\mathcal{H}(\underline{y},\underline{x}) = h(\underline{y}) + \varepsilon f(\underline{y},\underline{x})\ , \qquad \underline{y} \in \mathbf{R}^n\ ,\ \underline{x} \in \mathbf{T}^n\ .$$

The equations of motion are

$$\begin{aligned} \dot{\underline{x}} &= h_{\underline{y}}(\underline{y}) + \varepsilon f_{\underline{y}}(\underline{y},\underline{x}) \\ \dot{\underline{y}} &= -\varepsilon f_{\underline{x}}(\underline{y},\underline{x})\ . \end{aligned} \tag{7.37}$$

A KAM torus with rotation vector $\underline{\omega}$ is defined by the parametric equations

$$\begin{aligned} \underline{x}(\underline{\vartheta}) &= \underline{\vartheta} + \underline{u}(\underline{\vartheta}) \\ \underline{y}(\underline{\vartheta}) &= \underline{v}(\underline{\vartheta})\ , \end{aligned} \tag{7.38}$$

where $\underline{\vartheta} \in \mathbf{T}^n$ with $\dot{\underline{\vartheta}} = \underline{\omega}$ and $\underline{u}$, $\underline{v}$ are suitable vector functions. Let us introduce the operator $D \equiv \underline{\omega}\ \frac{\partial}{\partial\underline{\vartheta}}$. Inserting (7.38) in (7.37), one finds that $\underline{u}$ and $\underline{v}$ must satisfy the following quasi–linear partial differential equations on $\mathbf{T}^n$:

$$\begin{aligned} \underline{\omega} + D\underline{u} - h_{\underline{y}}(\underline{v}) - \varepsilon f_{\underline{y}}(\underline{v},\underline{\vartheta}+\underline{u}) &= \underline{0} \\ D\underline{v} + \varepsilon f_{\underline{x}}(\underline{v},\underline{\vartheta}+\underline{u}) &= \underline{0}\ . \end{aligned} \tag{7.39}$$

The KAM proof is obtained by solving (7.39) through a Newton iteration method, extending the procedure as it was described for finding the solution of (7.13).

7.2.2 The initial approximation and the estimate of the error term

The initial approximation $v \equiv v^{(0)}$ (see (7.14)) of the KAM theorem can be obtained taking advantage of the analyticity of the KAM surfaces with respect to the perturbing parameter in a neighborhood of the origin [139–141]). We consider the parametrization (7.10), where the function $u = u(\vartheta, t)$ depends parametrically on ε and therefore we denote it as $u = u(\vartheta, t; \varepsilon)$. Let us expand u in power series as

$$u(\vartheta, t; \varepsilon) = \sum_{k=1}^{\infty} u_k(\vartheta, t)\, \varepsilon^k \,. \tag{7.40}$$

In the case of the spin–orbit problem the coefficients u_k can be recursively computed as follows. Write equation (7.13) with the perturbation given by (7.2) as

$$D^2 u + \varepsilon \sum_{m\neq 0, m=N_1}^{N_2} W\left(\frac{m}{2}, \mathrm{e}\right) \sin(2\vartheta + 2u - mt) = 0 \,. \tag{7.41}$$

For u expanded as in (7.40), define the power series

$$e^{i(2\vartheta+2u)} \equiv \sum_{n=0}^{\infty} c_n(\vartheta, t)\, \varepsilon^n \,, \tag{7.42}$$

for some unknown complex coefficients c_n which can be determined as follows. Differentiating (7.42) with respect to ε and using the series expansion (7.40), one obtains

$$2i \sum_{k=1}^{\infty} k u_k \varepsilon^{k-1} \cdot \sum_{j=0}^{\infty} c_j \varepsilon^j = \sum_{n=1}^{\infty} n c_n \varepsilon^{n-1} \,.$$

Equating same powers of ε one obtains:

$$\begin{aligned} c_0(\vartheta, t) &\equiv e^{2i\vartheta} \\ c_n(\vartheta, t) &\equiv \frac{2i}{n} \sum_{k=1}^{n} k u_k c_{n-k} \,. \end{aligned} \tag{7.43}$$

Finally, (7.41) can be written as

$$D^2 u = -\frac{1}{2i} \sum_{n=1}^{\infty} \varepsilon^n \left[\sum_{m\neq 0, m=N_1}^{N_2} W\left(\frac{m}{2}, \mathrm{e}\right) (e^{-imt} c_{n-1} - e^{imt} \bar{c}_{n-1}) \right] ,$$

where the bar denotes complex conjugacy. A recursive relation defining the functions u_n is obtained comparing the terms of the same order in ε:

$$u_n(\vartheta, t) \equiv -\frac{1}{2i} D^{-2} \left[\sum_{m\neq 0, m=N_1}^{N_2} W\left(\frac{m}{2}, \mathrm{e}\right) (e^{-imt} c_{n-1} - e^{imt} \bar{c}_{n-1}) \right] . \tag{7.44}$$

Notice that u_n depends on the previous functions $u_1, \dots, u_{n-1}$. The initial approximation can be obtained as the finite truncation up to a suitable order k_0 (for some positive integer k_0) of the series expansion (7.40):

$$v^{(0)}(\vartheta, t; \varepsilon) \equiv \sum_{k=1}^{k_0} u_k(\vartheta, t)\, \varepsilon^k \ . \tag{7.45}$$

To give a concrete example, let us assume that the perturbing function in (7.2) is given by

$$\begin{aligned} f(x,t) \equiv\ & \frac{\mathrm{e}}{4}\cos(2x-t) - \left(\frac{1}{2} - \frac{5}{4}\mathrm{e}^2\right)\cos(2x-2t) \\ -\ & \frac{7}{4}\mathrm{e}\cos(2x-3t) - \frac{17}{4}\mathrm{e}^2\cos(2x-4t) \ . \end{aligned}$$

Then, the first two approximating functions $u_1(\vartheta,t)$ and $u_2(\vartheta,t)$ are given by the following expressions:

$$\begin{aligned} u_1(\vartheta,t) =\ & \frac{-\mathrm{e}}{2(2\omega-1)^2}\sin(2\vartheta-t) + \frac{(1-\frac{5}{2}\mathrm{e}^2)}{(2\omega-2)^2}\sin(2\vartheta-2t) + \\ +\ & \frac{7\mathrm{e}}{2(2\omega-3)^2}\sin(2\vartheta-3t) + \frac{17\mathrm{e}^2}{2(2\omega-4)^2}\sin(2\vartheta-4t) \end{aligned}$$

and

$$\begin{aligned} u_2(\vartheta,t) =\ & \left[-\frac{\mathrm{e}}{2(2\omega-1)^2} + \frac{4\mathrm{e}}{(2\omega-2)^2} - \frac{7\mathrm{e}}{2(2\omega-3)^2}\right]\sin t + \\ +\ & \frac{1}{4}\left[-\frac{7\mathrm{e}^2}{4(2\omega-1)^2} + \frac{17\mathrm{e}^2}{2(2\omega-2)^2} + \frac{7\mathrm{e}^2}{4(2\omega-3)^2} - \frac{17\mathrm{e}^2}{2(2\omega-4)^2}\right]\sin 2t \\ +\ & \frac{\mathrm{e}^2}{4(2\omega-1)^2}\,\frac{\sin(4\vartheta-2t)}{(4\omega-2)^2} \\ +\ & \left[-\frac{\mathrm{e}}{2(2\omega-2)^2} - \frac{\mathrm{e}}{2(2\omega-1)^2}\right]\frac{\sin(4\vartheta-3t)}{(4\omega-3)^2} + \\ +\ & \left[\frac{1-5\mathrm{e}^2}{(2\omega-2)^2} - \frac{7\mathrm{e}^2}{4}\left(\frac{1}{(2\omega-1)^2} + \frac{1}{(2\omega-3)^2}\right)\right]\frac{\sin(4\vartheta-4t)}{(4\omega-4)^2} + \\ +\ & \left[\frac{7\mathrm{e}}{2(2\omega-2)^2} + \frac{7\mathrm{e}}{2(2\omega-3)^2} + \right]\frac{\sin(4\vartheta-5t)}{(4\omega-5)^2} + \\ +\ & \left[\frac{17\mathrm{e}^2}{2(2\omega-2)^2} + \frac{49\mathrm{e}^2}{4(2\omega-3)^2} + \frac{17\mathrm{e}^2}{2(2\omega-4)^2}\right]\frac{\sin(4\vartheta-6t)}{(4\omega-6)^2} \ . \end{aligned}$$

To implement the KAM algorithm and to check the KAM condition, it is necessary to provide explicit estimates on some quantities, like the initial approximation, its derivative, the error term, etc. The most difficult task is the estimate of the error function $|\eta^{(0)}|_{\xi,\rho}$ (for some positive parameters ξ, ρ) associated to a given initial approximation $v^{(0)}$, which can be constructed by means of the recursive formulae (7.43), (7.44). The estimate of $\eta^{(0)}$ can be obtained through the following Lemma (see also [27]).

Lemma (Estimate of the error term). *Let* $v^{(0)}(\vartheta,t;\varepsilon) \equiv \sum_{k=1}^{k_0} u_k(\vartheta,t)\ \varepsilon^k$ *for some positive integer* k_0 *and let* $\eta \equiv \eta^{(0)}$ *satisfy* (7.14) *with* $v \equiv v^{(0)}$. *For some positive parameters* ξ, ρ, *let* $S^{(0)} \equiv \|v^{(0)}\|_{\xi,\rho}$, $U_k \equiv \|u_k\|_{\xi,\rho}$ *and* $\bar{F} \equiv \|f_x\|_{\xi,\rho}$. *Define recursively the sequences* $\{\alpha_j\}$, $\{\beta_j\}$ *as*

$$
\begin{aligned}
\alpha_0 &= 1 \\
\alpha_j &= \frac{2}{j}\sum_{k=1}^{j} kU_k\alpha_{j-k}\ , \qquad j \geq 1
\end{aligned}
$$

and

$$
\begin{aligned}
\beta_0 &= 1 \\
\beta_j &= -\frac{2}{j}\sum_{k=1}^{j} kU_k\beta_{j-k}\ , \qquad j \geq 1\ .
\end{aligned}
$$

Then, setting

$$
\begin{aligned}
a &= e^{2S^{(0)}} - \sum_{j=1}^{k_0-1} \alpha_j\rho^j \\
b &= e^{-2S^{(0)}} - \sum_{j=1}^{k_0-1} \beta_j\rho^j\ ,
\end{aligned}
$$

the error term is estimated as

$$
\|\eta^{(0)}\|_{\xi,\rho} = \bar{F}\sqrt{\frac{a^2+b^2}{2}}\ .
$$

We remark that in concrete applications the convergence of the KAM algorithm is improved as the order k_0 of the initial approximation (7.45) gets larger. Indeed, let us denote by $\varepsilon_{KAM}^{(k_0)} = \varepsilon_{KAM}^{(k_0)}(\omega)$ the lower bound provided by the KAM theorem on the persistence of the invariant torus with frequency ω, starting from the initial approximation (7.45) truncated at the order k_0. We report in Table 7.1 some results associated to (5.19) for the frequency $\omega = 1 + \frac{1}{2+\frac{\sqrt{5}-1}{2}}$; the results concern the values

Table 7.1. The threshold $\varepsilon_{KAM}^{(k_0)}(\omega)$ as a function of the order k_0 of the initial approximation.

k_0	$\varepsilon_{KAM}^{(k_0)}(\omega)$
1	$2 \cdot 10^{-5}$
5	$1.5 \cdot 10^{-3}$
10	$4.1 \cdot 10^{-3}$
15	$6 \cdot 10^{-3}$
20	$6.6 \cdot 10^{-3}$
25	$7.5 \cdot 10^{-3}$
30	$8.2 \cdot 10^{-3}$

$\varepsilon_{KAM}^{(k_0)}(\omega)$ as the order k_0 of the initial approximation increases (here we selected $\xi = 0.05$). We remark that the relative improvement of the threshold $\varepsilon_{KAM}^{(k_0)}(\omega)$ is higher as k_0 is small, while it gets smaller as k_0 increases.

7.2.3 Diophantine rotation numbers

One of the assumptions which is required to apply the KAM theorem is that the frequency of the motion must satisfy the diophantine condition (7.6). Moreover, we recall that the KAM estimates depend on the value of the diophantine constant (see, e.g., (7.26), (7.27), (7.28)) and a proper choice of the frequency certainly improves the performances of the theorem. In this section we review some results from number theory concerning the choice of diophantine numbers and the computation of the corresponding diophantine constants.

We start by introducing the *continued fraction* expansion of a positive real number α defined as the sequence of positive integer numbers $a_0, a_1, a_2, \ldots$, such that

$$\alpha \equiv a_0 + \frac{1}{a_1 + \frac{1}{a_2 + \frac{1}{a_3 + \ldots}}} , \qquad a_j \in \mathbf{Z}_+ . \tag{7.46}$$

Using standard notation, we shall write

$$\alpha \equiv [a_0; \, a_1, \, a_2, \, a_3, \, \ldots] .$$

A rational number has a finite continued fraction expansion, while irrationals have an infinite continued fraction expansion. For any irrational number α there exists an infinite *approximant* sequence of rational numbers, say $\{\frac{p_n}{q_n}\}_{n \in \mathbf{Z}_+}$, such that $\frac{p_n}{q_n}$ converges to α as n goes to infinity. Each $\frac{p_n}{q_n}$ can be obtained as the truncation to the order n of the continued fraction expansion (7.46):

$$\begin{aligned} \frac{p_0}{q_0} &= a_0 \\ \frac{p_1}{q_1} &= a_0 + \frac{1}{a_1} \\ \frac{p_2}{q_2} &= a_0 + \frac{1}{a_1 + \frac{1}{a_2}} \\ &\ldots \end{aligned}$$

For the golden number $\gamma = \frac{\sqrt{5}-1}{2}$, the rational approximants are given by the ratio of the Fibonacci's numbers:

$$\frac{0}{1}, \frac{1}{1}, \frac{1}{2}, \frac{2}{3}, \frac{3}{5}, \frac{5}{8}, \frac{8}{13}, \frac{13}{21}, \frac{21}{34}, \ldots$$

A bound on how close the rational numbers $\frac{p_n}{q_n}$ approximate α is given by the following inequalities:

$$\frac{1}{q_n(q_n + q_{n+1})} < \left| \alpha - \frac{p_n}{q_n} \right| \leq \frac{1}{q_n q_{n+1}} .$$

Definition. An *algebraic* number ω is a solution of a polynomial $P_n(z)$ of degree n with integer coefficients, say $c_0, \ldots, c_n$:

$$P_n(z) = c_n z^n + c_{n-1} z^{n-1} + \cdots + c_1 z + c_0 \ , \tag{7.47}$$

provided ω is not a solution of a polynomial of lower degree with integer coefficients. A *quadratic* number is an algebraic number of degree 2. An irrational number α is called a *noble number* if the terms of its continued fraction expansion (7.46) are definitely one, namely there exists an integer N such that $a_k = 1$ for all $k > N$. In this case we write

$$\alpha \equiv [a_0;\ a_1,\ \ldots,\ a_N,\ 1^\infty] \ ;$$

the number $[a_0;\ a_1,\ \ldots,\ a_N]$ is called the *head* of the noble number.

Noble numbers are a subset of the quadratic irrationals, which are in turn a subset of the algebraic irrationals. By a theorem due to Liouville one can show that an algebraic number is diophantine [104].

Theorem (Liouville). *Let ω be an algebraic number of degree n; then ω satisfies the diophantine condition* (7.7) *for some positive constant C and for $\tau = n - 1$.*

Proof. Let ω be a root of (7.47) so that we can write

$$P_n(z) = (z - \omega) P_{n-1}(z) \ , \tag{7.48}$$

for a suitable polynomial $P_{n-1}(z)$ of degree $n-1$. It is $P_{n-1}(\omega) \neq 0$, otherwise we could write $P_n(z) = (z-\omega)^2 P_{n-2}(z)$ for some polynomial $P_{n-2}(z)$. In this case $\frac{d}{dz} P_n(\omega) = 0$, in contrast to the assumption that ω is an algebraic number of degree n, being $\frac{d}{dz} P_n(z)$ a polynomial of degree $n-1$ with integer coefficients. Therefore there exists $\delta > 0$ such that $P_{n-1}(z) \neq 0$ for any $|z - \omega| \leq \delta$. If p, q are integer numbers such that $|\omega - \frac{p}{q}| \leq \delta$, from (7.48) we can write

$$\frac{p}{q} - \omega = \frac{P_n(\frac{p}{q})}{P_{n-1}(\frac{p}{q})} = \frac{c_0 q^n + c_1 p q^{n-1} + \cdots + c_n p^n}{q^n P_{n-1}(\frac{p}{q})} \ . \tag{7.49}$$

The numerator of the last expression in (7.49) is an integer greater or equal than one; let

$$M \equiv \sup_{|z-\omega| \leq \delta} |P_{n-1}(z)| \ .$$

Then we obtain

$$\left| \frac{p}{q} - \omega \right| \geq \frac{1}{M q^n} \ .$$

On the other hand, if $|\frac{p}{q} - \omega| > \delta$, then $|\frac{p}{q} - \omega| > \frac{\delta}{q^n}$, so that (7.7) is satisfied by defining

$$C \equiv \left(\min\left(\delta, \frac{1}{M} \right) \right)^{-1} \ .$$

□

We stress that there exist diophantine numbers which are not algebraic numbers. The set of diophantine numbers with constant C and exponent τ, say $D(C,\tau)$, has measure one as C tends to zero. For example, the measure $\mu(D(C,\tau)^c)$ of the complement $D(C,\tau)^c$ of the set $D(C,\tau)$ in the interval $[0,1]$ can be computed as follows. For any coprime integers m, n, one has

$$\mu(D(C,\tau)^c) = \sum_{n=1}^{\infty}\sum_{m=1}^{n} \frac{C}{n^{\tau+1}} = C\sum_{n=1}^{\infty}\frac{\phi(n)}{n^{\tau+1}} = C\frac{\zeta(\tau)}{\zeta(\tau+1)}\ ,$$

where $\phi(n)$ is the Euler function and $\zeta(\tau)$ is the Riemann zeta function. In conclusion, $\mu(D(C,\tau)^c)$ tends to zero as C tends to zero for any $\tau \geq 1$. The set of diophantine numbers is the union of the sets $D(C,\tau)$ for any positive C and τ.

7.2.4 Trapping diophantine numbers

For Hamiltonian systems like the spin–orbit problem, the KAM tori separate the phase space into invariant regions. One can make use of this property to trap periodic orbits between two KAM tori with suitable rotation numbers bounding the frequency of the periodic orbit from above and below. In this section we address the question concerning the choice of the bounding rotation numbers. In particular, the stability of the resonance of order $p:q$ (for some integers p, q with $q \neq 0$) can be inferred by proving the existence of a pair of invariant tori with frequency bounding the $p:q$ resonance from above and below. Having in mind an application of KAM theorem to the spin–orbit problem, we focus our attention on the 1:1 and 3:2 resonances. Let the golden ratio be $\gamma = \frac{\sqrt{5}-1}{2}$; a possible choice of trapping diophantine numbers for $p = q = 1$ is given by the sequences of noble numbers defined as

$$\begin{aligned}\Gamma_k &\equiv [0;\ 1,\ k-1,\ 1^\infty] \equiv 1 - \frac{1}{k+\gamma}\ ,\\ \Delta_k &\equiv [1;\ k,\ 1^\infty] \equiv 1 + \frac{1}{k+\gamma}\ , \qquad k \geq 2\ .\end{aligned} \tag{7.50}$$

Both Γ_k and Δ_k converge to one from below and above, respectively, and have the property that for all k, $|\Gamma_k - 1| = |\Delta_k - 1|$. Notice that Γ_k and Δ_k are noble algebraic numbers of degree two, since they are roots of the polynomials

$$\begin{aligned}P_{\Gamma_k}(x) &\equiv 4(k^4-2k^3-k^2+2k+1)x^2 - 4(2k^4-6k^3+k^2+5k+1)x\ +\\ &+\ 4k^4-16k^3+12k^2+8k-4\end{aligned}$$

and

$$\begin{aligned}P_{\Delta_k}(x) &\equiv 4(k^4-2k^3-k^2+2k+1)x^2 - 4(2k^4-2k^3-5k^2+3k+3)x\ +\\ &+\ 4k^4-12k^2+4\ .\end{aligned}$$

We remark that noble tori are conjectured to be the last surfaces to disappear in any interval of rotation numbers ([124, 125, 148], see also [62]). Numerical experiments on the standard and quadratic maps [124] show that noble tori are locally the most robust in the sense that

(*i*) for any critical (i.e., close to breakdown) noble surface of rotation number ω, there exists an interval around ω containing no other invariant tori;

(*ii*) let $\mathcal{T}_\varepsilon(\alpha)$ be a critical non–noble torus; then in any interval around α there always exists a non–critical noble.

Concerning the 3:2 resonance we can consider the trapping rotation numbers

$$\begin{aligned} \Gamma'_k &\equiv \frac{3}{2} - \frac{1}{k+\gamma} , \\ \Delta'_k &\equiv \frac{3}{2} + \frac{1}{k+\gamma} , \qquad\qquad k \geq 2 , \end{aligned} \tag{7.51}$$

converging to $\frac{3}{2}$ from below and above, respectively, and with $|\Gamma'_k - \frac{3}{2}| = |\Delta'_k - \frac{3}{2}|$ for any k. Notice that Γ'_k and Δ'_k are not necessarily noble numbers, but they are second–order algebraic numbers, since they are roots of the polynomials

$$\begin{aligned} P_{\Gamma'_k}(x) &\equiv 4(k^4 - 2k^3 - k^2 + 2k + 1)x^2 \ - \ 4(3k^4 - 8k^3 + 7k + 2)x \\ &+ \ 9k^4 - 30k^3 + 13k^2 + 20k - 1 \end{aligned}$$

and

$$\begin{aligned} P_{\Delta'_k}(x) &\equiv 4(k^4 - 2k^3 - k^2 + 2k + 1)x^2 \ - \ 4(3k^4 - 4k^3 - 6k^2 + 5k + 4)x \\ &+ \ 9k^4 - 6k^3 - 23k^2 + 8k + 11 \ . \end{aligned}$$

The computation of the diophantine constant C for the numbers Γ_k, Δ_k, Γ'_k, Δ'_k can be performed as follows.

Proposition. *Let* Γ_k, Δ_k, Γ'_k, Δ'_k *be as in* (7.50), (7.51); *then for any* $k \geq 2$ *the corresponding diophantine constants are, respectively,*

$$k+\gamma, \quad k+\gamma, \quad 4(k+\gamma), \quad 4(k+\gamma) \ .$$

Proof. Let us provide the details for the derivation of the diophantine constant associated to Δ_k; the computations for the other numbers follow easily. We want to show that

$$\left| \left(1 + \frac{1}{k+\gamma}\right) - \frac{p}{q} \right| \geq \frac{1}{(k+\gamma)q^2} \qquad \text{for all } p,\ q \in \mathbf{Z},\ \ q \neq 0 \ ,$$

which is equivalent to require that

$$\left| \frac{1}{k+\gamma} - \frac{p}{q} \right| \geq \frac{1}{(k+\gamma)q^2} \qquad \text{for all } p,\ q \in \mathbf{Z}\ ,\ \ q \neq 0 \ . \tag{7.52}$$

The rational approximants to $\frac{1}{k+\gamma}$ are given by

$$\left\{\frac{p_j}{q_j}\right\}_{j\geq 0} \equiv \left\{\frac{\alpha_j}{\alpha_j k+\alpha_{j-1}}\right\}_{j\geq 0},$$

where the α_j's are the Fibonacci's numbers defined via the recursive relation

$$\alpha_0=1,\ \alpha_1=1,\ \ldots,\ \alpha_{j+1}=\alpha_j+\alpha_{j-1} \qquad \text{for all } j\geq 1\ ;$$

then, it is sufficient to show (7.52) with $\frac{p}{q}$ replaced by the approximant $\frac{\alpha_j}{\alpha_j k+\alpha_{j-1}}$, namely

$$\left|\frac{1}{k+\gamma} - \frac{\alpha_j}{\alpha_j k+\alpha_{j-1}}\right| \geq \frac{1}{(k+\gamma)(\alpha_j k+\alpha_{j-1})^2} \qquad \text{for all } k\geq 2\ . \tag{7.53}$$

From (7.53) one gets the inequality

$$\left|1 - \frac{\alpha_j(k+\gamma)}{\alpha_j k+\alpha_{j-1}}\right| \geq \frac{1}{(\alpha_j k+\alpha_{j-1})^2},$$

which is equivalent to

$$\left|\gamma - \frac{\alpha_{j-1}}{\alpha_j}\right| \geq \frac{1}{\alpha_j^2(k+\frac{\alpha_{j-1}}{\alpha_j})}\ .$$

Since

$$k + \frac{\alpha_{j-1}}{\alpha_j} \geq 2 + \frac{\alpha_{j-1}}{\alpha_j},$$

it is sufficient to show that

$$\left|\gamma - \frac{\alpha_{j-1}}{\alpha_j}\right| \geq \frac{1}{\alpha_j^2(2+\frac{\alpha_{j-1}}{\alpha_j})}\ .$$

Defining A_j by the equality

$$\left|\gamma - \frac{\alpha_{j-1}}{\alpha_j}\right| \equiv \frac{1}{A_j\alpha_j^2},$$

it is readily seen that

$$A_j = \gamma + 1 + \frac{\alpha_{j-1}}{\alpha_j}\ ; \tag{7.54}$$

therefore we get that

$$\left|\gamma - \frac{\alpha_{j-1}}{\alpha_j}\right| = \frac{1}{\alpha_j^2(\gamma+1+\frac{\alpha_{j-1}}{\alpha_j})} \geq \frac{1}{\alpha_j^2(2+\frac{\alpha_{j-1}}{\alpha_j})},$$

since

$$2 + \frac{\alpha_{j-1}}{\alpha_j} \geq \gamma + 1 + \frac{\alpha_{j-1}}{\alpha_j}\ .$$

Applying the same procedure one proves that

$$\left|\Gamma_k - \frac{p}{q}\right| \geq \frac{1}{(k+\gamma)q^2} \qquad \text{for all } p,q \in \mathbf{Z}\ , \quad q \neq 0\ , \quad k \geq 2\ ,$$

where the sequence of rational approximants to Γ_k is given by

$$\left\{ \frac{\alpha_j(k-1)+\alpha_{j-1}}{\alpha_j k + \alpha_{j-1}} \right\} .$$

Analogous considerations hold for Γ'_k and Δ'_k. □

We remark that for the golden ratio equation (7.54) implies that the diophantine constant is equal to $C = \frac{3+\sqrt{5}}{2}$.

7.2.5 Computer–assisted proofs

The computation of the initial approximation and the control of the KAM algorithm usually require the use of a computer, due to the high number of operations involved. However, the computer introduces rounding–off and propagation errors. In order to leave unaltered the rigorous character of the result, one can keep track of the computer rounding–off errors through the application of the so–called *interval arithmetic* technique [60, 106], whose implementation is briefly explained as follows. The computer stores real numbers using a sign–exponent–fraction representation; the number of digits in the fraction and the exponent varies with the machine. The result of any elementary operation, i.e. sum, subtraction, multiplication and division, usually produces an approximation of the true result; other calculations, like exponent, square root, logarithm, etc., can be reduced to a sequence of elementary operations through a Taylor series expansion. The idea of the interval arithmetic technique is to represent any real number as an interval and to perform elementary operations on intervals, rather than on real numbers. For example, suppose we perform the sum of two numbers a and b, which are contained, respectively, within the intervals $[a_1, a_2]$ and $[b_1, b_2]$. Adding these two intervals one obtains $[c_1, c_2] \equiv [a_1 + b_1, a_2 + b_2]$. However, we have to consider that the end–points c_1, c_2 of the new interval are themselves produced by an elementary operation and therefore they are affected by rounding errors. Henceforth one needs to construct a new interval which gets rid of the fact that c_1 and c_2 are rounded. This can be done as follows. Let δ be the limiting precision of the machine (see, e.g., [159]). Then, multiply c_1 by $1 \mp \delta$ according to whether c_1 is positive or negative and let us call the final result $c_- \equiv down(c_1)$. Similarly, to get an upper bound of c_2 multiply it by $1 \pm \delta$ according to whether c_2 is positive or negative; let us call the final result $c_+ \equiv up(c_2)$. We finally get that $a + b \in [c_-, c_+]$. The subtraction can be treated in a similar way.

Concerning the multiplication (as well as the division), one needs to consider different cases according to the signs of the factors. More precisely, suppose we compute the multiplication $a \cdot b$, where a and b are represented by the intervals $[a_1, a_2]$ and $[b_1, b_2]$, while the result will be contained in $[c_-, c_+]$. We must distinguish the following cases:

(1) $a_1 \geq 0$ and $b_1 \geq 0$, then $c_- = down(a_1 b_1)$, $c_+ = up(a_2 b_2)$;
(2) $a_1 \geq 0$ and $b_2 \leq 0$, then $c_- = down(a_2 b_1)$, $c_+ = up(a_1 b_2)$;
(3) $a_1 \geq 0$ and $b_1 < 0$, $b_2 > 0$, then $c_- = down(a_2 b_1)$, $c_+ = up(a_2 b_2)$;
(4) $a_2 \leq 0$ and $b_1 \geq 0$, then $c_- = down(a_1 b_2)$, $c_+ = up(a_2 b_1)$;
(5) $a_2 \leq 0$ and $b_2 \leq 0$, then $c_- = down(a_2 b_2)$, $c_+ = up(a_1 b_1)$;
(6) $a_2 \leq 0$ and $b_1 < 0$, $b_2 > 0$, then $c_- = down(a_1 b_2)$, $c_+ = up(a_1 b_1)$;
(7) $a_1 < 0$, $a_2 > 0$ and $b_1 \geq 0$, then $c_- = down(a_1 b_2)$, $c_+ = up(a_2 b_2)$;
(8) $a_1 < 0$, $a_2 > 0$ and $b_2 \leq 0$, then $c_- = down(a_2 b_1)$, $c_+ = up(a_1 b_1)$;
(9) $a_1 < 0$, $a_2 > 0$ and $b_1 < 0$, $b_2 > 0$, then
(9a) let $\ell_- = down(a_1 b_2)$, $r_- = down(a_2 b_1)$; if $r_- < \ell_-$ then $\ell_- = r_-$;
(9b) let $\ell_+ = up(a_1 b_1)$, $r_+ = up(a_2 b_2)$; if $r_+ > \ell_+$ then $\ell_+ = r_+$;
set $b_1 = \ell_-$, $b_2 = \ell_+$.

A similar approach is used to deal with the division. Casting together the elementary operations on intervals one obtains the implementation of the interval arithmetic technique, where complex operations are reduced to a sequence of elementary operations by using their series expansion.

7.3 A survey of KAM results in Celestial Mechanics

7.3.1 Rotational tori in the spin–orbit problem

We consider the spin–orbit problem widely discussed in the previous sections and we aim to prove the existence of rotational invariant tori, trapping the synchronous resonance from above and below, thus providing a confinement property of the dynamics in the phase space. As a specific example we consider the Earth–Moon system. In writing the model (7.1)–(7.2) we have neglected all perturbations due to other celestial bodies as well as dissipative effects. Among the discarded contributions the most important term is due to the tidal torque generated by the non–rigidity of the satellite. For consistency, we expand the perturbing function in Fourier–Taylor series, neglecting all terms which are of the same order or less than the neglected tidal torque. Taking into account that the eccentricity of the Moon amounts to $\mathrm{e} = 0.0549$, one is led to consider the perturbing function (7.2) with $N_1 = 1$ and $N_2 = 7$. The corresponding Hamiltonian function reads as

$$\begin{aligned}\mathcal{H}(y,x,t) \equiv \frac{y^2}{2} - \varepsilon \Bigg[& \Big(-\frac{\mathrm{e}}{4} + \frac{\mathrm{e}^3}{32} \Big) \cos(2x-t) + \\ & + \Big(\frac{1}{2} - \frac{5}{4}\mathrm{e}^2 + \frac{13}{32}\mathrm{e}^4 \Big) \cos(2x-2t) + \Big(\frac{7}{4}\mathrm{e} - \frac{123}{32}\mathrm{e}^3 \Big) \cos(2x-3t) + \\ & + \Big(\frac{17}{4}\mathrm{e}^2 - \frac{115}{12}\mathrm{e}^4 \Big) \cos(2x-4t) + \Big(\frac{845}{96}\mathrm{e}^3 - \frac{32\,525}{1536}\mathrm{e}^5 \Big) \cos(2x-5t) + \\ & + \frac{533}{32}\mathrm{e}^4 \cos(2x-6t) + \frac{228\,347}{7680}\mathrm{e}^5 \cos(2x-7t) \Bigg] , \end{aligned} \tag{7.55}$$

where the physical value of the perturbing parameter amounts for the Moon to $\varepsilon \simeq 3.45 \cdot 10^{-4}$. The existence of two bounding tori with frequencies Γ_{40} and Δ_{40} (see (7.50)) has been proven in [23] by performing the following steps. Compute the initial approximation (7.45) up to the order $k_0 = 15$; apply the KAM theorem presented in Section 7.2; implement the interval arithmetic technique. Then, one gets [23] that the synchronous motion of the Moon is trapped in the region enclosed by the tori $\mathcal{T}(\Gamma_{40})$ and $\mathcal{T}(\Delta_{40})$, which is shown to be a subset of $\{(y,x,t) : (x,t) \in \mathbf{T}^2,\ 0.97 \le y \le 1.03\}$.

In a similar way one can prove the stability of the Mercury–Sun system. However, due to the bigger eccentricity of Mercury, being $\mathrm{e} = 0.2056$, the perturbing function contains a larger number of terms, so that the corresponding Hamiltonian is given by

$$\mathcal{H}(y,x,t) \equiv \frac{y^2}{2} - \frac{\varepsilon}{2} \sum_{m\neq 0, m=-11}^{3} W\left(\frac{m}{2}, \mathrm{e}\right) \cos(2x - mt) \ ,$$

with the coefficients $W(\frac{m}{2}, \mathrm{e})$ truncated to $O(\mathrm{e}^7)$. The stability of the observed 3:2 resonance is obtained for the true value of the perturbing parameter, i.e. $\varepsilon = 1.5 \cdot 10^{-4}$, by proving the existence of the tori with frequencies Γ'_{70} and Δ'_{70} (see (7.51)); the corresponding trapping region is contained in $\{(y,x,t) : (x,t) \in \mathbf{T}^2,\ 1.48 \le y \le 1.52\}$.

7.3.2 Librational invariant surfaces in the spin–orbit problem

The confinement of the motion associated to periodic orbits of the spin–orbit problem can also be obtained by constructing *librational* invariant surfaces. In the following we provide some details of the proof concerning the case of the 1:1 resonance (see [24]), whose outline is the following. The first task is to center the Hamiltonian on the 1:1 periodic orbit and to expand in Taylor series around the new origin. Next, diagonalize the quadratic terms to obtain a harmonic oscillator, perturbed by higher degree (time–dependent) terms. After introducing the action–angle variables associated to the harmonic oscillator, implement a Birkhoff normal form to reduce the size of the perturbation and then apply the KAM theorem to prove the existence of trapping librational tori.

According to the above strategy, we start by writing the Hamiltonian function as

$$\mathcal{H}_0(y,x,t) = \frac{y^2}{2} - \varepsilon a \, \cos(2x - 2t) - \frac{\varepsilon}{2} \sum_{m\neq 0,2,\ m=N_1}^{N_2} W\left(\frac{m}{2}, \mathrm{e}\right) \cos(2x - mt) \, , \qquad (7.56)$$

where $a \equiv \frac{1}{2} W(1, \mathrm{e})$. Perform the coordinate change $x' = 2x - 2t$, $y' = \frac{1}{2}(y-1)$, expand in Taylor series around the origin and diagonalize the time–independent quadratic terms by means of the symplectic transformation

$$\begin{aligned} p &= \alpha\, y' \\ q &= \beta\, x' \ , \end{aligned}$$

with $\alpha=\frac{\sqrt{2}}{(\varepsilon a)^{1/4}}$, $\beta=\frac{(\varepsilon a)^{1/4}}{\sqrt{2}}$. After these steps the Hamiltonian function becomes

$$\begin{aligned}\mathcal{H}_1(p,q,t) &= \frac{\omega}{2}(p^2+q^2)\ -\varepsilon a\left(\frac{q^4}{4!\beta^4}-\frac{q^6}{6!\beta^6}+\dots\right)\\ &- \frac{\mu}{2}\sum_{m\neq 0,-2}\tilde{W}\left(\frac{m+2}{2},\mathrm{e}\right)\left[\cos(mt)\left(1-\frac{q^2}{2\beta^2}+\frac{q^4}{4!\beta^4}+\dots\right)\right.\\ &+ \left.\sin(mt)\left(\frac{q}{\beta}-\frac{q^3}{3!\beta^3}+\frac{q^5}{5!\beta^5}+\dots\right)\right],\end{aligned}$$

where $\omega=2\sqrt{\varepsilon a}$ is the frequency of the harmonic oscillation, $\mu\equiv\varepsilon\mathrm{e}$, while the coefficients W have been rescaled as $\tilde{W}(\frac{m+2}{2},\mathrm{e})=\frac{1}{\mathrm{e}}W(\frac{m+2}{2},\mathrm{e})$. Introduce action–angle variables (I,φ) as

$$\begin{aligned}p &= \sqrt{2I}\ \cos\varphi\\ q &= \sqrt{2I}\ \sin\varphi\ ;\end{aligned}$$

the resulting Hamiltonian is given by

$$\begin{aligned}&\mathcal{H}_2(I,\varphi,t)=\omega I\ -\varepsilon a\left(\frac{I^2}{16\beta^4}-\frac{5I^3}{2\cdot 6!\beta^6}+\dots\right)\\ &-\ \varepsilon a\left[-\frac{I^2}{12\beta^4}\cos 2\varphi+\frac{I^2}{48\beta^4}\cos 4\varphi+\frac{I^3}{4\cdot 6!\beta^6}\ \cdot\right.\\ &\cdot\ \left.(15\cos 2\varphi-6\cos 4\varphi+\cos 6\varphi)+\dots\right]\\ &-\ \frac{\varepsilon\mathrm{e}}{2}\sum_{m\neq 0,-2}\tilde{W}(\frac{m+2}{2},\mathrm{e})\left\{\cos(mt)\left[1-\frac{I}{2!\beta^2}(1-\cos 2\varphi)+\frac{I^2}{8\cdot 3!\beta^4}\ \cdot\right.\right.\\ &\cdot\ \left.(3-4\cos 2\varphi+\cos 4\varphi)-\frac{I^3}{4\cdot 6!\beta^6}\ (10-15\cos 2\varphi+6\cos 4\varphi-\cos 6\varphi)+\dots\right]\\ &+\ \sin(mt)\left[\frac{\sqrt{2I}}{\beta}\sin\varphi-\frac{\sqrt{2}\ I^{3/2}}{12\beta^3}(3\sin\varphi-\sin 3\varphi)\right.\\ &+\ \left.\left.\frac{\sqrt{2}I^{5/2}}{4\cdot 5!\beta^5}\ (10\sin\varphi-5\sin 3\varphi+\sin 5\varphi)+\dots\right]\right\},\end{aligned}$$

which can be written in compact form as

$$\mathcal{H}_2(I,\varphi,t)\ =\ \omega I+\varepsilon\overline{h}(I)+\varepsilon\tilde{h}(I,\varphi)+\varepsilon\mathrm{e}f(I,\varphi,t)$$

with the obvious identification of the functions $\overline{h}$, h and f. A Birkhoff normal form can be implemented to reduce the size of the perturbation $R(I,\varphi,t)\equiv\tilde{h}(I,\varphi)+$

$ef(I,\varphi,t)$. After such reduction we write the Hamiltonian in the form

$$\mathcal{H}_k(I',\varphi',t) \;=\; h_k(I';\varepsilon)+\varepsilon^{k+1}R_k(I',\varphi',t;\varepsilon)\ ,$$

where the functions h_k and R_k can be explicitly determined. The application of (computer–assisted) KAM estimates [25] allows us to establish the existence of a librational invariant torus, which confines the synchronous resonance in the phase space.

As an example, we report the results for the Rhea–Saturn system, which is observed to move in a synchronous spin–orbit resonance; for this example the stability of the synchronous resonance can be established for the realistic values of the parameters.

Theorem [24]. *Consider the system described by the Hamiltonian* (7.56) *with* $N_1=-1$, $N_2=5$ *and let* $e=0.00098$. *If* $\varepsilon_{Rhea}=3.45\cdot 10^{-4}$ *is the physical value of the perturbing parameter, then there exists an invariant torus corresponding to a libration of* 1.95^o *for any* $\varepsilon\leq\varepsilon_{Rhea}$.

7.3.3 The spatial planetary three–body problem

The planetary problem concerns the study of two point–masses, say $\mathcal{P}_1$ and $\mathcal{P}_2$ with masses m_1 and m_2 of the same order of magnitude, orbiting around a central body, say $\mathcal{P}$ with mass M. It is therefore necessary to take into account the mutual interaction between $\mathcal{P}_1$ and $\mathcal{P}_2$, besides that with the central body. In order to write the Hamiltonian function, let us introduce the heliocentric positions of the planets, $\underline{r}_1,\underline{r}_2\in\mathbf{R}^3$, and the conjugated momenta referred to the center of mass, $\underline{v}_1,\underline{v}_2\in\mathbf{R}^3$. The Hamiltonian describing the motion of $\mathcal{P}_1$ and $\mathcal{P}_2$ can be decomposed as

$$\mathcal{H}=\mathcal{H}_0+\mathcal{H}_1\ , \tag{7.57}$$

where $\mathcal{H}_0$ is due to the decoupled Keplerian motions of the planets and $\mathcal{H}_1$ represents the interaction between $\mathcal{P}_1$ and $\mathcal{P}_2$. More precisely, one has

$$\mathcal{H}_0=\sum_{j=1}^{2}\frac{m_j+M}{2m_jM}\|\underline{v}_j\|^2-\mathcal{G}\,\frac{Mm_j}{\|\underline{r}_j\|}\ , \tag{7.58}$$

while the perturbation is given by

$$\mathcal{H}_1=\frac{\underline{v}_1\cdot\underline{v}_2}{M}-\mathcal{G}\,\frac{m_1m_2}{\|\underline{r}_1-\underline{r}_2\|}\ . \tag{7.59}$$

The preservation of the angular momentum allows us to state that the ascending nodes of the planets lie on the invariant plane perpendicular to the angular momentum and passing through the central body. The existence of invariant tori in the framework of the properly degenerate Hamiltonian (7.57), (7.58), (7.59) has been investigated in [3] under the assumption of planar motion and assuming that the ratio of the semimajor axes tends to zero. Invariant tori are shown to exist, provided that the planetary masses and the eccentricities are sufficiently small. The assumption that the ratio of the semimajor axes tends to zero has been removed in [155], where quantitative estimates have been worked out. The proper degeneracy of the Hamiltonian has been eliminated by a suitable normal form; after performing the reduction of the angular momentum, the perturbing function has been expanded using an adapted algebraic manipulator (see [110]). The result presented in [155] provides that, for sufficiently small planetary masses and eccentricities, one can apply Arnold's theorem on the existence of invariant tori, provided that the ratio α between the planetary semimajor axes satisfies $10^{-8} \leq \alpha \leq 0.8$ and that the mass ratio satisfies $0.01 \leq \frac{m_1}{m_2} \leq 100$.

The specific case of the Sun–Jupiter–Saturn planetary problem has been studied in [120]. After the Jacobi reduction of the nodes [120], the problem turns out to be described by a Hamiltonian function with four degrees of freedom, which is expanded up to the second order in the masses and averaged over the fast angles. The resulting two–degrees–of–freedom Hamiltonian describes the slow motion of the orbital parameters, and precisely of the eccentricities. The existence of invariant tori in a suitable neighborhood of an elliptic point is obtained as follows. After expressing the perturbing function in Poincaré variables, an expansion up to the order 6 in the eccentricities is performed. The computation of the Birkhoff normal form and a computer–assisted KAM theorem yield the existence of two invariant surfaces trapping the *secular* motions of Jupiter and Saturn for the astronomical values of the parameters. This approach was later extended [121] to include the description of the fast variables, like the semimajor axes and the mean longitudes of the planets. A preliminary average over the fast angles was performed without eliminating the terms with degree greater or equal than 2 with respect to the fast actions. The canonical transformations involving the secular coordinates can be adapted to produce a good initial approximation of an invariant torus for the reduced Hamiltonian of the planetary three–body problem. Afterwards the Kolmogorov normal form was constructed (so that the Hamiltonian is reduced to a harmonic oscillator plus higher–order terms) and it was numerically shown to be convergent. The numerical results on the convergence of the Kolmogorov normal form have been obtained for a planetary solar system composed by two planets with masses equal to those of Jupiter and Saturn.

7.3.4 The circular, planar, restricted three–body problem

We consider the motion of a small body ($\mathcal{P}_2$), say an asteroid, under the influence of two primaries, say the Sun ($\mathcal{P}_1$) and Jupiter ($\mathcal{P}_3$) in the framework of the circular, planar, restricted three–body problem (see Section 4.1). The Sun–Jupiter–asteroid problem was selected in [31] as a test–bench for KAM theory, which provided estimates on the mass–ratio very far from the astronomical observations; in particular, the existence of invariant tori was obtained for mass–ratios less than 10^{-333} by applying Arnold's theorem and 10^{-48} using Moser's theorem. We recall that the perturbative parameter ε coincides with the Jupiter–Sun mass ratio, which amounts to about $\varepsilon = \varepsilon_J \equiv 0.954 \cdot 10^{-3}$. The small body was chosen as the asteroid 12 Victoria, whose orbital elements are:

$$a_V \simeq 2.335 \text{ AU} , \qquad e_V \simeq 0.220 , \qquad i_V \simeq 8.362^o ,$$

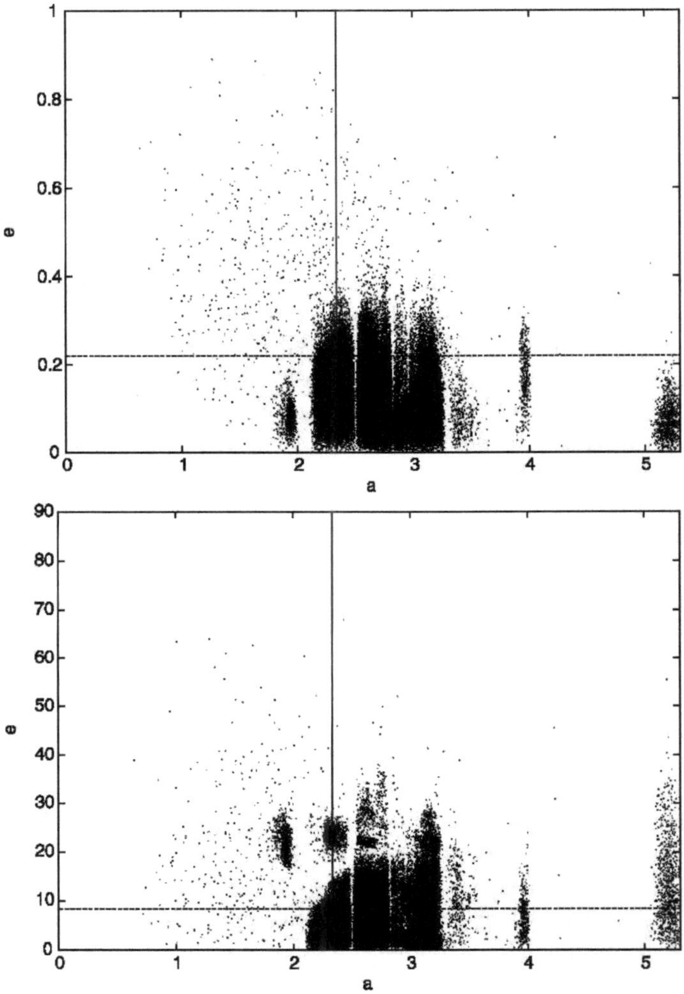

Fig. 7.2. Orbital elements of the numbered asteroids. Top: semimajor axis versus eccentricity. Bottom: semimajor axis versus inclination. The internal lines locate the position of the asteroid 12 Victoria (reprinted from [30]).

where a_V is the semimajor axis of the asteroid, e_V is the eccentricity, i_V is the inclination with respect to the ecliptic plane. Figure 7.2 shows that 12 Victoria is a typical object of the asteroidal belt[2], since the semimajor axes of most asteroids lie within the interval $1.8 \leq a \leq 3.5$ AU, while the eccentricity is usually within $0 \leq \mathrm{e} \leq 0.35$.

The model presented above does not include many effects, most notably the eccentricity of Jupiter, the mutual inclinations, the influence of other planets, as well as dissipative effects. For consistency, the perturbing function, representing

[2] The elements of the numbered asteroids are provided by the JPL's DASTCOM database at http://ssd.jpl.nasa.gov/?sb_elem

the influence of Jupiter on the asteroid, has been expanded in the eccentricity and semimajor axes ratio, and truncated to discard all terms which are of the same order of magnitude or less than the maximum contribution due to the effects we have neglected. Indeed, in the Sun–Jupiter–Victoria model the biggest neglected contribution is due to the eccentricity of the orbit of Jupiter, which has been assumed to be zero in the present model. According to this criterion we obtain the following Hamiltonian:

$$\mathcal{H}(L,G,\ell,g) = -\frac{1}{2L^2} - G - \varepsilon\mathcal{R}(L,G,\ell,g) , \tag{7.60}$$

where (L, G) are the Delaunay action variables, ℓ is the mean anomaly, g is the difference between the argument of perihelion and the true anomaly of Jupiter (see Chapter 4) and the perturbing function is given by

$$\begin{aligned}
\mathcal{R}(L,G,\ell,g) &\equiv 1 + \frac{L^4}{4} + \frac{9}{64}L^8 + \frac{3}{8}L^4 \mathrm{e}^2 - \left(\frac{1}{2} + \frac{9}{16}L^4\right) L^4 \mathrm{e} \cos\ell \\
&+ \left(\frac{3}{8}L^6 + \frac{15}{64}L^{10}\right) \cos(\ell+g) - \left(\frac{9}{4} + \frac{5}{4}L^4\right) L^4 \mathrm{e} \cos(\ell+2g) \\
&+ \left(\frac{3}{4}L^4 + \frac{5}{16}L^8\right) \cos(2\ell+2g) + \frac{3}{4}L^4 \mathrm{e} \cos(3\ell+2g) \\
&+ \left(\frac{5}{8}L^6 + \frac{35}{128}L^{10}\right) \cos(3\ell+3g) + \frac{35}{64}L^8 \cos(4\ell+4g) \\
&+ \frac{63}{128}L^{10} \cos(5\ell+5g) ,
\end{aligned}$$

where $\mathrm{e} = \sqrt{1 - \frac{G^2}{L^2}}$. Let us write (7.60) as

$$\mathcal{H}(L,G,\ell,g;\varepsilon) = \mathcal{H}_0(L,G) + \varepsilon\mathcal{R}(L,G,\ell,g) ,$$

where $\mathcal{H}_0(L,G) \equiv -\frac{1}{2L^2} - G$. The KAM theorem described in Section 7.2 cannot be applied, since the integrable part $\mathcal{H}_0$ is degenerate. However, it is possible to apply a different version of the theorem, which requires the *isoenergetic non–degeneracy condition* due to Arnold [6]:

$$C_E(L,G) \equiv \det \begin{pmatrix} \mathcal{H}_0'' & \mathcal{H}_0' \\ \mathcal{H}_0' & 0 \end{pmatrix} \neq 0 \qquad \text{for all } 0 < G < L ,$$

where $\mathcal{H}_0'$ and $\mathcal{H}_0''$ denote, respectively, the Jacobian vector and the Hessian matrix associated to $\mathcal{H}_0$. A straightforward computation shows that $C_E(L,G) = \frac{3}{L^4}$. To fix the energy level we proceed as follows (see [31]). From the physical value of the asteroid 12 Victoria, using normalized units one gets that $L_{\mathrm{V}} \simeq 0.670$, $G_{\mathrm{V}} \simeq 0.654$. Let

$$E_{\mathrm{V}}^{(0)} = -\frac{1}{2L_{\mathrm{V}}^2} - G_{\mathrm{V}} \simeq -1.768 , \quad E_{\mathrm{V}}^{(1)} \equiv \langle \mathcal{R}(L_{\mathrm{V}}, G_{\mathrm{V}}, \ell, g) \rangle \simeq -1.060 .$$

We define the energy level through the expression

$$E_{\rm V}^* = E_{\rm V}^{(0)} + \varepsilon_J E_{\rm V}^{(1)} \simeq -1.769 \ ,$$

where ε_J denotes the observed Jupiter–Sun mass–ratio. The existence of two invariant tori, bounding from above and below the observed values $L_{\rm V}$ and $G_{\rm V}$, is proven on the level set $\mathcal{H}^{-1}(E_{\rm V}^*)$. Setting $\tilde{L}_\pm = L_V \pm 0.001$, the bounding frequencies are computed as

$$\tilde{\omega}_\pm = \left(\frac{1}{\tilde{L}_\pm^3}, -1\right) \equiv (\tilde{\alpha}_\pm, -1) \ .$$

Since we need *diophantine* numbers, we proceed to compute the continued fraction expansion of $\tilde{\alpha}_\pm$ up to the order 5 and then we add a tail of ones to obtain the following diophantine numbers:

$$\begin{aligned} \alpha_- &\equiv [3; 3, 4, 2, 1^\infty] = 3.30976937631389\ldots \ , \\ \alpha_+ &\equiv [3; 2, 1, 17, 5, 1^\infty] = 3.33955990647860\ldots \ . \end{aligned}$$

Next we introduce the frequencies

$$\omega_\pm \equiv (\alpha_\pm, -1) \ ,$$

which satisfy the diophantine condition (7.7) with $\tau = 1$ and with diophantine constants respectively equal to

$$C_- = 138.42 \ , \qquad C_+ = 30.09 \ .$$

The stability of the asteroid 12 Victoria is finally obtained by proving the persistence of the unperturbed KAM tori $\mathcal{T}_0^\pm \equiv \{(L_\pm, G_\pm)\} \times \mathbf{T}^2$ for a value of the perturbing parameter ε greater or equal than the Jupiter–Sun mass ratio.

Theorem [31]. *For $|\varepsilon| \leq 10^{-3}$ the unperturbed tori $\mathcal{T}_0^\pm$ can be analytically continued into invariant KAM tori $\mathcal{T}_\varepsilon^\pm$ for the perturbed system on the energy level $\mathcal{H}^{-1}(E_{\rm V}^*)$ keeping fixed the ratio of the frequencies.*

Since the orbital elements are related to the Delaunay action variables, the theorem guarantees that the semimajor axis and the eccentricity stay close to the unperturbed values within an interval of order ε (see [31] for full details on the KAM isoenergetic, computer–assisted proof).

7.4 Greene's method for the breakdown threshold

There exist different techniques which allow us to evaluate numerically the breakdown threshold of an invariant surface (see, e.g., [82, 109, 145]). One of the most accepted methods, which has been partially rigorously proved [54, 63, 127], was developed by J. Greene in [82]. His method is based on the conjecture that the breakdown of an invariant surface is closely related to the stability character of the approximating periodic orbits [92]. The key role of the periodic orbits had already been stressed by H. Poincarè in [149], who formulated the following conjecture:

"... here is a fact that I have not been able to prove rigorously, but that seems to me very reasonable. Given equations of the form (13) [Hamilton's equations] *and a particular solution of these equations, one can always find a periodic solution (whose period, it is true, can be very long) such that the difference between the two solutions may be as small as one wishes for as long as one wishes".*

Greene's algorithm for computing the breakdown threshold was originally formulated for the standard mapping, but we present it here for the spin–orbit problem, which has been assumed as a model problem throughout this chapter. Let us reduce the analysis of the differential equation (7.1) to the study of the discrete mapping obtained integrating (7.1) through an area–preserving leapfrog method:

$$\begin{aligned} y_{j+1} &= y_j - \varepsilon f_x(x_j, t_j) h \\ x_{j+1} &= x_j + y_{j+1} h \ , \end{aligned} \tag{7.61}$$

where $t_{j+1} = t_j + h$ and $h \geq 0$ denotes the integration step, $y_j \in \mathbf{R}$, $x_j \in \mathbf{T}$, $t_j \in \mathbf{T}$. We say that a periodic orbit has length q (for some positive integer q), if it closes after q iterations. We shall consider the periodic orbits which exist for all values of the parameter ε down to $\varepsilon = 0$. Analogously, we consider rotational KAM tori with the same property. In the integrable limit the rotation number is given by $\omega \equiv y_0$; if the frequency of motion is rational, say $\omega = \frac{p}{q}$ for some positive integers p and q with $q \neq 0$, then the second of (7.61) implies that

$$p = \sum_{j=1}^{q} y_j = \sum_{j=1}^{q} \frac{x_j - x_{j-1}}{h} = \frac{x_q - x_0}{h} \ .$$

If the frequency ω is irrational, the periodic orbits with frequency equal to its rational approximants $\frac{p_j}{q_j}$ are those which nearly approach the torus with rotation number ω (see Figure 7.3).

In order to determine the linear stability of a periodic orbit, we compute the tangent space trajectory $(\partial y_j, \partial x_j)$ at (y_j, x_j), which is related to the initial conditions $(\partial y_0, \partial x_0)$ at (y_0, x_0) by

$$\begin{pmatrix} \partial y_j \\ \partial x_j \end{pmatrix} = M \begin{pmatrix} \partial y_0 \\ \partial x_0 \end{pmatrix} \ ,$$

where the matrix M is the product of the Jacobian of (7.61) along a full cycle of the periodic orbit:

$$M = \prod_{i=1}^{q} \begin{pmatrix} 1 & -\varepsilon f_{xx}(x_j, t_j) h \\ h & 1 - \varepsilon f_{xx}(x_j, t_j) h^2 \end{pmatrix} \ .$$

The eigenvalues of M are the associated Floquet multipliers (compare with Appendix D); by the area–preservation of the mapping it is $\det(M) = 1$ and denoting by $\mathrm{tr}(M)$ the trace of M, the eigenvalues are the solutions of the equation

$$\lambda^2 - \mathrm{tr}(M)\lambda + 1 = 0 \ .$$

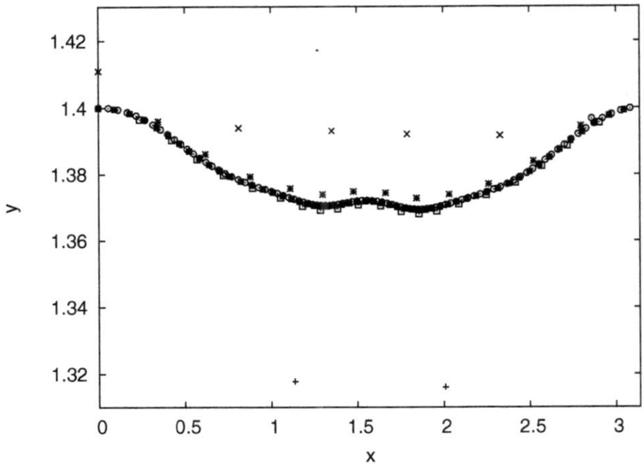

Fig. 7.3. Periodic orbits corresponding to the equations of motion associated to (7.55) approaching the torus with rotation number $\omega = 1 + \frac{1}{2+\frac{\sqrt{5}-1}{2}}$ for $\varepsilon = 0.03$ on a Poincaré section at times 2π. The graph shows the periodic orbits with frequencies $4/3$ $+$, $7/5$ $\times$, $18/13$ $*$, $29/21$ $\square$, $76/55$ $\blacksquare$, $123/89$ $\circ$.

Let us introduce a quantity, called the *residue,* by means of the relation (see [82]):

$$R \equiv \frac{1}{4}(2 - \mathrm{tr}(M)) ,$$

where the factors 2 and 4 are introduced for convenience. The eigenvalues of M are related to the residue R by

$$\lambda = 1 - 2R \pm 2\sqrt{R^2 - R} .$$

When $0 < R < 1$ the eigenvalues are complex conjugates with modulus one and the orbit is stable, otherwise when $R < 0$ or $R > 1$ the periodic orbit is unstable. Due to a theorem by Poincaré, for each rational frequency the number of orbits with positive or negative residue is the same. The positive residue orbits are stable for low values of ε. The residue gets larger as the perturbing parameter increases, until it becomes greater than one, thus showing the instability of the associated periodic orbit.

According to [82], we define the *mean residue* of a periodic orbit of period p/q as the quantity

$$f\left(\frac{p}{q}; \varepsilon\right) \equiv (4|R|)^{1/q} .$$

The definition of the mean residue for irrational frequencies ω is obtained as follows: if $\omega \equiv [a_0; a_1, \ldots, a_N, \ldots]$, then

$$f(\omega; \varepsilon) = \lim_{N\to\infty} f(\omega_N; \varepsilon) ,$$

where $\omega_N \equiv [a_0; a_1, \ldots, a_N]$. If ω is a noble number, say $\omega \equiv [a_0; a_1, \ldots, a_N, 1, 1, 1, \ldots]$, let $\varepsilon = \varepsilon_c(\omega)$ be such that

$$f(\omega, \varepsilon_c(\omega)) = 1 ;$$

then the corresponding residue converges to

$$R \equiv R(\omega; \varepsilon_c(\omega)) = \frac{1}{4}$$

(this assertion justifies the factor 4 introduced in the definition of the mean residue). Greene's method is based on the conjecture that a KAM rotational torus with frequency ω exists if and only if

$$f(\omega; \varepsilon) < 1$$

(see [63] for a partial proof of this statement). In Table 7.2 we consider the first few frequencies of the periodic orbits approaching the torus with frequency equal to the golden ratio. For each periodic orbit of period $\frac{p}{q}$ we report the value of the perturbing parameter $\varepsilon = \varepsilon_c(\frac{p}{q})$ at which the corresponding residue becomes bigger than $\frac{1}{4}$. As $\frac{p}{q}$ increases, the limit of the values $\varepsilon_c(\frac{p}{q})$ provides the breakdown threshold $\varepsilon_c(\omega)$ of the torus with frequency ω.

Table 7.2. Critical values $\varepsilon_c(\frac{p}{q})$ of the perturbing parameter for some periodic orbits approaching the torus with frequency equal to the golden ratio.

$\frac{p}{q}$	$\varepsilon_c(\frac{p}{q})$	$\frac{p}{q}$	$\varepsilon_c(\frac{p}{q})$
$\frac{1}{2}$	0.103	$\frac{13}{21}$	0.144
$\frac{2}{3}$	0.124	$\frac{21}{34}$	0.139
$\frac{3}{5}$	0.158	$\frac{34}{55}$	0.146
$\frac{5}{8}$	0.112	$\frac{55}{89}$	0.145
$\frac{8}{13}$	0.151	$\frac{89}{144}$	0.144

The efficiency of Greene's method strongly depends on the computational speed for the determination of the periodic orbits approaching the invariant surface. In the particular case of the spin–orbit discretized system (7.61), one can get advantage from the fact that the mapping (7.61) including the time variation $t_{j+1} = t_j + h$, herewith denoted as S, can be decomposed as the product of two involutions:

$$S = I_2 \, I_1 \, ,$$

where $I_1^2 = I_2^2 = 1$. In particular I_1 is given by

$$\begin{aligned} y_{j+1} &= y_j - \varepsilon f_x(x_j, t_j) h \\ x_{j+1} &= -x_j \\ t_{j+1} &= -t_j \, , \end{aligned}$$

while I_2 takes the form

$$\begin{aligned} y_{j+1} &= y_j \\ x_{j+1} &= -x_j + h y_j \\ t_{j+1} &= -t_j + h \ . \end{aligned}$$

The periodic orbits can be found as fixed points of one of these involutions. This decomposition of the original mapping significantly reduces the computational time for the determination of the periodic orbits, thus making easier the implementation of Greene's method.

7.5 Low–dimensional tori

For a nearly–integrable system with $m+n$ degrees of freedom, we consider the case when the unperturbed Hamiltonian is not integrable in the whole phase space, but rather on some surface foliated by invariant tori whose dimension is less than $m+n$. The proof of the existence of low–dimensional tori is based on Kolmogorov's approach under the requirement that the system satisfies two conditions, namely that it is *isotropic* and *reducible.* The theory of low–dimensional tori is very wide and heavily depends on the properties of the main frequencies of motion. Here, we just aim to give an idea of the problem, referring to [101,119] for complete details. We start by providing the definitions of isotropic and reducible systems.

Definition. Consider an n–dimensional manifold W endowed with a symplectic non–degenerate 2–form; a submanifold U of W is called *isotropic* if the 2–form restricted to U vanishes.

Definition. Consider a nearly–integrable Hamiltonian $\mathcal{H} = \mathcal{H}_0 + \varepsilon\mathcal{H}_1$ with $m+n$ degrees of freedom. An invariant torus for $\mathcal{H}$ with frequency ω is called *reducible,* if in its neighborhood there exists a set of coordinates $(\underline{I}, \underline{\varphi}, \underline{z}) \in \mathbf{R}^n \times \mathbf{T}^n \times \mathbf{R}^{2m}$, such that the unperturbed Hamiltonian takes the form

$$\mathcal{H}_0(\underline{I}, \underline{\varphi}, \underline{z}) = h(\underline{I}) + \frac{1}{2} A(\underline{I})\underline{z} \cdot \underline{z} + \mathcal{R}_3(\underline{I}, \underline{\varphi}, \underline{z}) \ , \qquad (7.62)$$

where h is a function only of $\underline{I}$, $A(\underline{I})$ is a $2m \times 2m$ symmetric matrix and $\mathcal{R}_3(\underline{I}, \underline{\varphi}, \underline{z})$ is $O(|z|^3)$.

Hamilton's equations associated to (7.62) are given by

$$\begin{aligned} \dot{\underline{z}} &= \Omega(\underline{I})\underline{z} + O(|\underline{z}|^2) \\ \dot{\underline{I}} &= O(|\underline{z}|^3) \\ \dot{\underline{\varphi}} &= \underline{\omega}(\underline{I}) + O(|\underline{z}|^2) \ , \end{aligned}$$

where $\Omega(\underline{I}) \equiv J A(\underline{I})$, J being the standard symplectic matrix, and $\underline{\omega}(\underline{I}) \equiv \frac{\partial h(\underline{I})}{\partial \underline{I}}$. The KAM theorem for low–dimensional tori states that, under suitable conditions on $\Omega(\underline{I})$ and on $\omega(\underline{I})$, one can prove the existence of isotropic, reducible,

n–dimensional invariant tori on which a quasi–periodic motion takes place. The invariant tori are *elliptic* if the eigenvalues of $\Omega(\underline{I})$ are purely imaginary, while they are *hyperbolic* if $\Omega(\underline{I})$ has no purely imaginary eigenvalues. As already mentioned, the proofs of the existence of low–dimensional tori may vary according to the assumptions on the frequencies Ω, $\underline{\omega}$ and we refer to the specialized literature for further details (see, e.g., [2]). Here we just mention how a parametrization in the style of (7.10) can be found for lower–dimensional tori. To see how it works, let us consider a concrete example, and precisely the four–dimensional standard map described by the equations

$$\begin{aligned} y_{n+1} &= y_n + \varepsilon f_1(x_n, z_n, \lambda) \\ x_{n+1} &= x_n + y_{n+1} \\ w_{n+1} &= w_n + \varepsilon f_2(x_n, z_n, \lambda) \\ z_{n+1} &= z_n + w_{n+1} \ , \end{aligned} \tag{7.63}$$

where $(y_n, w_n) \in \mathbf{R}^2$, $(x_n, z_n) \in \mathbf{T}^2$, $\varepsilon > 0$ is the perturbing parameter and $\lambda > 0$ is the coupling parameter. From (7.63) it follows that

$$\begin{aligned} x_{n+1} - 2x_n + x_{n-1} = \varepsilon f_1(x_n, z_n, \lambda) \\ z_{n+1} - 2z_n + z_{n-1} = \varepsilon f_2(x_n, z_n, \lambda) \ . \end{aligned}$$

Let us parametrize a one–dimensional invariant torus with frequency ω by means of the equations

$$\begin{aligned} x_n &= \vartheta + u_1(\vartheta; \varepsilon, \lambda) \\ z_n &= \vartheta + u_2(\vartheta; \varepsilon, \lambda) \ , \end{aligned}$$

where $\vartheta_{n+1} = \vartheta_n + \omega$. One finds that the unknown functions u_1 and u_2 must satisfy the equations

$$\begin{aligned} u_1(\vartheta+\omega) - 2u_1(\vartheta) + u_1(\vartheta-\omega) &= \varepsilon f_1(\vartheta + u_1(\vartheta;\varepsilon,\lambda), \vartheta + u_2(\vartheta;\varepsilon,\lambda), \lambda) \\ u_2(\vartheta+\omega) - 2u_2(\vartheta) + u_2(\vartheta-\omega) &= \varepsilon f_2(\vartheta + u_1(\vartheta;\varepsilon,\lambda), \vartheta + u_2(\vartheta;\varepsilon,\lambda), \lambda) \ , \end{aligned}$$

whose solution describes the low–dimensional torus with frequency ω (see [100]).

Within the spatial three–body problem the existence of low–dimensional tori has been investigated in [99]. In particular, the three–body model studied in [99] admits four degrees of freedom after having performed the reduction of the nodes. Solutions with two or three rationally independent frequencies have been proved, provided the mutual inclinations i_i, i_2 satisfy the condition (see [99])

$$\cos^2(i_1 + i_2) < \frac{3}{5} \ .$$

The existence of quasi–periodic motions with a number of frequencies less than the number of degrees of freedom has been studied also in [113]; in particular, the solutions of the planar three–body problem such that the mean value of the difference of the perihelia is zero have been investigated. The planetary planar $(N+1)$–body problem has been analyzed in [16] and [17], where the existence of N–dimensional elliptic (i.e. linearly stable) tori is shown. Around the elliptic tori there exists a set of positive measure of maximal tori. The proof is based on an elliptic KAM theorem under suitable non–degeneracy conditions (i.e., the so–called Melnikov conditions).

7.6 A dissipative KAM theorem

Let us consider the dissipative spin–orbit equation that we write in compact form as (compare with (5.21))

$$\ddot{x} + \eta(\dot{x} - \nu) + \varepsilon f_x(x,t) = 0 \ , \tag{7.64}$$

where $f_x(x,t) \equiv \varepsilon(\frac{a}{r})^3 \sin(2x-2f)$, $\eta \equiv K_d \bar{L}(\mathrm{e})$, $\nu \equiv \frac{\bar{N}(\mathrm{e})}{\bar{L}(\mathrm{e})}$. We immediately remark that for $\eta \neq 0$ and $\varepsilon = 0$ the torus $\mathcal{T}_0 \equiv \{y = \nu\} \times \{(\vartheta,\tau) \in \mathbf{T}^2\}$ is a global attractor and the flow on $\mathcal{T}_0$ is given by $(\vartheta,\tau) \to (\vartheta + \nu t, \tau + t)$. This is easily seen from the fact that the solution of (7.64) for $\varepsilon = 0$ is given by

$$x(t) = x_0 + \nu(t - t_0) + \frac{1 - e^{-\eta(t-t_0)}}{\eta}\,(v_0 - \nu) \ ,$$

where $x_0 \equiv x(t_0)$ and $v_0 \equiv \dot{x}(t_0)$. An invariant attractor with frequency ω is parametrized by

$$x(t) = \vartheta + u(\vartheta,t) \ , \tag{7.65}$$

where $u = u(\vartheta,t)$ is a real analytic function for $(\vartheta,t) \in \mathbf{T}^2$ and $\dot{\vartheta} = \omega$. The existence of the invariant attractor with frequency ω for (7.64) is provided by the following

Theorem [32]. *Assume that ω is diophantine; then, there exists $\varepsilon_0 > 0$ such that for $0 \le \varepsilon \le \varepsilon_0$ and for any $0 \le \eta < 1$, there exists a function $u = u(\vartheta,t)$ with $\langle u \rangle = 0$ and $1 + u_\vartheta \neq 0$, such that (7.65) is a solution of (7.64) provided*

$$\nu = \omega\ (1 + \langle (u_\vartheta)^2 \rangle) \ . \tag{7.66}$$

The proof of the theorem is based on the following ideas (we refer to [32] for full details). Let us start by introducing the operator $\partial_\omega \equiv \omega \frac{\partial}{\partial\vartheta} + \frac{\partial}{\partial t}$, so that $\dot{x} = \omega + \partial_\omega u$ and $\ddot{x} = \partial_\omega^2 u$. The solution (7.65) is quasi–periodic if the function u satisfies

$$\partial_\omega^2 u + \eta \partial_\omega u + \varepsilon f_x(\vartheta + u, t) + \gamma = 0 \ , \qquad \gamma \equiv \eta(\omega - \nu) \ . \tag{7.67}$$

The unknowns u, γ must satisfy the compatibility condition

$$\eta\omega\ \langle (u_\vartheta)^2 \rangle + \gamma = 0 \ , \tag{7.68}$$

which is equivalent to (7.66). The proof of the existence of the quasi–periodic attractor is perturbative in ε, but uniform in η; the conservative KAM torus bifurcates in the attractor as far as $\eta \neq 0$. For the spin–orbit problem, one has to keep in mind that in place of η and ν one should consider the dissipative constant K_d and the eccentricity e. As a consequence, the theorem is stated for any $0 \leq K_d < 1$ and besides the existence of a function $u = u(\vartheta, t)$, one needs to find a function $\mathrm{e} = \mathrm{e}(K_d, \omega, \varepsilon) = \nu_{\mathrm{e}}^{-1}(\omega) + O(\varepsilon^2)$ to satisfy the compatibility condition (7.66).

Coming back to equation (7.67), let us introduce the operators

$$D_\eta u \equiv \partial_\omega u + \eta u \ , \qquad \Delta_\eta u \equiv D_\eta \partial_\omega u = \partial_\omega D_\eta u \ .$$

Then, (7.67) becomes

$$\mathcal{F}_\eta(u;\gamma) \equiv \Delta_\eta u + \varepsilon f_x(\vartheta + u, t) + \gamma = 0 \ .$$

In particular, if $u = \sum_{(n,m)\in\mathbf{Z}^2} \hat{u}_{n,m} e^{i(n\vartheta + mt)}$, then

$$\begin{aligned} \partial_\omega u &= \sum_{(n,m)\in\mathbf{Z}^2} i(\omega n + m)\hat{u}_{n,m} e^{i(n\vartheta+mt)} \\ D_\eta u &= \sum_{(n,m)\in\mathbf{Z}^2} [i(\omega n + m) + \eta]\hat{u}_{n,m} e^{i(n\vartheta+mt)} \ ; \end{aligned}$$

being $|i(\omega n + m) + \eta| \geq |\eta| > 0$, then D_η is invertible with

$$D_\eta^{-1} u = \sum_{(n,m)\in\mathbf{Z}^2} \frac{\hat{u}_{n,m} e^{i(n\vartheta+mt)}}{i(\omega n + m) + \eta} \ .$$

Having introduced the norm $\|u\|_\xi \equiv \sum_{(n,m)\in\mathbf{Z}^2} |\hat{u}_{n,m}| e^{(|n|+|m|)\xi}$, one can state the following

Theorem. *Let $0 < \xi < \bar{\xi} \leq 1$, $0 \leq \eta < 1$; let ω be diophantine and define M such that*

$$\|\varepsilon f_{xxx}\|_{\bar{\xi}} \leq M \ .$$

Assume that there exists an approximate solution $v = v(\vartheta, t; \eta)$, $\beta = \beta(\eta)$ such that v_ϑ is bounded and invertible; let the error function $\chi = \chi(\vartheta, t; \eta) \equiv \mathcal{F}_\eta(v;\beta)$ satisfy a smallness requirement of the form

$$\mathcal{D}\ \|\chi\|_\xi \leq 1 \ ,$$

where $\mathcal{D}$ depends upon ξ, M, as well as upon the norms of v and of its derivatives. Then, there exist $u = u(\vartheta, t; \eta) \in C^\infty$ and $\gamma = \gamma(\eta) \in C^\infty$, which solve $\mathcal{F}_\eta(u;\gamma) = 0$.

The proof is constructive and the solution is obtained as the limit of a sequence of approximate solutions (v_j, β_j), quadratically converging to the solution (u, γ). We sketch here the proof as a sequence of five main steps, referring to [32] for complete details.

Step 1. Establish some properties of the operators D_η, Δ_η as well as of their derivatives and inverse functions, providing formulae of the form

$$\|D_\eta^{-s}\,\partial_\vartheta^p u\|_{\xi-\delta} \le \sigma_{p,s}(\delta)\,\|u\|_\xi \ ,$$

for some $0<\delta<\xi$ and for $p, s \in \mathbf{Z}_+$, where

$$\sigma_{p,s}(\delta) \equiv \sup_{(n,m)\in\mathbf{Z}^2\setminus\{\underline{0}\}} \Big(|i(\omega n+m)+\eta|^{-s}|n|^p e^{-\delta(|n|+|m|)} \Big) \ ,$$

which can be bounded as

$$\sigma_{p,s}(\delta) \le \left(\frac{s\tau+p}{e}\right)^{s\tau+p} C^s\delta^{-(s\tau+p)} \ .$$

It turns out that $\langle(1+u_\vartheta)\mathcal{F}_\eta(u;\gamma)\rangle = \eta\omega\langle(u_\vartheta)^2\rangle+\gamma$; if $\mathcal{F}_\eta(u;\gamma)=0$ one finds the compatibility condition (7.68).

Step 2. Given an approximate solution (v,β) of $\mathcal{F}_\eta(u;\gamma)=0$, a quadratically smaller approximation (v',β') is found by a Newton iteration scheme. More precisely, starting from

$$\chi \equiv \mathcal{F}_\eta(v;\beta) = \Delta_\eta v + \varepsilon f_x(\vartheta+v,t)+\beta \ ,$$

one looks for a solution

$$v' = v+\tilde v \ , \qquad \beta' = \beta+\tilde\beta \ ,$$

such that $\tilde v, \tilde\beta = O(\|\chi\|)$, $\mathcal{F}_\eta(v';\beta') = O(\|\chi\|^2)$. In order to find $\tilde v$ and $\tilde\beta$, setting $V\equiv 1+v_\vartheta$ let us introduce the quantities

$$Q_1 \equiv \varepsilon[f_x(\vartheta+v+\tilde v,t) - f_x(\vartheta+v,t) - f_{xx}(\vartheta+v,t)\tilde v] \ , \qquad Q_2 \equiv V^{-1}\chi_\vartheta\,\tilde v \ ;$$

it follows that

$$\mathcal{F}_\eta(v';\beta') \equiv \mathcal{F}_\eta(v+\tilde v;\beta+\tilde\beta) = \chi+\tilde\beta+A_{\eta,v}\tilde v+Q_1+Q_2$$

with $A_{\eta,v}\tilde v \equiv V^{-1}\,D_\eta\Big(V^2 D_0(V^{-1}\tilde v)\Big)$. One can find explicit expressions for $\tilde v$, $\tilde\beta$, such that they satisfy the relation

$$\chi+\tilde\beta+A_{\eta,v}\tilde v = 0 \ ;$$

the latter equation provides $\chi' \equiv \mathcal{F}_\eta(v+\tilde v,\beta+\tilde\beta) = Q_1+Q_2$, so that the new error term is quadratically smaller.

Step 3. Given the estimates on the norms of v_θ, $\tilde v$, $\tilde v_\theta$, $\tilde\beta$, a KAM algorithm is implemented to compute an estimate on the norm of the error function χ' of the form

$$\|\chi'\|_{\xi-\delta} \le C_1\delta^{-s}\|\chi\|_\xi^2 \ ,$$

for some C_1, $s>0$.

Step 4. Implement a KAM algorithm which provides that under smallness conditions on the parameters there exists a sequence (v_j, β_j) of approximate solutions, which converges to the true solution:

$$(u, \gamma) \equiv \lim_{j\to\infty} (v_j, \beta_j) ,$$

where (u, γ) satisfy $\mathcal{F}_\eta(u; \gamma) = 0$.

Step 5. A local uniqueness is shown by proving that if there exists a solution $\xi(t) = \vartheta + w(\vartheta, t)$ with $\dot{\vartheta} = \omega$ and $\langle w \rangle = 0$, then $w \equiv u$, while ν coincides with (7.66).

7.7 Converse KAM

Converse KAM theory provides upper bounds on the perturbing parameter ensuring the non–existence of invariant tori. Following [126, 128, 129] (see also [6]) we adopt the Lagrangian formulation as follows. As in the previous sections, we are concerned with applications to the spin–orbit model; therefore we introduce a one–dimensional, time–dependent Lagrangian function of the form $\mathcal{L} = \mathcal{L}(x, y, t)$, where $x \in \mathbf{T}$, $y \in \mathbf{R}$. We assume that the Lagrangian function satisfies the so–called *Legendre condition,* which requires that $\frac{\partial^2 \mathcal{L}}{\partial \dot{x}^2}$ is everywhere positive. A function $x = x(t)$ is an orbit for $\mathcal{L}$ if for any $t_0 < t_1$ and for any variation $\delta x = \delta x(t)$ such that $\delta x(t_0) = \delta x(t_1) = 0$, the variation $\delta\mathcal{A}$ of the action is zero, where

$$\mathcal{A}(x) \equiv \int_{t_0}^{t_1} \mathcal{L}(x(t), \dot{x}(t), t)\ dt\ . \tag{7.69}$$

A trajectory $x = x(t)$ has *minimal action* if for any $t_0 < t_1$ and $\tilde{x}(t)$ such that $\tilde{x}(t_0) = x(t_0)$, $\tilde{x}(t_1) = x(t_1)$, then $\mathcal{A}(x) \leq \mathcal{A}(\tilde{x})$. The minimal action is *non–degenerate* if for any $t_0 < t_1$, then $\delta^2\mathcal{A}$ is positive definite for any variation δx such that $\delta x(t_0) = \delta x(t_1) = 0$.

The Legendre transformation allows us to introduce the Hamiltonian function $\mathcal{H} = \mathcal{H}(y, x, t)$ associated to $\mathcal{L}$, where $y \in \mathbf{R}$ is the momentum associated to x. A *Lagrangian graph* is described by a C^1–generating function $\mathcal{S} = \mathcal{S}(x, t)$ such that $y = \mathcal{S}_x(x, t)$, $T = \mathcal{S}_t(x, t)$, where T is the the variable conjugated to the time in the extended phase space. We now give a characterization of Lagrangian graphs and rotational tori.

Proposition [129]. *An invariant rotational two–dimensional torus for* $\mathcal{H}_1(y, x, T, t) \equiv \mathcal{H}(y, x, t) + T$ *with* $\mathcal{H}_{yy}$ *positive definite is a Lagrangian graph.*

Moreover, we have the following

Lemma [129]. *If* Σ *is an invariant surface for the Hamiltonian* $\mathcal{H}_1(y, x, T, t) \equiv \mathcal{H}(y, x, t) + T$ *such that locally* $y = \mathcal{S}_x(x, t)$, *then* Σ *is a Lagrangian graph.*

In order to introduce a converse KAM criterion, we need the following theorem due to K. Weierstrass (see [129]).

Theorem. *If Σ is an invariant Lagrangian graph for a Lagrangian system satisfying the Legendre condition, then any orbit on Σ has a non–degenerate minimal action.*

From the Weierstrass theorem it follows that if the orbit segment $x = x(t)$ for $t \in [t_0, t_1]$ is not a non–degenerate minimum for $\mathcal{A}$, then it is not contained in any invariant Lagrangian graph. In practice, one should compute the quantity $\delta^2\mathcal{A}$ for some variation δx with $\delta x(t_0) = \delta x(t_1) = 0$ and check whether it fails to be positive definite. This method allows us to give an elementary analytical estimate, which can be explicitly computed. Following [40], let us consider the spin–orbit equation (5.18) that we write as

$$\ddot{x} + \varepsilon \sum_{m=1}^{N} \alpha_m(\mathrm{e}) \sin(2x - mt) = 0 \tag{7.70}$$

for some $N > 0$; the coefficients $\alpha_m(\mathrm{e})$ are trivially related to the coefficients $W(\frac{m}{2}, \mathrm{e})$ in (5.18). We apply the criterion based on the Weierstrass theorem to the model described by (7.70). The Lagrangian function associated to (7.70) has the form

$$\mathcal{L}(x, \dot{x}, t) = \frac{1}{2}\dot{x}^2 + \frac{\varepsilon}{2} \sum_{m=1}^{N} \alpha_m(\mathrm{e}) \cos(2x - mt) \ .$$

The second variation of the action is given by

$$\delta^2\mathcal{A} = \int_{t_0}^{t_1} \left[\delta\dot{x}^2 - 2\varepsilon \sum_{m=1}^{N} \alpha_m(\mathrm{e}) \cos(2x - mt)\delta x^2 \right] dt \ .$$

Consider the deviation $\delta x(t) = \cos\frac{t}{4\tau}$ such that $\delta x(\pm 2\pi\tau) = 0$; notice that $\int_0^{2\pi\tau} \delta x^2 = \pi\tau$, $\int_0^{2\pi\tau} \delta\dot{x}^2 = \frac{\pi}{16\tau}$. Writing (7.70) as

$$\ddot{x} = g(x, t) \equiv -\varepsilon \sum_{m=1}^{N} \alpha_m(\mathrm{e}) \sin(2x - mt)$$

and assuming the initial conditions $x(0) = 0$, $\dot{x}(0) = v_0$, the solution of (7.70) can be written in integral form as

$$x(t) = v_0 t + \int_0^t (t - s) g(x(s), s)\, ds \ .$$

Let G be a bound on $g(x, t)$, i.e. $|g(x, t)| \leq \varepsilon \sum_{m=1}^{N} |\alpha_m(\mathrm{e})| \equiv G$; as a first approximation we can use the inequality

$$|x(t) - v_0 t| \leq \frac{G}{2} t^2 \ .$$

Since $\cos\vartheta \geq 1 - \frac{1}{2}\vartheta^2$, we obtain

$$\cos(2x - mt) \geq 1 - \frac{1}{2}(|m - 2v_0|t + Gt^2)^2 \ .$$

Therefore the second variation of the action for the variation $\delta x(t) = \cos \frac{t}{4\tau}$, $-2\pi\tau \leq t \leq 2\pi\tau$, is bounded by

$$\delta^2 \mathcal{A} \leq \frac{\pi}{8\tau} - 4\varepsilon \sum_{m=1}^{N} |\alpha_m(\mathrm{e})| \int_0^{2\pi\tau} \left[1 - \frac{1}{2}(|m - 2v_0|t + Gt^2)^2\right] \delta x^2 \, dt$$
$$\leq \frac{\pi}{8\tau} - 4G\pi\tau + 2\varepsilon \sum_{m=1}^{N} |\alpha_m(\mathrm{e})| \int_0^{\frac{\pi}{2}} \Big[|m - 2v_0|^2 (4\tau)^3 \vartheta^2 \cos^2\vartheta$$
$$+ G^2 (4\tau)^5 \vartheta^4 \cos^2\vartheta + 2|m - 2v_0| G (4\tau)^4 \vartheta^3 \cos^2\vartheta\Big] \, d\vartheta \ .$$

Let us define the quantity

$$I_n \equiv 2 \int_0^{\frac{\pi}{2}} \vartheta^n \cos^2\vartheta \, d\vartheta \ ;$$

then, one obtains

$$\frac{\delta^2 \mathcal{A}}{\tau} \leq \frac{\pi}{8\tau^2} - 4G\pi + \frac{\varepsilon}{\tau} \sum_{m=1}^{N} |\alpha_m(\mathrm{e})| \cdot \Big[|m - 2v_0|^2 (4\tau)^3 I_2$$
$$+ 2|m - 2v_0| G (4\tau)^4 I_3 + G^2 (4\tau)^5 I_4\Big] \equiv \Phi(\varepsilon, v_0, \tau) \ . \tag{7.71}$$

The non–existence criterion is fulfilled whenever one can find $\tau > 0$ such that $\Phi(\varepsilon, v_0, \tau) < 0$, so that $\delta^2 \mathcal{A} < 0$. Denote by ε_{NE} the value of the perturbing parameter at which this condition first occurs. As concrete examples we consider the orbital eccentricity of the Moon (e = 0.0549) and of Mercury (e = 0.2056); moreover we consider $v_0 = 1$ and $v_0 = 1.5$, corresponding, respectively, to the 1:1 and 3:2 resonance. The results of the implementation of the Weierstrass criterion based on the estimate (7.71) are provided in Table 7.3, where $N = 7$ has been taken in (7.70) (see [40]). Though the estimates are rather crude and could be further refined, they show how to find by simple explicit computations the regions of non–existence of rotational invariant tori.

Table 7.3. The non–existence criterion based on the Weierstrass theorem provides the following values associated to the Moon with eccentricity e = 0.0549 and to Mercury with eccentricity e = 0.2056 (reprinted, with permission, from [40], Copyright 2007, American Institute of Physics).

	Moon	Mercury
$v_0 = 1$	$\varepsilon_{NE} \simeq 0.15$	$\varepsilon_{NE} \simeq 0.82$
$v_0 = 1.5$	$\varepsilon_{NE} \simeq 0.77$	$\varepsilon_{NE} \simeq 0.58$

7.7.1 Conjugate points criterion

A method of investigating the non–existence of invariant tori has been formulated in [129] for conservative systems, based on the following

Definition. Let $(x, y) : [t_0, t_1] \to \mathbf{T} \times \mathbf{R}$ be an orbit; the times t_0 and t_1 are said to be *conjugate,* if there exists a non–zero tangent orbit $(\delta x, \delta y)$, such that $\delta x(t_0) = \delta x(t_1) = 0$.

We also introduce the twist property as follows. Let us write (7.70) as

$$\begin{aligned} \dot{x} &= y \\ \dot{y} &= -\varepsilon \sum_{m=1}^{N} \alpha_m(e) \sin(2x - mt) \ . \end{aligned} \tag{7.72}$$

We say that (7.72) satisfies the *twist property* if there exists a constant $A > 0$ such that

$$\frac{\partial \dot{x}}{\partial y} \geq A$$

(in our case $A = 1$). A result due to K. Jacobi shows that minimizing orbits (with respect to the action (7.69)) have no conjugate points. This leads to the following non–existence criterion, which can be formulated to encompass also the dissipative context [40].

Conjugate points criterion: The existence of conjugate points implies that the orbit does not belong to any rotational invariant torus, otherwise the forward orbit starting from the initial *vertical* vector $(0, 1)$ at $t = t_0$ is prevented from crossing the tangent to the torus and the twist property implies that $\delta x(t) > 0$ for all $t > t_0$.

For the conservative case with time–reversal symmetry and initial conditions on the symmetry line $x = 0$, the backward trajectory and the backward tangent orbit are determined by reflecting the forward ones. We can conclude that the times $\pm t$ are conjugate whenever the tangent orbit of the *horizontal* vector $(\delta x(0), \delta y(0)) = (1, 0)$ satisfies $\delta x(t) = 0$. This remark considerably decreases the computational time, also due to the fact that close to a suitable symmetry line the rotation of the tangent orbits is strongest and it is convenient to select orbit segments which straddle it symmetrically.

The dissipative case does not admit time–reversal symmetry and it is necessary to integrate backward and forward orbits. However, one can choose $t_0 = 0$ and avoid backward integration, thus integrating just forwards from the vertical vector $(\delta x(0), \delta y(0)) = (0, 1)$ and then looking for a change of sign of $\delta x(t)$. We report in Figure 7.4 an application of the conjugate points criterion for the dissipative spin–orbit problem and for different values of the eccentricity. A grid of 500×500 points over $y(0) \in [0.2, 2]$ and $\varepsilon \in [0, 0.1]$ has been computed.

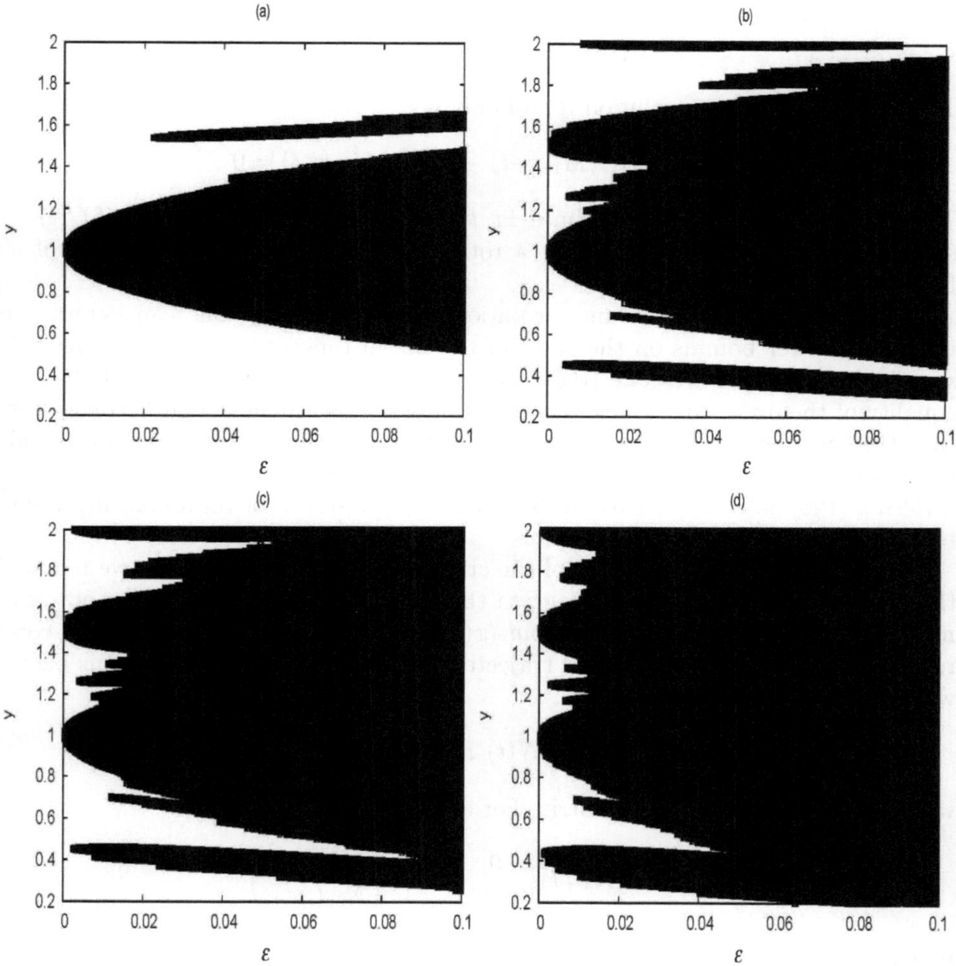

Fig. 7.4. The black region denotes the non–existence of rotational invariant tori for $K_d = 10^{-3}$. (*a*) e $= 0.001$, (*b*) e $= 0.0549$, (*c*) e $= 0.1$, (*d*) e $= 0.2056$.

7.7.2 Cone-crossing criterion

Without using time–reversal symmetry and without taking initial conditions on a symmetry line, the conjugate points criterion with $t_0 = -t_1$ can be applied, provided one computes the slope of an initial tangent vector, say $(\delta x(0), \delta y(0))$, such that $\delta x(\pm t_1) = 0$ simultaneously. To this end one can compute the monodromy matrix M at times $\pm t$ by integrating the equations

$$\dot{M} = F(x, y, t)M \ ,$$

where $M(0)$ equals the identity matrix and $F(x, y, t)$ denotes the Jacobian of the vector field. Then, the initial condition $(\delta x(0), \delta y(0)) \equiv (\xi, \eta)$ satisfies the relations

$$M_{11}(t)\xi + M_{12}(t)\eta = 0 ,$$
$$M_{11}(-t)\xi + M_{12}(-t)\eta = 0 .$$

There exists a non–zero solution if and only if

$$C(t) \equiv M_{11}(t)M_{12}(-t) - M_{12}(t)M_{11}(-t) = 0 .$$

Therefore we conclude that the times $\pm t$, $t > 0$, are conjugate if and only if $C(t) = 0$. A result by Birkhoff states that a rotational invariant torus is a graph of a Lipschitz function.

If the initial condition is on a rotational invariant torus, one can determine upper and lower bounds on the slope of the initial tangent vector, providing the so–called local *Lipschitz cone* [165]. The condition $C(t) = 0$ corresponds to the equality of the upper and lower bounds; for larger t the upper bound becomes less than the lower bound. However, this is in contrast with the existence of a rotational invariant torus through that initial point, thus yielding the so–called *cone–crossing criterion* [128] as a method to establish the non–existence of rotational invariant tori.

The practical implementation of the criterion is the following. First we remark that it is more convenient to integrate the equation for the inverse monodromy matrix $N(t) = M(t)^{-1}$. Starting from $(x(0), y(0))$, let $(x(\pm t), y(\pm t))$ be the corresponding forward and backward trajectories; then integrate the equations backwards and forwards in time

$$\dot{N}(t) = -N(t)\ F(x(t), y(t), t)$$

with $N(0)$ being the identity matrix. For any $t > 0$, let

$$w^{\pm}(t) = N(\mp t)\begin{pmatrix} 0 \\ \pm 1 \end{pmatrix} = \begin{pmatrix} \pm N_{12}(\mp t) \\ \pm N_{22}(\mp t) \end{pmatrix}$$

be tangent vectors at $(x(0), y(0))$, which give a local Lipschitz cone through the initial condition. Let $C(t) = w^-(t) \wedge w^+(t)$; then $C(0) = 0$ and $\dot{C}(0) > 0$. Finally, if there exists a time $t' > 0$ such that $C(t') \leq 0$, then the orbit starting from $(x(0), y(0))$ does not belong to an invariant rotational torus.

7.7.3 Tangent orbit indicator

Based on the conjugate points criterion, we introduce an indicator of chaos, which can be used as a complementary tool to Lyapunov exponents, frequency analysis, FLIs, etc. (see Chapter 2). We start by remarking that through the change of sign of $\delta x(t)$ we can distinguish between rotational tori, librational tori and chaos. Starting from a horizontal tangent vector, for a librational torus the δx–component oscillates around zero (a linear increase is observed when starting from the vertical tangent vector). The results are shown in Figure 7.5(*a*,*b*), obtained by integrating (7.70) through a fourth–order symplectic Yoshida's method [175] shortly recalled in Appendix F. Notice that the first crossing occurs at $t = 3.39$. A similar behavior is observed for the chain of islands of Figure 7.5(*c*,*d*). Oscillations with large

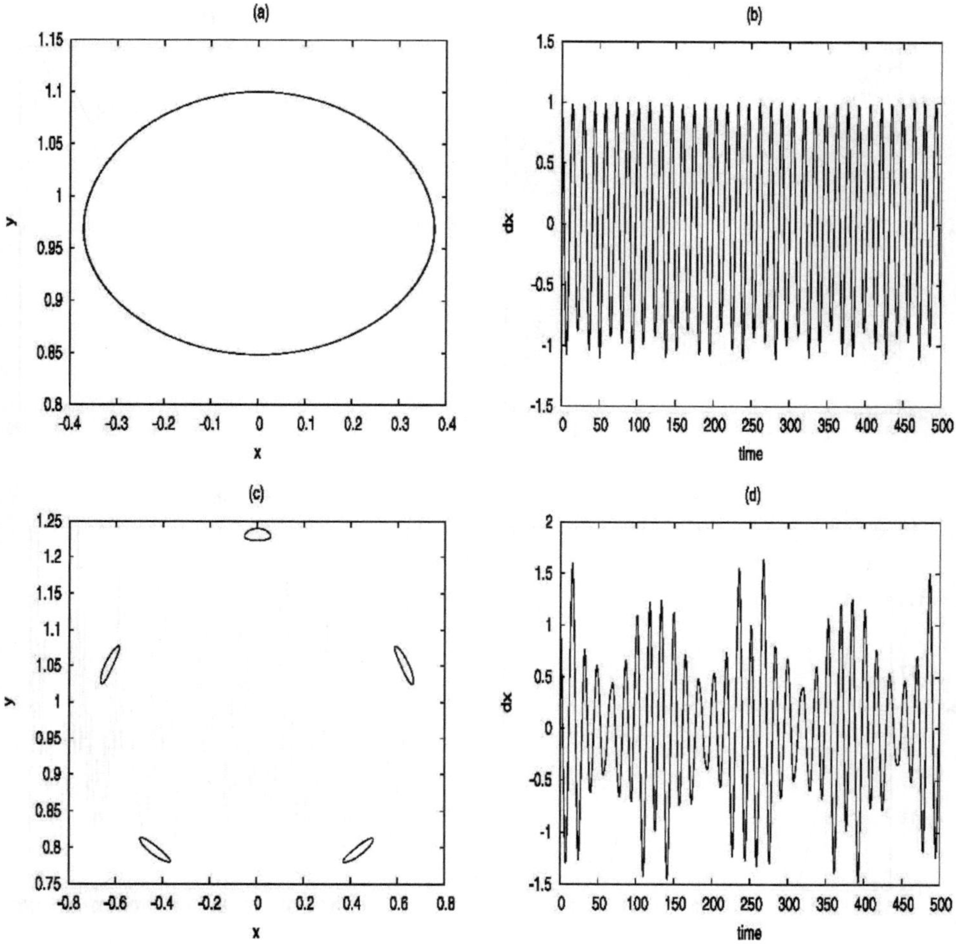

Fig. 7.5. Analysis of (7.70) with $\varepsilon = 0.1$, e $= 0.0549$ (after [40]). The left column shows the Poincaré section on the plane $t = 0$; the right column shows the implementation of the conjugate points method from the horizontal tangent vector. (a, b) refer to an example of a librational invariant torus for the initial conditions $x = 0$, $y = 1.1$. (c, d) refer to an example with a chain of islands for the initial conditions $x = 0$, $y = 1.24$. (e, f) refer to an example of chaotic motion for the initial conditions $x = 0$, $y = 1.3$. (g, h) refer to an example of a rotational invariant torus for the initial conditions $x = 0$, $y = 1.8$ (reprinted, with permission, from [40], Copyright 2007, American Institute of Physics).

amplitudes are observed for chaotic motions as shown in Figure 7.5(e,f). Finally, rotational invariant tori are characterized by positive oscillations of δx far from zero (see Figure 7.5(g,h)).

This scenario leads to the introduction of the so–called *tangent orbit indicator* by computing the average of $\delta x(t)$ over a finite interval of time. The resulting value characterizes the dynamics as follows: a zero value denotes a librational regime, a

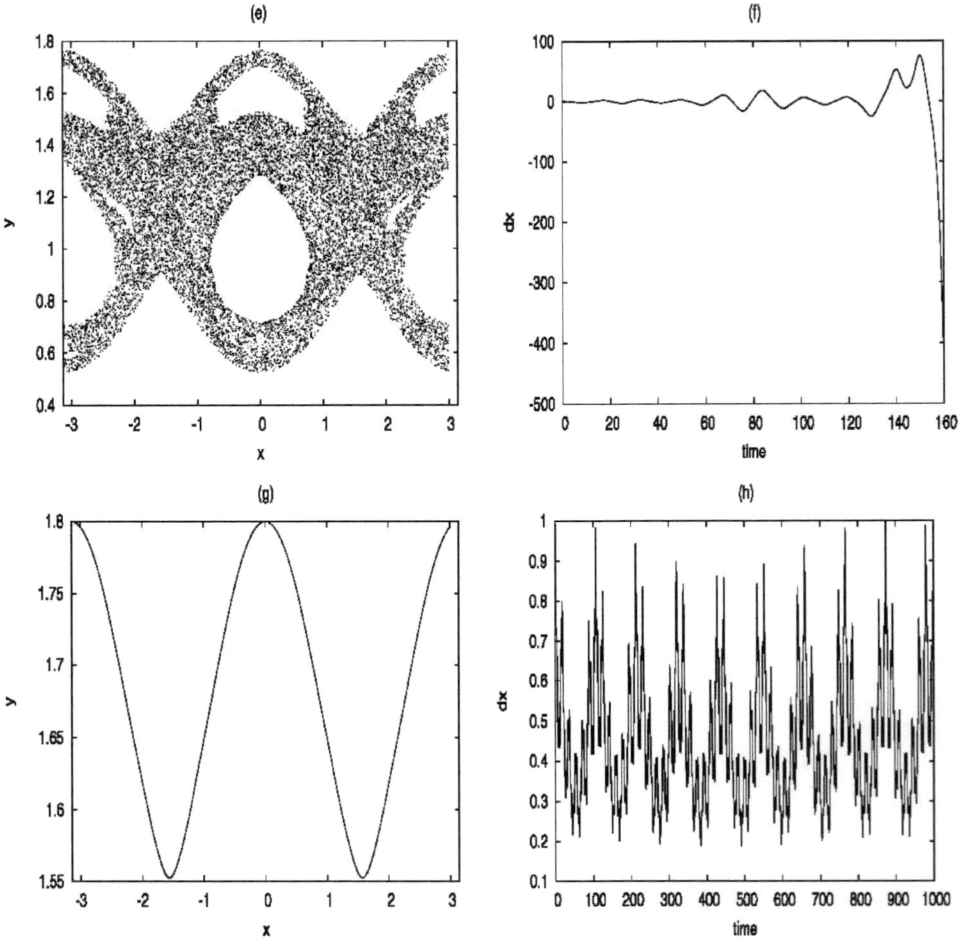

Fig. 7.5. (continued).

moderate value is associated to rotational tori, high values correspond to chaotic motions.

As an example we report in Figure 7.6 (top panels) the computation of the tangent orbit indicators with horizontal initial tangent vector over a grid of 500×500 initial conditions in x and y for the equation (7.70). Figure 7.6 (bottom panels) provides the tangent orbit indicator in the plane y–ε for a fixed x_0. A black color denotes tangent orbit indicators close to zero, grey stands for moderate values, while white corresponds to large values. The results are in full agreement with those obtained implementing other techniques, like frequency analysis or the computation of the FLIs introduced in Chapter 2 (see [37]).

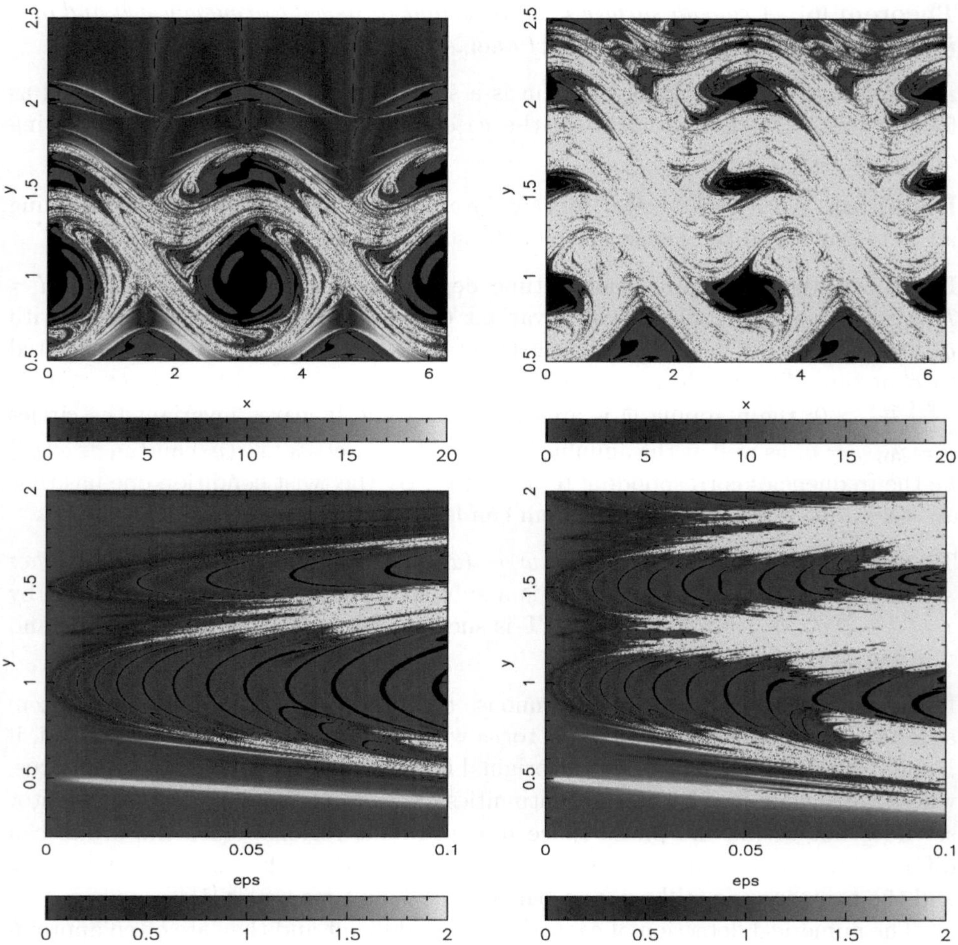

Fig. 7.6. Tangent orbit indicator associated to (7.70) for $\varepsilon = 0.1$ from the initial horizontal tangent vector. Top left: graph in the plane x–y with e $= 0.0549$; top right: graph in the plane x–y with e $= 0.2056$; bottom left: graph in the plane ε–y with e $= 0.0549$; bottom right: graph in the plane ε–y with e $= 0.2056$. (Reprinted, with permission, from [40], Copyright 2007, American Institute of Physics.)

7.8 Cantori

Let $\mathcal{L} = \mathcal{L}(\underline{x}, \underline{X})$ be a Lagrangian function with $\underline{x} \in \mathbf{T}^n$ and $\underline{X} \equiv \dot{\underline{x}} \in \mathbf{R}^n$. For a function $v = v(\underline{\vartheta})$, let $D_{\underline{\omega}}$ be the operator defined as $D_{\underline{\omega}} v = \underline{\omega} \cdot \frac{\partial v}{\partial \underline{\vartheta}}$. An n–dimensional torus is described by the equations $\underline{x} = \underline{x}(\underline{\vartheta})$, $\underline{X} = D_{\underline{\omega}} \underline{x}(\underline{\vartheta})$; let a variation be described as $\underline{x}(\underline{\vartheta}) + \delta \underline{x}(\underline{\vartheta})$, $D_{\underline{\omega}} \underline{x}(\underline{\vartheta}) + D_{\underline{\omega}} \delta \underline{x}(\underline{\vartheta})$. Let us introduce the functional

$$\mathcal{A}_{\underline{\omega}} \equiv \frac{1}{(2\pi)^n} \int_{\mathbf{T}^n} \mathcal{L}(\underline{x}(\underline{\vartheta}), D_{\underline{\omega}} \underline{x}(\underline{\vartheta}))\, d\underline{\vartheta} \; .$$

A variational principle can be stated as follows

Theorem [6]. *A smooth surface is an invariant torus with frequency* $\underline{\omega}$ *if and only if it is a stationary point of the functional* $\mathcal{A}_{\underline{\omega}}$.

A solution of the variational problem is a so–called *cantorus,* which is defined as follows (see [43]). Let us consider the case $n = 2$. We introduce the following definition (see [7,132]).

Definition. An *Aubry–Mather set* is an invariant set, which is obtained embedding a Cantor subset in the phase space of the standard two–dimensional torus.

Let us consider a one–dimensional, time–dependent Hamiltonian of the form $\mathcal{H} = \mathcal{H}(y, x, t)$. Assume it admits two invariant tori described by $y = y_0$, $y = y_1$ with $y_0 < y_1$. Denote by $\underline{\Phi} = \underline{\Phi}(y, x) \equiv (\Phi_1(y, x), \Phi_2(y, x))$ the Poincaré map associated to $\mathcal{H}$ at times 2π, which we assume to satisfy the so–called twist condition namely $\frac{\partial \Phi_2(y,x)}{\partial y} > 0$; the mapping $\underline{\Phi}$ is area preserving and it leaves invariant the circles $y = y_0$, $y = y_1$ as well as the annulus between them. Let $\omega_0 \equiv \omega(y_0)$ and $\omega_1 \equiv \omega(y_1)$ be the frequencies corresponding to y_0 and y_1. By the twist condition one has that $\omega_0 < \omega_1$. The Aubry–Mather theorem can be stated as follows.

Theorem [6]. *For any irrational* $\omega \in (\omega_0, \omega_1)$, *there exists an Aubry–Mather set* with rotation number ω, which is a subset of a closed curve parametrized by $x = \vartheta + u(\vartheta)$, $y = v(\vartheta)$, where $\vartheta \in \mathbf{T}$ is such that $\vartheta' = \vartheta + \omega$, u is monotone and u, v are 2π–periodic.

If the functions u and v are continuous, then the original Hamiltonian system admits a two–dimensional invariant torus with frequency ω. On the other hand, if u and v are discontinuous, then the original Hamiltonian system admits a cantorus, whose gaps coincide with the discontinuities of u and v. We remark that a Cantor set does not divide the phase space into invariant regions, since the orbits can diffuse through the gaps of the Cantor set. However, the leakage cannot be easy and the cantorus can still act as a barrier over long time scales [153].

The numerical detection of cantori is rather difficult and they are often approximated by high–order periodic orbits [49,85]. In very peculiar examples, an analytic expression of the cantori can be given. This is the case of the conservative sawtooth map, which is described by the equations

$$\begin{aligned} y_{n+1} &= y_n + \lambda f(x_n) \\ x_{n+1} &= x_n + y_{n+1} \ , \end{aligned} \tag{7.73}$$

where $x_n \in \mathbf{T}$, $y_n \in \mathbf{R}$, $\lambda \in \mathbf{R}_+$ denotes the perturbing parameter and the perturbation f on the covering $\mathbf{R}$ of $\mathbf{T}$ is defined as

$$\begin{aligned} f(x) &\equiv \mathrm{mod}(x, 1) - \frac{1}{2} \qquad && \text{if } 0 < \mathrm{mod}(x, 1) < 1 \\ f(x) &\equiv 0 && \text{if } x \in \mathbf{Z} \ . \end{aligned} \tag{7.74}$$

The mapping (7.73) is area–preserving; for $\lambda > 0$ there do not exist invariant circles and the phase space is filled by cantori and periodic orbits. Since $x_{n+1} - x_n = y_{n+1}$,

$x_n - x_{n-1} = y_n$, one obtains

$$x_{n+1} - 2x_n + x_{n-1} = \lambda f(x_n) .$$

Let us parametrize a solution with frequency $\omega \in \mathbf{R}$ as

$$x(\vartheta) = \vartheta + u(\vartheta) , \qquad \vartheta \in \mathbf{T} , \tag{7.75}$$

where $\vartheta' = \vartheta + \omega$. Then, the function u must satisfy the equation

$$u(\vartheta + \omega) - 2u(\vartheta) + u(\vartheta - \omega) = \lambda\ f(\vartheta + u(\vartheta)) .$$

We can determine $u(\vartheta)$ by expanding it as

$$u(\vartheta) = \sum_{n=-\infty}^{\infty} a_n f(\vartheta + n\omega) \tag{7.76}$$

for some coefficients a_n which are given by

$$a_n = -\alpha\ \rho^{-|n|} ,$$

with

$$\alpha \equiv \left(1 + \frac{4}{\lambda}\right)^{-1/2} , \qquad \rho = 1 + \frac{\lambda}{2} + \left(\lambda + \frac{\lambda^2}{4}\right)^{1/2} .$$

In fact, inserting the series expansion (7.76) in (7.73) we obtain

$$\sum_j (a_{j-1} - 2a_j + a_{j+1}) f(\vartheta + j\omega) = \lambda f(\vartheta + u(\vartheta)) .$$

Being $f(\vartheta + u(\vartheta)) = \vartheta + u(\vartheta) - \frac{1}{2} = f(\vartheta) + u(\vartheta)$, one finds the following recursive relations

$$\begin{aligned} a_{-1} - 2a_0 + a_1 &= \lambda(1 + a_0) && j = 0 \\ a_{j-1} - 2a_j + a_{j+1} &= \lambda a_j && j \neq 0 . \end{aligned} \tag{7.77}$$

Let us write $a_j = -\alpha\rho^{-|j|}$; from the first of (7.77) for $j = 0$ one has $-\alpha\rho^{-1} + 2\alpha - \alpha\rho^{-1} = \lambda(1 - \alpha)$, namely

$$\alpha = \frac{\lambda\rho}{2\rho - 2 + \lambda\rho} . \tag{7.78}$$

Equation (7.77) for $j \neq 0$ implies that $-\alpha\rho^{-|j-1|} + 2\alpha\rho^{-|j|} - \alpha\rho^{-|j+1|} = -\lambda\alpha\rho^{-|j|}$, namely $\rho^2 - (2 + \lambda)\rho + 1 = 0$ with solution

$$\rho = 1 + \frac{\lambda}{2} + \left(\lambda + \frac{\lambda^2}{4}\right)^{1/2} .$$

Replacing this expression for ρ in (7.78) one obtains $\alpha = (1 + \frac{4}{\lambda})^{-1/2}$.

Taking advantage of the solution parametrized as in (7.75) with u given by (7.76), one can prove the existence of cantori for the sawtooth map through the following proposition as stated in [42].

Proposition. *Let ω be irrational, let*

$$\tilde{M}_\omega \equiv \{(x(\vartheta), x(\vartheta + \omega)) \ : \ \vartheta \in \mathbf{R}\}$$

and let $M_\omega \equiv \tilde{M}_\omega / \mathbf{Z}$. Then, M_ω is a Cantor set.

A proof of the existence of cantori in the dissipative sawtooth map, defined by the equations

$$\begin{aligned} y_{n+1} &= b y_n + c + \lambda f(x_n) \\ x_{n+1} &= x_n + y_{n+1} \ , \end{aligned}$$

for $b \in \mathbf{R}$, $c \in \mathbf{R}$ and $f(x)$ as in (7.74) is provided in [39].

8 Long–time stability

Hamiltonian systems with (strictly) more than two degrees of freedom do not admit a confinement of the phase space by invariant tori. For example, for a three–dimensional Hamiltonian system, the phase space has dimension 6, the constant energy level is five–dimensional and therefore the three–dimensional KAM tori do not separate the phase space by invariant regions. As a consequence the motion can diffuse through the invariant tori and can reach arbitrarily far regions of the phase space. This phenomenon is known as *Arnold's diffusion* (Section 8.1) and a theorem due to N.N. Nekhoroshev allows us to state that it takes place at least on exponentially long times (Section 8.2, see also [75]).

In this chapter we focus on the main ideas at the basis of Nekhoroshev's theorem and we also formulate it in the neighborhood of elliptic equilibria (Section 8.3). The theorem has been widely used in Celestial Mechanics, in particular in connection with the stability of the three–body problem (Section 8.4) and of the Lagrangian solutions (Section 8.5).

8.1 Arnold's diffusion

We consider an n–dimensional Hamiltonian system described by the following real–analytic Hamiltonian function (in action–angle coordinates):

$$\mathcal{H}(\underline{y},\underline{x}) = h(\underline{y}) + \varepsilon f(\underline{y},\underline{x}) \ , \qquad \underline{y} \in Y \ , \quad \underline{x} \in \mathbf{T}^n \ , \tag{8.1}$$

where Y is an open subset of $\mathbf{R}^n$. The phase–space associated to (8.1) is $2n$–dimensional; fixing an energy level we obtain a $(2n-1)$–dimensional manifold. In this framework, the n–dimensional invariant tori divide the constant energy level only if $n = 2$. On the other hand, if $n > 2$ the separation property is no longer valid as we showed above in the case of a three–dimensional Hamiltonian system. Indeed this property is not valid already if we consider a non–autonomous two–dimensional Hamiltonian system. In all the cases when the separation property does not apply, invariant tori form the majority of the solutions, but the resonances generate gaps between the invariant tori, so that some orbits can leak to arbitrarily far regions of the phase space. An example of such a phenomenon was given for the first time in [4] and it is now known as *Arnold's diffusion.* More precisely, the example provided in [4] is based on the Hamiltonian function

$$\mathcal{H}(y_1, y_2, x_1, x_2, t) = \frac{1}{2}(y_1^2 + y_2^2) + \varepsilon[(\cos x_1 - 1)(1 + \mu \sin x_2 + \mu \cos t)] \ ,$$

where y_1, $y_2 \in \mathbf{R}$, x_1, $x_2 \in \mathbf{T}$ and ε, μ are positive parameters. For suitable values of the parameters with μ exponentially small with respect to ε and for given values of the second action, say $y_2^{(a)}$ and $y_2^{(b)}$ with $0 < y_2^{(a)} < y_2^{(b)}$, there exists an orbit connecting regions where the values of the action y_2 are far from each other, say $y_2 < y_2^{(a)}$ and $y_2 > y_2^{(b)}$; the time for such diffusion can be exponentially long with respect to ε. This phenomenon is obtained by constructing unstable tori, called *whiskered* tori. A chain of heteroclinic intersections provides the device to transport the trajectories; such a transition chain makes it possible to pass from the neighborhood of a torus to the neighborhood of another torus. However, practical difficulties in constructing the transition chain are due to the fact that the elements of the chain must be sufficiently close and that the intersection angle of the manifolds of successive tori is typically exponentially small. The literature on this topic is extensive and we refer the reader to it for complete details; we just mention that analytical proofs and examples of Arnold's diffusion can be found in [45,46,102,174], while numerical investigations have been performed in [87,111,164].

8.2 Nekhoroshev's theorem

A theorem developed by A.A. Nekhoroshev [143] proves that Arnold's diffusion eventually takes place at least on exponential times. In particular, under suitable applicability conditions, the theorem allows us to state that the actions remain confined over an exponentially long time (see [137] for a connection to KAM theory). The original version of the theorem was formulated under a suitable assumption on the Hamiltonian (called the *steepness* condition); following [150] we present the theorem under the convex and quasi–convex hypotheses which are introduced as follows. With reference to (8.1) with $n \geq 2$, we define a complex neighborhood of $Y \times \mathbf{T}^n$ with radii r_0 and s_0 as the complex set $V_{r_0}Y \times W_{s_0}\mathbf{T}^n$, where $V_{r_0}Y$ denotes the complex neighborhood of radius r_0 around Y with respect to the Euclidean norm $\|\cdot\|$ and $W_{s_0}\mathbf{T}^n$ is the complex strip of width s_0 around $\mathbf{T}^n$: $W_{s_0}\mathbf{T}^n \equiv \{\underline{x} \in \mathbf{C}^n \ : \ \max_{1\leq j\leq n} |\mathrm{Im}\ x_j| < s_0\}$. Let $U_{r_0}Y \equiv V_{r_0}Y \cap \mathbf{R}^n$ be the real neighborhood of Y. For an analytic function $u = u(\underline{y}, \underline{x})$ on $V_{r_0}Y \times W_{s_0}\mathbf{T}^n$ with Fourier expansion $u(\underline{y}, \underline{x}) = \sum_{\underline{k}\in\mathbf{Z}^n} \hat{u}_{\underline{k}}(\underline{y}) e^{i\underline{k}\cdot\underline{x}}$, let us define the norm

$$\|u\|_{Y,r_0,s_0} \equiv \sup_{\underline{y}\in V_{r_0}Y} \sum_{\underline{k}\in\mathbf{Z}^n} |\hat{u}_{\underline{k}}(\underline{y})|\ e^{(|k_1|+\cdots+|k_n|)s_0}\ .$$

Let F be an upper bound on $\|f\|_{Y,r_0,s_0}$ and let M be an upper bound on the Hessian matrix $Q = Q(\underline{y})$ of the unperturbed Hamiltonian: $\sup_{\underline{y}\in V_{r_0}Y} \|Q(\underline{y})\| \leq M$. Finally, let $\underline{\omega}(\underline{y}) \equiv \frac{\partial h(\underline{y})}{\partial \underline{y}}$ be the frequency vector.

Definition. Given a positive real parameter m, the unperturbed Hamiltonian is said to be m–convex if

$$(Q(\underline{y})\underline{v}, \underline{v}) \geq m\|\underline{v}\|^2 \qquad \text{for all } \underline{v} \in \mathbf{R}^n\ , \quad \text{for all } \underline{y} \in U_{r_0}Y\ .$$

Given positive real parameters m, ℓ, the unperturbed Hamiltonian is said to be m,ℓ–quasi–convex if for any $\underline{y} \in U_{r_0}Y$ one of the following inequalities holds for any $\underline{v} \in \mathbf{R}^n$:

$$|(\underline{\omega}(\underline{y}), \underline{v})| > \ell\|\underline{v}\| \ , \qquad (Q(\underline{y})\underline{v}, \underline{v}) \geq m\|\underline{v}\|^2 \ . \tag{8.2}$$

We remark that quasi–convex Hamiltonians satisfy the isoenergetic non–degeneracy condition (7.5). We can finally formulate Nekhoroshev's theorem, providing the explicit estimates for the confinement of the actions and for the stability time as given in [150].

Theorem (Nekhoroshev). *Let us consider the Hamiltonian function* (8.1) *and assume that the unperturbed Hamiltonian satisfies the quasi–convexity hypothesis* (8.2). *Let* $r_0 \leq \frac{4\ell}{m}$, $A \equiv \frac{11M}{m}$, $\varepsilon_0 \equiv \frac{mr_0^2}{2^{10}A^{2n}}$; *if for* $s_0 > 0$, *one has* $\|f\|_{Y,r_0,s_0}\varepsilon \leq \varepsilon_0$, *then for any initial condition* $(\underline{y}_0, \underline{x}_0) \in Y \times \mathbf{T}^n$ *the following estimates hold:*

$$\|\underline{y}(t) - \underline{y}_0\| \leq \frac{r_0}{A}\left(\frac{\varepsilon}{\varepsilon_0}\right)^{\frac{1}{2n}} \qquad \text{for } |t| \leq \frac{A^2 s_0}{\Omega_0}\ e^{\frac{s_0}{6}(\frac{\varepsilon_0}{\varepsilon})^{\frac{1}{2n}}} \ ,$$

where $\Omega_0 \equiv \sup_{\|\underline{y}-\underline{y}_0\| \leq \frac{r_0}{A}} \|\underline{\omega}(\underline{y})\|$.

The proof is based on three main steps: the construction of a suitable normal form, the use of the convexity and quasi–convexity assumptions, and a careful analysis of the geography of the resonances. We provide here the main ideas of the proof, referring the reader to [143, 150] for complete details.

Step 1: Normal form. Let Λ be a sublattice of $\mathbf{Z}^n$ and for $K \geq 0$ set $\mathbf{Z}^n_K \equiv \{\underline{k} \in \mathbf{Z}^n \ : \ |\underline{k}| \leq K\}$; we say that a subset Y_0 of Y is α, *K–non–resonant modulo* Λ, if for any $\underline{y} \in Y_0$ and for any $\underline{k} \in \mathbf{Z}^n_K \backslash \Lambda$, one has

$$|\underline{k} \cdot \underline{\omega}(\underline{y})| \geq \alpha \ .$$

If Λ contains just the zero, then Y_0 is said to be *completely* α, K–non–resonant. In the domain Y_0 one looks for a suitable change of coordinates so to obtain a Λ–resonant normal form, where the new Hamiltonian is given by the sum of three terms:

(*i*) the unperturbed Hamiltonian;
(*ii*) a resonant term $g_\Lambda = g_\Lambda(\underline{y}, \underline{x})$ which can be expanded in Fourier series as

$$g_\Lambda(\underline{y}, \underline{x}) = \sum_{\underline{k} \in \Lambda} g_{\underline{k}}(\underline{y})\ e^{i\underline{k}\cdot\underline{x}} \ ,$$

containing only the resonant terms belonging to the lattice Λ;
(*iii*) a remainder function $f_K = f_K(\underline{y}, \underline{x})$, which can be made exponentially small by a suitable choice of K; in particular, for some $r < r_0$, $s < s_0$ it is

$$\|f_K\|_{Y_0,r,s} \leq \varepsilon F e^{\frac{-Ks}{6}} \ ,$$

where we assume that $Ks \geq 6$.

Step 2: Using the convexity and quasi–convexity. With reference to the previous step, let the normal form Hamiltonian be given by

$$\mathcal{H}_\Lambda(\underline{y},\underline{x}) = h(\underline{y}) + \varepsilon g_\Lambda(\underline{y},\underline{x}) + f_K(\underline{y},\underline{x}) \ .$$

The evolution of the actions associated to $\mathcal{H}_\Lambda$ belongs almost completely to the *plane of fast drift,* namely the hyperplane generated by Λ and passing through the initial condition $\underline{y}_0$. Indeed, one finds that

$$\dot{\underline{y}} = \varepsilon \sum_{\underline{k}\in\Lambda} \Big(- i g_{\underline{k}}(\underline{y}) e^{i\underline{k}\cdot\underline{x}} \Big) \cdot \underline{k} + O(\varepsilon e^{\frac{-Ks}{6}}) \ .$$

Therefore in the non–resonant domain $\dot{\underline{y}}$ is almost parallel to Λ and the distance between $\underline{y}(t)$ and the plane of fast drift is small for exponential times.

In the proximity of the resonances we use the convexity and the quasi–convexity to prove that the orbits stay bounded for exponentially long times. To this end, let R_Λ be the *resonant manifold* defined as

$$R_\Lambda \equiv \{\underline{\omega} \in \mathbf{R}^n \ : \ \underline{k}\cdot\underline{\omega} = 0 \qquad \text{for all } \underline{k} \in \Lambda\} \ .$$

For $\delta > 0$ let D be a set δ–close to the resonant manifold R_Λ, namely D is composed by the points $\underline{y} \in \mathbf{R}^n$ such that $\min_{\underline{\nu}\in R_\Lambda} \|\underline{\omega}(\underline{y}) - \underline{\nu}\| \leq \delta$. If the unperturbed Hamiltonian is m–convex, the Λ–resonant normal form ensures that for any initial condition in a real neighborhood of D, the corresponding orbit is bounded for finite times. A similar result holds in the quasi–convex case. If we now assume that D is also α, K–non–resonant modulo Λ, then one can prove the following result.

Proposition. *Let the unperturbed Hamiltonian be ℓ, m–quasi–convex; then, if ε is sufficiently small, any orbit with initial conditions $(\underline{y}_0, \underline{x}_0)$ in $D \times \mathbf{T}^n$ satisfies*

$$\|\underline{y}(t) - \underline{y}_0\| \leq r \qquad \text{for all } |t| \leq C_2 e^{\frac{Ks}{6}} \ ,$$

for a suitable $r > 0$ and a positive constant C_2.

We remark that the quasi–convexity implies that the plane of fast drift and the inverse image of the resonant manifold intersect transversally (see, e.g., [11]). Due to the complementary dimensions, their intersection is a point, where the unperturbed Hamiltonian has an extremum; around such a point the surfaces corresponding to a constant unperturbed Hamiltonian level provide an elliptic structure on the plane of fast drift.

Step 3: The geography of the resonances. The results of steps 1 and 2 must be applied to encompass the whole phase space $Y \times \mathbf{T}^n$. A suitable covering is first constructed in the frequency space and then in the actions. The geography of the resonances can be explored by introducing the following sets.

- The *family of maximal K–lattices* $\mathcal{M}_K$ is the set of lattices Λ in $\mathbf{Z}^n$ generated by vectors in $\mathbf{Z}^n_K$ and not properly contained in other lattices of the same dimension.

- Let $0 < \lambda_1 < \lambda_2 < \dots < \lambda_n < \dots$ be a sequence of real numbers; let Λ be a maximal K–lattice of dimension d and let $\delta_\Lambda \equiv \frac{\lambda_d}{|\Lambda|}$, where $|\Lambda|$ is the volume of the parallelepiped built from a basis vector of Λ. The *resonant zone* of order d associated to Λ with size δ_Λ is the set

$$\mathcal{Z}_\Lambda \equiv \{\underline{\omega} \in \mathbf{R}^n \ : \ \min_{\underline{\nu} \in R_\Lambda} \|\underline{\omega} - \underline{\nu}\| < \delta\} \ .$$

- A *resonant region* of order d is the union of all resonant zones of order d:

$$\mathcal{Z}_d^* \equiv \cup_{\dim(\Lambda)=d} \mathcal{Z}_\Lambda \ .$$

  In particular, one finds that $\mathcal{Z}_{n+1}^*$ is the empty set: $\mathcal{Z}_{n+1}^* = \emptyset$.
- A *resonant block* of order d is obtained by removing from a resonant zone of order d a resonant region of order $d+1$:

$$\mathcal{B}_\Lambda \equiv \mathcal{Z}_\Lambda \backslash \mathcal{Z}_{d+1}^* \ .$$

  The block $\mathcal{B}_\emptyset$ denotes the completely non–resonant frequency space.

From $\mathbf{R}^n = \mathcal{Z}_\emptyset$ it follows that the frequency space can be covered by resonant blocks as

$$\mathbf{R}^n = \mathcal{B}_\emptyset \cup (\cup_{\dim(\Lambda)=1} \mathcal{B}_\Lambda) \cup .. \cup (\cup_{\dim(\Lambda)=n} \mathcal{B}_\Lambda) \ .$$

The following result provides the conditions under which $\mathcal{B}_\Lambda$ is α, K–non–resonant modulo Λ.

Geometric Lemma. *Assume that $\lambda_{j+1} \geq C_3 \lambda_j$, for $j = 1, \dots, n$ and for some positive constant C_3. Each block $\mathcal{B}_\Lambda$ is α_Λ, K–non–resonant modulo Λ with*

$$\alpha_\Lambda \geq C_4 \delta_\Lambda \ ,$$

for some positive constant C_4; moreover, $\mathcal{B}_\emptyset$ is completely $\alpha_\emptyset$, K–non–resonant modulo Λ with $\alpha_\emptyset = \lambda_1$.

The transformation from the frequency to the action space is provided by the following result.

Covering Lemma. *Let*

$$D_\Lambda \equiv \{\underline{y} \in Y \ : \ \underline{\omega}(\underline{y}) \in \mathcal{B}_\Lambda\} \ ;$$

let r_Λ, α_Λ be positive parameters. Then, there exists a covering of the action space Y through resonant blocks D_Λ, $\Lambda \in \mathcal{M}_K$, which are α_Λ, K–non–resonant modulo Λ, though being δ_Λ–close to exact resonances.

The proof of the theorem can be obtained by combining the three steps outlined above. In particular, by the covering lemma there exists a covering of Y by suitable resonant blocks D_Λ, $\Lambda \in \mathcal{M}_K$. The stability estimates can now be extended to all blocks simultaneously for an exponentially long stability time.

We remark that there exist several versions of the Nekhoroshev's theorem, according to the specific dynamical context. For example, when dealing with completely non–resonant domains there is a sharp decrease in the stability radius. On the other side, in the proximity of a resonance, one could get much bigger stability times.

8.3 Nekhoroshev's estimates around elliptic equilibria

We discuss the formulation of Nekhoroshev's theorem in a neighborhood of an elliptic equilibrium position of a Hamiltonian system with n degrees of freedom (see [64, 151]). Assume that the natural frequencies $(\omega_1, \dots, \omega_n)$ are non–resonant up to the order $\ell \geq 4$, namely there does not exist any vector $\underline{k} \in \mathbf{R}^n \backslash \{\underline{0}\}$ with $|\sum_{j=1}^n k_j| \leq \ell$, such that $\underline{k} \cdot \underline{\omega} = 0$. Under this condition, there exists a symplectic coordinate transformation by implementing a Birkhoff normal form such that the original Hamiltonian takes the form

$$\mathcal{H}(\underline{x}, \underline{y}) = \underline{\omega} \cdot \underline{I} + \frac{1}{2} A \underline{I} \cdot \underline{I} + B(\underline{I}) + R(\underline{x}, \underline{y}) \ , \qquad (\underline{x}, \underline{y}) \in \mathbf{R}^{2n} \ , \tag{8.3}$$

where $I_j = \frac{1}{2}(x_j^2 + y_j^2)$, the coefficients of the matrices A and the term B are the so–called *Birkhoff's invariants* with B being of order 3 at least in $\underline{I}$, while R is of order $\ell+1$ in the variables x_j, y_j. We refer to the integrable part as the Hamiltonian $h_0 = h_0(\underline{I})$ which is obtained from (8.3) neglecting the remainder R. Then, in a small neighborhood of the origin, h_0 is convex if and only if the matrix A is positive definite. Let $|\underline{I}| = |I_1| + \dots + |I_n|$.

Theorem. *Let $\delta > 0$; if A is positive definite, then for any orbit associated to* (8.3) *with initial condition $|\underline{I}_0| < \delta^2$ sufficiently small, one obtains*

$$|\underline{I}(t) - \underline{I}_0| < C_5 \delta^{2 + \frac{\ell - 3}{2n}} \qquad \text{for } |t| < \frac{1}{|\underline{\omega}|} e^{C_6 \delta^{-\frac{\ell-3}{2n}}} \ ,$$

for suitable positive constants C_5 and C_6.

It is readily seen that rescaling the variables as

$$\underline{x} = \delta \underline{\tilde{x}} \ , \quad \underline{y} = \delta \underline{\tilde{y}} \ , \quad \tilde{I}_j = \frac{1}{2}(\tilde{x}_j^2 + \tilde{y}_j^2) \ ,$$

the region $|\underline{I}_0| < \delta^2$ is transformed to $|\underline{\tilde{I}}_0| < 1$; as a consequence the Hamiltonian (8.3) becomes

$$\tilde{\mathcal{H}}(\underline{\tilde{I}}, \underline{\tilde{x}}, \underline{\tilde{y}}) = \frac{1}{\delta^2} \ \underline{\omega} \cdot \underline{\tilde{I}} + \frac{1}{2} A \underline{\tilde{I}} \cdot \underline{\tilde{I}} + \delta^{-4} B(\delta^2 \underline{\tilde{I}}) + \delta^{\ell - 3} \tilde{R}(\underline{\tilde{x}}, \underline{\tilde{y}}) \ . \tag{8.4}$$

Setting $\varepsilon \equiv \delta^{\ell-3}$ one has the following

Theorem. *If A is positive definite and ε is sufficiently small, then for any orbit associated to* (8.4) *with initial condition $|\underline{\tilde{I}}_0| < 1$, one obtains*

$$|\underline{\tilde{I}}(t) - \underline{\tilde{I}}_0| < C_7 \varepsilon^{\frac{1}{2n}} \qquad \text{for } |t| < \frac{\delta^2}{|\underline{\omega}|} e^{C_8 \varepsilon^{-\frac{1}{2n}}} \ ,$$

for suitable positive constants C_7 and C_8.

8.4 Effective estimates in the three–body problem

The stability estimates provided by Nekhoroshev's theorem are particularly relevant in Celestial Mechanics. In fact, they can be used to provide bounds on the elliptic elements for an exponentially long time, possibly comparable with the age of the solar system, namely 5 billion years (see [108, 168] for a long–time integration of the equations of motion). Effective estimates have been developed for the triangular Lagrangian points ([34, 38, 77, 160], see also [12, 86]), the resonant D'Alembert problem [15], and the perturbed Euler rigid body [13]. Here we discuss two applications in the context of the three–body problem. First we consider the planar, circular, restricted three–body problem; being described by a two–dimensional Hamiltonian function, diffusion does not take place, but the relevant result is a bound on the semimajor axis and eccentricity for astronomical times. The second example concerns the stability of the Lagrangian points in the circular, restricted, spatial, three–body problem; the procedure developed in [38] has been successfully extended in [77, 160], providing physically relevant estimates within the asteroidal belt.

8.4.1 Exponential stability of a three–body problem

Following [34], we consider the planar, circular, restricted three–body problem. Let $\mathcal{P}_1$, $\mathcal{P}_3$ be the primaries with masses m_1, m_3, moving on circular orbits around the common barycenter. Let $\mathcal{P}_2$ be a small body of negligible mass. We assume that the motion of the three bodies takes place on the same plane. Using Delaunay action–angle variables (L, G, ℓ, g), the Hamiltonian takes the form:

$$\mathcal{H}_0(L, G, \ell, g) \quad = \quad -\frac{1}{2L^2} - G + \varepsilon R_0(L, G, \ell, g) \ , \tag{8.5}$$

where $(L, G) \in \mathbf{R}^2$, $(\ell, g) \in \mathbf{T}^2$ and where we assume that the perturbing function is given by

$$\begin{aligned} R_0(L, G, \ell, g) &= L^4 \left[\frac{1}{4} + \frac{9}{64} L^4 + \frac{3}{8} \left(1 - \frac{G^2}{L^2} \right) \right] \\ &+ L^4 \left(\frac{3}{4} + \frac{5}{16} L^4 \right) \cos(2\ell + 2g) - L^4 \sqrt{1 - \frac{G^2}{L^2}} \left(\frac{1}{2} + \frac{9}{16} L^4 \right) \cos \ell \\ &+ \frac{3}{4} L^4 \sqrt{1 - \frac{G^2}{L^2}} \cos(3\ell + 2g) - L^4 \sqrt{1 - \frac{G^2}{L^2}} \left(\frac{9}{4} + \frac{5}{4} L^4 \right) \cos(\ell + 2g) \\ &+ L^6 \left(\frac{3}{8} + \frac{15}{64} L^4 \right) \cos(\ell + g) + L^6 \left(\frac{5}{8} + \frac{35}{128} L^4 \right) \cos(3\ell + 3g) \\ &+ \frac{35}{64} L^8 \cos(4\ell + 4g) + \frac{63}{128} L^{10} \cos(5\ell + 5g) \ . \end{aligned} \tag{8.6}$$

In order to obtain optimal Nekhoroshev stability estimates, it is convenient to apply perturbation theory to reduce the perturbing function in (8.5) to higher orders in ε. This can be achieved through a canonical transformation $(L, G, \ell, g) \to$

(L', G', ℓ', g') which conjugates (8.5) to a Hamiltonian of the form

$$\mathcal{H}_n(L', G', \ell', g') \; = \; h_n(L', G') + \varepsilon^{n+1} R_n(L', G', \ell', g') \; , \tag{8.7}$$

where n is the *order* of the normal form, h_n is the new unperturbed Hamiltonian, while R_n is the new perturbing function. For $n = 1$, let $\Phi^{(1)}(L', G', \ell, g)$ be the generating function providing the transformation

$$\begin{aligned} L &= L' + \varepsilon \frac{\partial \Phi^{(1)}}{\partial \ell} \\ G &= G' + \varepsilon \frac{\partial \Phi^{(1)}}{\partial g} \\ \ell' &= \ell + \varepsilon \frac{\partial \Phi^{(1)}}{\partial L'} \\ g' &= g + \varepsilon \frac{\partial \Phi^{(1)}}{\partial G'} \; . \end{aligned} \tag{8.8}$$

Inserting (8.8) in (8.5) and imposing that the terms of the first order in ε do not depend on the angle variables, one gets that the new unperturbed Hamiltonian is given by

$$h_1(L', G') = h_0(L', G') + \varepsilon \overline{R}_0(L', G') \; ,$$

where $h_0(L', G') \equiv -\frac{1}{2L'^2} - G'$, while $\overline{R}_0(L', G')$ denotes the average of $R_0(L', G', \ell, g)$ over the angles:

$$\overline{R}_0(L', G') \; \equiv \; \frac{1}{(2\pi)^2} \int_{\mathbf{T}^2} R_0(L', G', \ell, g) \; d\ell \; dg \; .$$

Expand $R_0(L, G, \ell, g)$ in Fourier series as

$$R_0(L, G, \ell, g) \; = \; \sum_{(m_1, m_2) \in \mathcal{R}} \hat{R}^{(0)}_{(m_1, m_2)}(L, G) \; e^{i(m_1 \ell + m_2 g)} \; ,$$

where $\mathcal{R}$ is the set of Fourier indexes corresponding to (8.6):

$$\mathcal{R} \equiv \{(0,0), \pm(1,0), \pm(1,1), \pm(1,2), \pm(2,2), \pm(3,2), \pm(3,3), \pm(4,4), \pm(5,5)\} \; .$$

The generating function $\Phi^{(1)}$ is finally given by

$$\Phi^{(1)}(L', G', \ell, g) = - \sum_{(m_1, m_2) \in \mathcal{R} \setminus \{(0,0)\}} \frac{\hat{R}^{(0)}_{(m_1, m_2)}(L', G')}{i(m_1 \omega_1 + m_2 \omega_2)} \; e^{i(m_1 \ell + m_2 g)} \; ,$$

where (ω_1, ω_2) are the frequencies

$$\omega_1 = \omega_1(L, G) \equiv \frac{\partial h_0(L, G)}{\partial L} \; , \qquad \omega_2 = \omega_2(L, G) \equiv \frac{\partial h_0(L, G)}{\partial G} \; .$$

The new perturbing function $R_1 = R_1(L', G', \ell', g')$ is obtained from

$$\begin{aligned} R_1(l', G', \ell, g) &= \frac{1}{2}\frac{\partial^2 h}{\partial L^2}\left(\frac{\partial \Phi^{(1)}}{\partial \ell}\right)^2 + \frac{1}{2}\frac{\partial^2 h}{\partial G^2}\left(\frac{\partial \Phi^{(1)}}{\partial g}\right)^2 \\ &+ \frac{\partial^2 h}{\partial L \partial G}\left(\frac{\partial \Phi^{(1)}}{\partial \ell}\right)\left(\frac{\partial \Phi^{(1)}}{\partial g}\right) \\ &+ \frac{\partial R_0}{\partial L}\frac{\partial \Phi^{(1)}}{\partial \ell} + \frac{\partial R_0}{\partial G}\frac{\partial \Phi^{(1)}}{\partial g} , \end{aligned}$$

with (ℓ, g) expressed in terms of (ℓ', g') through (8.8).

A higher–order normal form provides a new Hamiltonian of the form (8.7), where the generating function is given by

$$\Phi^{(n)}(L', G', \ell, g) = - \sum_{(m_1,m_2)\in\mathcal{R}_n\setminus\{(0,0)\}} \frac{\hat{R}^{(n-1)}_{(m_1,m_2)}(L', G')}{i(m_1\omega_1 + m_2\omega_2)} e^{i(m_1\ell + m_2 g)}$$

for a suitable index set $\mathcal{R}_n$, being R_{n-1} the perturbing function at the order $n-1$ and $\hat{R}^{(n-1)}_{(m_1,m_2)}$ are its Fourier coefficients. The new unperturbed Hamiltonian takes the form

$$h_n(L', G') = h_{n-1}(L', G') + \varepsilon^n \overline{R}_{n-1}(L', G') ,$$

where h_{n-1} is the unperturbed Hamiltonian at the order $n-1$ and $\overline{R}_{n-1}$ is the average of R_{n-1}. The new perturbing function is given by

$$\begin{aligned} R_n(L', G', \ell, g) &= \sum_{k=2}^{n+1} \frac{1}{k!}\frac{\partial^k h_0}{\partial L^k} \sum^{*} \prod_{j=1}^{k} \frac{\partial \Phi^{(p_j)}}{\partial \ell} \\ &+ \sum_{1\le|a_1+a_2|\le n} \frac{1}{a_1!}\frac{1}{a_2!}\frac{\partial^{a_1+a_2} R_0}{\partial^{a_1} L \partial^{a_2} G} \sum^{**} \prod_{j=1}^{a_1} \frac{\partial \Phi^{(p_j^{(1)})}}{\partial \ell} \prod_{j=1}^{a_2} \frac{\partial \Phi^{(p_j^{(2)})}}{\partial g} , \end{aligned}$$

where $*$ means that the sum is computed over all integers $p_1, \ldots, p_n \geq 1$ such that $\sum_{j=1}^k p_j = n+1$ and $**$ means that the sum is performed over all integers $p_1^{(1)}, \ldots, p_{a_1}^{(1)} \geq 1$, $p_1^{(2)}, \ldots, p_{a_2}^{(2)} \geq 1$ such that $\sum_{j=1}^{a_1} p_j^{(1)} + \sum_{j=1}^{a_2} p_j^{(2)} = n$. Again, (ℓ, g) must be expressed in terms of (ℓ', g') by means of (8.8) with $\Phi^{(1)}$ replaced by $\Phi^{(n)}$.

In [34] the stability estimates developed in [150] have been applied in the framework of the Hamiltonian (8.5)–(8.6) in order to investigate the dynamics of the small body Ceres under the gravitational influence of the Sun and Jupiter. Normalizing to unity the mass of the Sun and the semimajor axis of Jupiter, the elliptic elements of Ceres are

$$a = 0.5319 , \qquad \mathrm{e} = 0.0766 ,$$

which correspond to the Delaunay variables

$$L_0 = 0.7293 , \qquad G_0 = 0.7272 . \tag{8.9}$$

Perturbation theory has been implemented up to the order $n = 4$; the corresponding generating function contains 440 Fourier coefficients. After the normal form, the fundamental frequencies are given by

$$\underline{\omega}' = (\omega_1', \omega_2') = (-2.5131, -1) \ ,$$

which satisfy the α, K–non–resonance condition with

$$\alpha = 2.6225 \cdot 10^{-2} \ , \qquad K = 120 \ .$$

The value of the parameter K was chosen in order to optimize the results, while α was computed numerically by means of the non–resonant condition. We report in Table 8.1 the results obtained as the parameters vary; in particular, the stability time and the perturbing parameter are given as a function of K, α, r_0, s_0.

The advantage of performing normal forms before implementing Nekhoroshev's theorem is shown in Table 8.2, which provides the results at several orders n of the normalization procedure. The integer n denotes the order of the normal form ($n = 0$ corresponds to the direct application of the stability estimates given in [150] without performing a preliminary normal form). Table 8.2 reports the variation of the action variables up to a time $|t| \leq t_{max}$ for any $\varepsilon \leq \varepsilon_0$.

Table 8.1. Stability times $t_{\max}$ as a function of the parameters K, α, r_0, s_0 and setting $\varepsilon = \varepsilon_0$ (reprinted from [34]).

K	α	r_0	s_0	ε_0	$t_{\max}$ (years)
12	$1.5669 \cdot 10^{-1}$	$6.1190 \cdot 10^{-4}$	0.5	$4.6036 \cdot 10^{-9}$	$1.02 \cdot 10^{4}$
20	$1.0827 \cdot 10^{-1}$	$2.5369 \cdot 10^{-4}$	0.3	$2.7285 \cdot 10^{-9}$	$1.47 \cdot 10^{4}$
24	$1.0827 \cdot 10^{-1}$	$2.1141 \cdot 10^{-4}$	0.3	$1.8948 \cdot 10^{-9}$	$2.16 \cdot 10^{4}$
30	$4.8418 \cdot 10^{-2}$	$7.5632 \cdot 10^{-5}$	0.2	$3.8651 \cdot 10^{-10}$	$3.29 \cdot 10^{4}$
67	$4.8418 \cdot 10^{-2}$	$3.3865 \cdot 10^{-5}$	0.2	$7.7491 \cdot 10^{-11}$	$2.52 \cdot 10^{4}$
75	$1.1436 \cdot 10^{-2}$	$7.1457 \cdot 10^{-6}$	0.08	$4.9066 \cdot 10^{-12}$	$1.39 \cdot 10^{5}$
100	$1.1436 \cdot 10^{-2}$	$5.3593 \cdot 10^{-6}$	0.5	$3.5314 \cdot 10^{-13}$	$1.78 \cdot 10^{9}$
120	$1.1436 \cdot 10^{-2}$	$4.4661 \cdot 10^{-6}$	0.5	$2.4524 \cdot 10^{-13}$	$1.13 \cdot 10^{10}$
150	$1.1436 \cdot 10^{-2}$	$3.5729 \cdot 10^{-6}$	0.5	$1.5695 \cdot 10^{-13}$	$1.72 \cdot 10^{11}$
300	$2.6724 \cdot 10^{-3}$	$4.1744 \cdot 10^{-7}$	0.02	$1.7911 \cdot 10^{-14}$	$5.97 \cdot 10^{5}$

Table 8.2. Variation of the action variables up to $|t| \leq t_{max}$ for any $\varepsilon \leq \varepsilon_0$ (reprinted from [34]).

n	Action variation	Stability time t_{max} (years)	ε_0
0	$4.47 \cdot 10^{-6}$	$1.13 \cdot 10^{10}$	$2.45 \cdot 10^{-13}$
1	$2.00 \cdot 10^{-7}$	$1.13 \cdot 10^{10}$	$5.00 \cdot 10^{-8}$
2	$2.81 \cdot 10^{-10}$	$1.16 \cdot 10^{10}$	$7.00 \cdot 10^{-7}$
3	$2.49 \cdot 10^{-14}$	$8.45 \cdot 10^{9}$	$8.50 \cdot 10^{-7}$
4	$7.29 \cdot 10^{-18}$	$4.93 \cdot 10^{9}$	$1.00 \cdot 10^{-6}$

Using the normal form at the order $n = 4$ we can provide the following result [150]:

Let (L_0, G_0) be as in (8.9) and fix the analyticity parameters as $r_0 = 10^{-6}$ and $s_0 = 0.5$. Then, the Hessian matrix Q associated to the normal form at the fourth order $h_4(L', G')$ in the new set of variables (L', G') is bounded by

$$\sup_{(L'_0, G'_0) \in V_{r_0} Y} \|Q(L'_0, G'_0)\| \leq M = 1.50 \cdot 10^{13} .$$

For $\varepsilon \leq \varepsilon_0 \equiv 10^{-6}$, the action variables satisfy

$$\|L'(t) - L'_0\|, \ \|G'(t) - G'_0\| < 7.29 \cdot 10^{-18} \qquad \text{for all } |t| \leq T \equiv 4.93 \cdot 10^9 \text{ years} .$$

To pull back the estimates to the original Delaunay variables, we remark that

$$\|L(t) - L_0\| \leq \|L(t) - L'(t)\| + \|L'(t) - L'_0\| + \|L'_0 - L_0\| \tag{8.10}$$

(similarly for G), which takes into account also the displacement generated by the canonical transformation. The first and third terms of the right–hand side of (8.10) can be estimated through (8.8) (or the equivalent formulae at order n). We finally obtain that (see [34])

$$\|L(t) - L_0\| \leq 1.61 \cdot 10^{-7}, \qquad \|G(t) - G_0\| \leq 1.55 \cdot 10^{-7} ,$$

which provide a confinement of the action variables for a time comparable to the age of the solar system. The perturbing parameter ε should be taken less than 10^{-6}, while the present value of the Jupiter–Sun mass ratio amounts to about 10^{-3}. However, we stress that, as shown in Table 8.2, a higher–order normal form computation could provide results in better agreement with the astronomical parameters.

8.5 Effective stability of the Lagrangian points

As an application of the exponential estimates provided by Nekhoroshev's theorem to higher–dimensional Hamiltonian systems, we consider the stability of the triangular Lagrangian points in the circular, restricted, spatial, three–body problem. With reference to chapter 4 we consider the motion of a body of negligible mass around two primaries with masses μ and $1 - \mu$ (in suitable normalized units, see chapter 4). Let (ξ, η, ζ) be the coordinates of the small body in a synodic reference frame with origin in the barycenter of the primaries and rotating with their angular velocity. Denoting by (p_ξ, p_η, p_ζ) the corresponding momenta, the Hamiltonian takes the form

$$\begin{aligned} \mathcal{H}(p_\xi, p_\eta, p_\zeta, \xi, \eta, \zeta) &= \frac{1}{2}(p_\xi^2 + p_\eta^2 + p_\zeta^2) + \eta p_\xi - \xi p_\eta \\ &- \frac{1-\mu}{\sqrt{(\xi - \mu)^2 + \eta^2 + \zeta^2}} - \frac{\mu}{\sqrt{(\xi + 1 - \mu)^2 + \eta^2 + \zeta^2}} \end{aligned}$$

(notice that the primaries are now located at $(\mu, 0, 0)$ and $(-1+\mu, 0, 0)$). Let us now assume a reference frame $(\mathcal{O}, x, y, z)$ centered in $L_4(\mu - \frac{1}{2}, \frac{\sqrt{3}}{2}, 0)$; the transformed Hamiltonian is

$$\begin{aligned}\mathcal{H}(p_x, p_y, p_z, x, y, z) &= \frac{1}{2}(p_x^2 + p_y^2 + p_z^2) + yp_x - xp_y - x\left(\mu - \frac{1}{2}\right) - y\frac{\sqrt{3}}{2} \\ &- \frac{1-\mu}{r_1} - \frac{\mu}{r_2} ,\end{aligned}$$

where $r_1^2 = 1 - x + \sqrt{3}y + x^2 + y^2 + z^2$, $r_2^2 = 1 + x + \sqrt{3}y + x^2 + y^2 + z^2$. Expanding the Hamiltonian around the equilibrium position, we can write

$$\mathcal{H}(p_x, p_y, p_z, x, y, z) = \sum_{k \geq 2} \mathcal{H}_k(p_x, p_y, p_z, x, y, z) ,$$

where, setting $a \equiv -\frac{3\sqrt{3}}{4}(1 - 2\mu)$, one has

$$\mathcal{H}_2(p_x, p_y, p_z, x, y, z) = \frac{1}{2}(p_x^2 + p_y^2) + yp_x - xp_y + \frac{x^2}{8} - \frac{5}{8}y^2 - axy + \frac{1}{2}(p_z^2 + z^2) ,$$

while

$$\mathcal{H}_k(p_x, p_y, p_z, x, y, z) = (1-\mu)r^k P_k\left(\frac{x - \sqrt{3}y}{2r}\right) + \mu r^k P_k\left(\frac{-x - \sqrt{3}y}{2r}\right) ,$$

being $r^2 = x^2 + y^2 + z^2$, with P_k denoting the Legendre polynomial of order k. Next, we diagonalize the quadratic part introducing a new set of variables $(x_1, x_2, x_3, y_1, y_2, y_3)$ defined as follows. Let the characteristic equation be written as

$$(\omega^2 - 1)\left(\omega^4 - \omega^2 + \frac{27}{16} - a^2\right) = 0 ,$$

where the solution $\omega = 1$ corresponds to the vertical component (p_z, z), while the remaining roots

$$\omega_1^2 = \frac{1}{2} + \frac{1}{2}\sqrt{1 - \frac{27}{4} + 4a^2} , \qquad \omega_2^2 = \frac{1}{2} - \frac{1}{2}\sqrt{1 - \frac{27}{4} + 4a^2}$$

are real and distinct provided μ satisfies the inequality $27\mu(1-\mu) < 1$. Define the vectors

$$\underline{e}_j = \begin{pmatrix} a \\ -\frac{3}{4} - \omega_j^2 \\ \frac{3}{4} - \omega_j^2 \\ a \end{pmatrix} , \qquad \underline{f}_j = \begin{pmatrix} 2\omega_j \\ 0 \\ a\omega_j \\ (\frac{5}{4} - \omega_j^2)\omega_j \end{pmatrix} ;$$

setting $d \equiv \sqrt{\omega_j(2\omega_j^4 + \frac{1}{2}\omega_j^2 - \frac{3}{4})}$, let

$$\underline{e}'_j = \frac{\underline{e}_j}{d} , \qquad \underline{f}'_j = \frac{\underline{f}_j}{d} .$$

Then, the coordinate change is defined as

$$\begin{pmatrix} x \\ y \\ p_x \\ p_y \end{pmatrix} = C \begin{pmatrix} x_1 \\ x_2 \\ y_1 \\ y_2 \end{pmatrix} ,$$

where C is the matrix whose columns coincide with $\underline{e}'_1$, $\underline{e}'_2$, $\underline{f}'_1$, $\underline{f}'_2$. Notice that[1] $C^T \tilde{\mathcal{H}}_2 C = diag(\omega_1, \omega_2, \omega_1, \omega_2)$, being $\tilde{\mathcal{H}}_2$ the Hessian matrix of $\mathcal{H}_2$. We complete the change of coordinates by setting $x_3 = z$ and $y_3 = p_z$. Finally, we write the transformed Hamiltonian as

$$\begin{aligned} \mathcal{H}_{new}(x_1, x_2, x_3, y_1, y_2, y_3) &= \mathcal{H}_{2,new}(x_1, x_2, x_3, y_1, y_2, y_3) \\ &+ \mathcal{H}_{3,new}(x_1, x_2, x_3, y_1, y_2, y_3) + \ldots , \end{aligned}$$

with

$$\mathcal{H}_{2,new}(x_1, x_2, x_3, y_1, y_2, y_3) \equiv \frac{1}{2} \sum_{k=1}^{3} \omega_k (y_k^2 + x_k^2) ,$$

while $\mathcal{H}_{3,new}$ denotes higher–order terms. Thus, the quadratic part admits the first integrals

$$I_k = \frac{1}{2}(y_k^2 + x_k^2) , \qquad k = 1, 2, 3 ;$$

let us look for first integrals of the perturbed system in the form

$$\Phi^{(k)} \equiv I_k + \Phi_3^{(k)} + \Phi_4^{(k)} + \ldots ,$$

where $\Phi_j^{(k)}$ is a homogeneous polynomial of degree j in I_1, I_2, I_3. Let $\Phi^{(k,r)}$, $r \geq 3$, be the truncation of the integral, defined as

$$\Phi^{(k,r)} = I_k + \Phi_3^{(k)} + \cdots + \Phi_r^{(k)} ,$$

where the terms $\Phi_j^{(k)}$, $j = 3, \ldots, r$, can be explicitly constructed or they can be recursively estimated. For $\rho > 0$ and $\underline{R} \in \mathbf{R}^3_+$, define the domain

$$\Delta_{\rho \underline{R}} = \{(x, y) \in \mathbf{R}^6 \ : \ y_k^2 + x_k^2 \leq \rho^2 R_k^2 \ , \ 1 \leq k \leq 3\} .$$

Assume that the initial condition at $t = 0$ lies within the domain $\Delta_{\rho_0 \underline{R}}$ for some $\rho_0 > 0$ and look for the time t_{max} such that the solution is confined within the domain $\Delta_{\rho \underline{R}}$ with $\rho > \rho_0$ up to a maximal time, say $t \leq t_{max}$. By trivial inequalities and by the definition of the domain, one has

$$\begin{aligned} |I_k(t) - I_k(0)| &\leq |I_k(t) - \Phi^{(k,r)}(t)| + |\Phi^{(k,r)}(t) - \Phi^{(k,r)}(0)| + |\Phi^{(k,r)}(0) - I_k(0)| \\ &\leq \frac{1}{2} R_k^2 (\rho^2 - \rho_0^2) . \end{aligned}$$

[1] C^T is the transposed matrix of C and $diag(a_1, a_2, a_3, a_4)$ is the 4×4 matrix with diagonal elements a_1, a_2, a_3, a_4, while all other elements are zero.

We now introduce the following norm: given a complex polynomial $f(x,y) = \sum_{j,k} f_{jk} x^j y^k$ with $f_{jk} \in \mathbf{C}$, setting $\underline{R} = (R_1, \dots, R_n) \in \mathbf{R}^n$, we obtain

$$\|f\|_{\underline{R}} \equiv \sum_{j,k} |f_{jk}| \underline{R}^{j+k} \ ;$$

for any $(x,y) \in \Delta_{\rho \underline{R}}$ and for any homogeneous function $f = f(x,y)$ of order s, we define

$$|f(x,y)| \leq \|f\|_{\underline{R}} \ \rho^s \ .$$

Then, one has

$$|\Phi^{(k,r)}(0) - I_k(0)| = |\Phi_3^{(k)}(0) + \dots + \Phi_r^{(k)}(0)| \leq \eta_r^{(k)}(\rho_0) \equiv \sum_{j=3}^{r} \|\Phi_j^{(k)}\|_{\underline{R}} \rho_0^j \ .$$

Similarly one gets

$$|\Phi^{(k,r)}(t) - I_k(t)| \leq \eta_r^{(k)}(\rho) \ .$$

Finally, one has:

$$|\Phi^{(k,r)}(t) - \Phi^{(k,r)}(0)| \leq R^{(k,r)}(\rho_0, \rho) \equiv \frac{1}{2} R_k^2 (\rho^2 - \rho_0^2) - \eta_r^{(k)}(\rho) - \eta_r^{(k)}(\rho_0) \ .$$

On the other hand, since

$$|\Phi^{(k,r)}(t) - \Phi^{(k,r)}(0)| \leq |\dot{\Phi}^{(k,r)}| \ |t| \leq F^{(k,r)}(\rho) \ |t| \ ,$$

where $F^{(k,r)}$ is an upper bound on $|\dot{\Phi}^{(k,r)}|$, then the stability time can be computed as

$$|t| \leq \min_r \ \frac{R^{(k,r)}(\rho_0, \rho)}{F^{(k,r)}(\rho)} \ .$$

Based on the above strategy, a number of results provide estimates of the stability times and of the regions of stability for the Sun–Jupiter–asteroid problem (see [34, 38, 76, 77, 101, 160]). Physically relevant stability regions have been found for a time interval of the order of the age of the solar system. The analytical estimates of [38] have been improved in [77] and [160], and the resulting stability region includes a few asteroids.

Exponential stability estimates have also been performed in [114] for the planar, elliptic, restricted three–body problem. According to [89] the study of this problem is reduced to the analysis of a suitable four–dimensional symplectic mapping. A Birkhoff normal form is implemented and first integrals are explicitly constructed. A Nekhoroshev stability domain is computed around Jupiter's Lagrangian points for a time span comparable to the age of the solar system.

9 Determination of periodic orbits

Periodic orbits play a very important role in many problems of Celestial Mechanics; for example, their study provides interesting information on spin–orbit and orbital resonances (see [96, 136]). From the dynamical point of view periodic orbits can be used to approximate quasi–periodic trajectories; more precisely, a truncation of the continued fraction expansion of an irrational frequency yields a sequence of rational numbers, which correspond to periodic orbits eventually approximating a quasi–periodic torus.

We present some results on the existence of periodic orbits through a constructive version of the implicit function theorem, both in a conservative and in a dissipative setting (Section 9.1). Then we review classical methods for computing periodic orbits, like the Lindstedt–Poincarè (Section 9.2) and the KBM (Section 9.3) techniques. We conclude with a discussion of Lyapunov's theorem on the determination of families of periodic orbits (Section 9.4) and an application to the J_2–problem.

9.1 Existence of periodic orbits

The existence of periodic orbits can be proved through the implementation of an implicit function theorem, which yields a constructive algorithm to find suitable approximations of the solution [29, 149]. We discuss the existence of periodic orbits in the conservative and in the dissipative setting, with concrete reference to the specific sample provided by the spin–orbit problem (see Section 5.5.1).

9.1.1 Existence of periodic orbits (conservative setting)

Let us write the spin–orbit equation of motion (5.16) in the form

$$\ddot{x} - \varepsilon g(x,t) \;=\; 0 \;, \tag{9.1}$$

where $g(x,t) \equiv -(\frac{a}{r})^3 \sin(2x - 2f)$. Equation (9.1) can also be written as

$$\begin{aligned} \dot{x} &= y \\ \dot{y} &= \varepsilon g(x,t) \;. \end{aligned} \tag{9.2}$$

Here ε represents the equatorial ellipticity and it can be assumed that $\varepsilon < 1$. A spin–orbit resonance of order $p : q$ is a periodic solution of (9.2) with period

$T = 2\pi q$ ($q \in \mathbf{Z}_+$), such that

$$\begin{aligned} x(t + 2\pi q) &= x(t) + 2\pi p \\ y(t + 2\pi q) &= y(t) \ . \end{aligned} \tag{9.3}$$

From (9.2) one obtains

$$\begin{aligned} x(t) &= x(0) + y(0)t + \varepsilon \int_0^t \int_0^\tau g(x(s), s)\ ds\ d\tau = x(0) + \int_0^t y(s)\ ds \\ y(t) &= y(0) + \varepsilon \int_0^t g(x(s), s)\ ds \ . \end{aligned} \tag{9.4}$$

Using the periodicity conditions (9.3) one gets

$$\begin{aligned} \int_0^{2\pi q} y(s)ds - 2\pi p &= 0 \\ \int_0^{2\pi q} g(x(s), s)ds &= 0 \ . \end{aligned} \tag{9.5}$$

Let us expand the solution in powers of ε as

$$\begin{aligned} x(t) &\equiv \overline{x} + \overline{y}t + \varepsilon x_1(t) + \varepsilon^2 x_2(t) + \dots \\ y(t) &\equiv \overline{y} + \varepsilon y_1(t) + \varepsilon^2 y_2(t) + \dots \ , \end{aligned}$$

where $x(0) = \overline{x}$ and $y(0) = \overline{y}$ are suitable initial conditions, while $x_j(t)$, $y_j(t)$, $j \geq 1$, are unknown corrections to higher orders in ε. Let us expand also the initial conditions in powers of ε as

$$\begin{aligned} \overline{x} &= \overline{x}_0 + \varepsilon \overline{x}_1 + \varepsilon^2 \overline{x}_2 + \dots \\ \overline{y} &= \overline{y}_0 + \varepsilon \overline{y}_1 + \varepsilon^2 \overline{y}_2 + \dots \ , \end{aligned} \tag{9.6}$$

for some unknown terms $\overline{x}_0$, $\overline{y}_0$, $\overline{x}_1$, $\overline{y}_1$, ... Equating in (9.2) the same orders in ε and using (9.6), one obtains

$$\begin{aligned} \overline{y} + \varepsilon \dot{x}_1(t) + \dots &= \overline{y} + \varepsilon y_1(t) + \dots \\ \varepsilon \dot{y}_1(t) + \dots &= \varepsilon g(\overline{x} + \overline{y}t, t) + \dots \ , \end{aligned}$$

which yield

$$\begin{aligned} \dot{x}_1(t) &= y_1(t) \\ \dot{y}_1(t) &= g(\overline{x}_0 + \overline{y}_0 t, t) \ , \end{aligned}$$

namely

$$\begin{aligned} x_1(t) &= x_1(t; \overline{x}, \overline{y}) = \int_0^t y_1(s)\ ds \\ y_1(t) &= y_1(t; \overline{x}, \overline{y}) = \int_0^t g(\overline{x}_0 + \overline{y}_0 s, s)\ ds \ . \end{aligned} \tag{9.7}$$

Notice that $x_1(t)$ and $y_1(t)$ can be computed explicitly. Concerning the initial data, using the second of (9.4) and the periodicity conditions (9.5) one obtains

$$\int_0^{2\pi q} \left[\overline{y}_0 + \varepsilon \overline{y}_1 + \varepsilon \int_0^t g(\overline{x}_0 + \overline{y}_0 s, s)\ ds \right]\ dt \ = 2\pi p\ .$$

Therefore, $\overline{y}_0$ and $\overline{y}_1$ are given by

$$\begin{aligned} \overline{y}_0 &= \frac{p}{q} \\ \overline{y}_1 &= -\frac{1}{2\pi q} \int_0^{2\pi q} \int_0^t g(\overline{x}_0 + \overline{y}_0 s, s)\ ds\ dt\ . \end{aligned} \tag{9.8}$$

In a similar way, $\overline{x}_0$ and $\overline{x}_1$ are obtained using

$$\int_0^{2\pi q} g(\overline{x}_0 + \overline{y}_0 s + \varepsilon(\overline{x}_1 + \overline{y}_1 s + x_1(s)), s)\ ds \ = \ 0\ ;$$

expanding in series of ε, the quantity $\overline{x}_0$ is determined as the solution of

$$\int_0^{2\pi q} g(\overline{x}_0 + \overline{y}_0 s, s)\ ds \ = \ 0\ , \tag{9.9}$$

while $\overline{x}_1$ is given by

$$\overline{x}_1 = -\frac{1}{\int_0^{2\pi q} g_x^0 dt} \left[\overline{y}_1 \int_0^{2\pi q} g_x^0\ t\ dt + \int_0^{2\pi q} g_x^0\ x_1(t)\ dt \right]\ , \tag{9.10}$$

where $g_x^0 = g_x(\overline{x}_0 + \overline{y}_0 t, t)$.

9.1.2 Computation of the libration in longitude

Applying the results of Section 9.1.1, we can implement the above formulae to compute the *libration in longitude* of the Moon, which measures the displacement from the synchronous resonance corresponding to $p = q = 1$. The initial data and the first–order corrections are computed through (9.7), (9.8), (9.9), (9.10):

$$\begin{aligned} \overline{x}_0 &= 0 \\ \overline{y}_0 &= 1 \\ x_1(t) &= 0.232086\, t - 0.218318 \sin(t) - 6.36124 \cdot 10^{-3} \sin(2t) \\ &- 3.21314 \cdot 10^{-4} \sin(3t) - 1.89137 \cdot 10^{-5} \sin(4t) \\ &- 1.18628 \cdot 10^{-6} \sin(5t) \\ y_1(t) &= 0.232086 - 0.218318 \cos(t) - 0.0127225 \cos(2t) \\ &- 9.63942 \cdot 10^{-4} \cos(3t) - 7.56548 \cdot 10^{-5} \cos(4t) \\ &- 5.93138 \cdot 10^{-6} \cos(5t) \\ \overline{x}_1 &= 0 \\ \overline{y}_1 &= -0.232086\ , \end{aligned}$$

where for the Moon we used $e = 0.0549$, $\varepsilon = 3.45 \cdot 10^{-4}$. To the first order, the solution corresponding to the synchronous periodic orbit is given by

$$\begin{aligned} x(t) &= \overline{x}_0 + \overline{y}_0 t + \varepsilon x_1(t) = (1 + 8.00697 \cdot 10^{-5})t \\ &- 7.53196 \cdot 10^{-5} \sin(t) - 2.19463 \cdot 10^{-6} \sin(2t) \\ &- 1.10853 \cdot 10^{-7} \sin(3t) - 6.52523 \cdot 10^{-9} \sin(4t) \\ &- 4.09265 \cdot 10^{-10} \sin(5t) \\ y(t) &= \overline{y}_0 + \varepsilon y_1(t) = 1 - 7.53196 \cdot 10^{-5} \cos(t) - 4.38926 \cdot 10^{-6} \cos(2t) \\ &- 3.3256 \cdot 10^{-7} \cos(3t) - 2.61009 \cdot 10^{-8} \cos(4t) \\ &- 2.04633 \cdot 10^{-9} \cos(5t) \ . \end{aligned} \tag{9.11}$$

We remark that having set to unity the angular velocity of rotation, the time t coincides with the Moon's longitude. For $\varepsilon = 0$, the equations of motion can be solved as

$$\begin{aligned} x(t) &= \overline{x}_0 + \overline{y}_0 t = \overline{x}_0 + t \\ y(t) &= \overline{y}_0 = 1 \ ; \end{aligned}$$

since $\overline{x}_0 = 0$, the difference between $x(t)$ and t is zero and therefore the direction on the equatorial plane joining the barycenter of the Moon with the Earth does not vary with time. When adding the perturbation due to the non–spherical structure of the Moon, the function $x(t)$ varies by a quantity of order ε, which provides a measure of the *libration in longitude*. The computation to the first order as in (9.11) gives a displacement of the quantity $x(t) - t$ of the order of $8 \cdot 10^{-5}$ in agreement with the astronomical data.

9.1.3 Existence of periodic orbits (dissipative setting)

We consider the dissipative spin–orbit problem described in Section 5.5.3, whose equation of motion (5.21) can be written in compact form as

$$\dot{\underline{z}} = \underline{G}(\underline{z}, t; \mu) \ ,$$

where $\underline{z} = (x, y)$, while $\underline{G}$ is a periodic two–dimensional vector function, depending parametrically on the dissipative constant μ. Assume that for $\mu = 0$ (conservative case) we know a T–periodic solution of the form

$$\underline{z}(t) = \underline{\varphi}(t)$$

with $\underline{\varphi}(T) = \underline{\varphi}(0)$. For μ sufficiently small, there still exists a periodic solution of the dissipative problem with period T [149]; this result is based on the implicit function theorem under quite general hypotheses as we are going to describe. For the dissipative spin–orbit problem we assume for simplicity that the dissipative constant and the perturbing parameter are related by $\mu = \mu_0 \varepsilon$ for a suitable quantity

$\mu_0 < 1$. Then equation (5.21) becomes

$$\begin{aligned}\dot{x} &= y \\ \dot{y} &= \varepsilon g(x, y, t) ,\end{aligned} \tag{9.12}$$

with $g(x, y, t) = -(\frac{a}{r})^3 \sin(2x - 2f) - \mu_0(y - \eta)$. We denote by $\bar{x}$ and $\bar{y}$ the initial conditions and by $(x(t; \bar{x}, \bar{y}), y(t; \bar{x}, \bar{y}))$ the solution at time t with initial conditions $(\bar{x}, \bar{y})$. By (9.12) we obtain

$$\begin{aligned}x(t) &\equiv x(t; \bar{x}, \bar{y}) = \bar{x} + \bar{y}t + \varepsilon \int_0^t \int_0^\tau g(x(s; \bar{x}, \bar{y}), y(s; \bar{x}, \bar{y}), s) ds d\tau \\ y(t) &\equiv y(t; \bar{x}, \bar{y}) = \bar{y} + \varepsilon \int_0^t g(x(s; \bar{x}, \bar{y}), y(s; \bar{x}, \bar{y}), s) ds .\end{aligned} \tag{9.13}$$

A spin–orbit resonance of order $p : q$ satisfies the periodicity conditions (9.3) which, together with (9.13), are equivalent to find solutions of the equations

$$\begin{aligned}F_1(\bar{x}, \bar{y}) &= 0 \\ F_2(\bar{x}, \bar{y}) &= 0 ,\end{aligned} \tag{9.14}$$

where

$$\begin{aligned}F_1(\bar{x}, \bar{y}) &\equiv 2\pi(q\bar{y} - p) + \varepsilon \int_0^{2\pi q} \int_0^\tau g(x(s; \bar{x}, \bar{y}), y(s; \bar{x}, \bar{y}), s) ds d\tau \\ F_2(\bar{x}, \bar{y}) &\equiv \int_0^{2\pi q} g(y(s; \bar{x}, \bar{y}), x(s; \bar{x}, \bar{y}), s) ds .\end{aligned} \tag{9.15}$$

Expanding (9.15) to the first order in ε, one obtains

$$\begin{aligned}F_1(\bar{x}, \bar{y}) &= 2\pi(q\bar{y} - p) + \varepsilon \Phi_1(\bar{x}, \bar{y}) \\ F_2(\bar{x}, \bar{y}) &= \int_0^{2\pi q} g(\bar{x} + \bar{y}s, \bar{y}, s) ds + \varepsilon \Phi_2(\bar{x}, \bar{y})\end{aligned} \tag{9.16}$$

for suitable functions $\Phi_1(\bar{x}, \bar{y})$, $\Phi_2(\bar{x}, \bar{y})$. Let us expand the initial conditions as $\bar{x} = \bar{x}_0 + \varepsilon \bar{x}_1 + \varepsilon^2 \bar{x}_2 + \ldots$, $\bar{y} = \bar{y}_0 + \varepsilon \bar{y}_1 + \varepsilon^2 \bar{y}_2 + \ldots$ Then we find $\bar{y}_0 = \frac{p}{q}$, while $\bar{x}_0$ is determined as a non–degenerate critical point of the function

$$\Psi_{p,q}(x) \equiv \frac{1}{2} \int_0^{2\pi q} \left(\frac{a}{r(t)} \right)^3 \cos(2x + 2\frac{p}{q}t - 2f(t)) \, dt ,$$

so that using (9.16) one obtains

$$\begin{aligned}F_1(\bar{x}_0, \bar{y}_0) &= \varepsilon \Phi_1(\bar{x}_0, \bar{y}_0) \\ F_2(\bar{x}_0, \bar{y}_0) &= -2\pi q \mu_0 \left(\frac{p}{q} - \eta \right) + \varepsilon \Phi_2(\bar{x}_0, \bar{y}_0) .\end{aligned}$$

Let us evaluate the Jacobian J of (9.16) at $(\bar{x}_0, \bar{y}_0)$ and let us denote the result by $J_0 + \varepsilon J_1$; then J_0 is non–degenerate, since

$$J_0 = \begin{pmatrix} 2\pi q & 0 \\ \Phi_{p,q}(\bar{x}_0, \bar{y}_0; \mu_0) & \frac{d^2}{dx^2}\Psi_{p,q}(\bar{x}_0) \end{pmatrix}$$

for a suitable function $\Phi_{p,q} = \Phi_{p,q}(\bar{x}_0, \bar{y}_0; \mu_0)$. Let M be the inverse of the Jacobian J evaluated at $(\bar{x}_0, \bar{y}_0)$; let $\rho > 0$ and denote by $\bar{B}_\rho(\bar{x}_0, \bar{y}_0)$ the closed ball of radius ρ around $(\bar{x}_0, \bar{y}_0)$. Let A be a compact subset of $\mathbf{R}$ and let $0 < \alpha < 1$, $R > 0$ be real parameters. The implicit function theorem can be applied provided the following conditions are satisfied (I_2 is the 2×2 identity matrix):

$$\begin{aligned} \sup_{\bar{B}_\rho(\bar{x}_0,\bar{y}_0)\times A} \|I_2 - M\ J\| &\le \alpha \\ \sup_A |\underline{F}(\bar{x}_0, \bar{y}_0)| \cdot \sup_A \|M\| &\le (1-\alpha)\ R\ ; \end{aligned}$$

the above inequalities turn out to be smallness conditions on the parameters. Under these conditions the implicit function theorem guarantees that for ε sufficiently small there exists a solution $(x(\varepsilon), y(\varepsilon)) \in \bar{B}_\rho(\bar{x}_0, \bar{y}_0)$ of the system

$$\begin{aligned} F_1(x(\varepsilon), y(\varepsilon)) &= 0 \\ F_2(x(\varepsilon), y(\varepsilon)) &= 0\ , \end{aligned}$$

providing a fixed point of (9.14) with the required periodicity conditions.

9.1.4 Normal form around a periodic orbit

The dynamics in a neighborhood of the periodic orbits determined as in Section 9.1.1 can be studied through the development of a suitable normal form, which turns out to be useful in a number of samples in Celestial Mechanics. We briefly sketch the procedure referring to equations (9.2), whose associated Hamiltonian function takes the form

$$\mathcal{H}_1(y, x, t) = \frac{y^2}{2} - \varepsilon V(x,t)\ , \qquad y \in \mathbf{R}\ , \quad (x,t) \in \mathbf{T}\ ,$$

where $y = \dot{x}$ is the variable conjugated to x and $V_x(x,t) = g(x,t)$. Let $(\tilde{x}(t), \tilde{y}(t))$ be a periodic orbit of order $p : q$ with periodicity conditions (9.3). We assume to know the periodic orbit for example through its series expansion as explained in Section 9.1.1. In the proximity of the periodic orbit, let γ be a positive, small parameter, measuring the distance from the periodic orbit and let $(\gamma\xi(t), \gamma\eta(t))$ be a small displacement such that we can write the solution in the form

$$\begin{aligned} x(t) &= \tilde{x}(t) + \gamma\xi(t) \\ y(t) &= \tilde{y}(t) + \gamma\eta(t)\ . \end{aligned} \tag{9.17}$$

Inserting (9.17) in (9.2), one obtains

$$\begin{aligned} \dot{x}(t) &= \dot{\tilde{x}}(t) + \gamma\dot{\xi}(t) = \tilde{y}(t) + \gamma\eta(t) \\ \dot{y}(t) &= \dot{\tilde{y}}(t) + \gamma\dot{\eta}(t) = \varepsilon g(\tilde{x}(t) + \gamma\xi(t), t)\ , \end{aligned}$$

where we can expand g in Taylor series around $\gamma = 0$ as

$$g(\tilde{x}(t) + \gamma\xi(t), t) = g(\tilde{x}(t), t) + \gamma g_x(\tilde{x}(t), t)\xi + \frac{1}{2}\gamma^2 g_{xx}(\tilde{x}(t), t)\xi^2 + \dots$$

Since $(\tilde{x}(t), \tilde{y}(t))$ is a solution of the equations of motion, one gets

$$\begin{aligned} \dot{\xi} &= \eta \\ \dot{\eta} &= \varepsilon g_x(\tilde{x}(t), t)\xi + \frac{\varepsilon}{2}\gamma g_{xx}(\tilde{x}(t), t)\xi^2 + \dots , \end{aligned} \tag{9.18}$$

whose associated Hamiltonian is

$$\mathcal{H}_2(\eta, \xi, t) = \frac{\eta^2}{2} - \frac{\varepsilon}{2} g(\tilde{x}(t), t)\xi^2 - \frac{\varepsilon}{6} g_x(\tilde{x}(t), t)\gamma\xi^3 + \dots$$

Defining

$$\underline{u} \equiv \begin{pmatrix} \xi \\ \eta \end{pmatrix} , \quad Q(t) \equiv \begin{pmatrix} 0 & 1 \\ \varepsilon g_x(\tilde{x}(t), t) & 0 \end{pmatrix} , \quad R_2(\underline{u}, t) \equiv \begin{pmatrix} 0 \\ \frac{\varepsilon}{2} g_{xx}(\tilde{x}(t), t)\xi^2 + \dots \end{pmatrix} ,$$

we can write (9.18) in the form

$$\dot{\underline{u}} = Q(t)\underline{u} + \gamma R_2(\underline{u}, t) . \tag{9.19}$$

Floquet theory (see Appendix D) can be implemented to eliminate the time–dependence in the linear part. Through a symplectic, periodic change of variables one can reduce (9.19) to the form

$$\dot{\underline{v}} = A\underline{v} + \gamma S_2(\underline{v}(t), t) , \tag{9.20}$$

where $\underline{v} \equiv (v_1, v_2) \in \mathbf{R}^2$, A is a constant matrix and S_2 is a suitable function. We can assume that A takes the form

$$A \equiv \begin{pmatrix} 0 & \omega \\ -\omega & 0 \end{pmatrix} ,$$

so that the linear part reduces to

$$\begin{aligned} \dot{v}_1 &= \omega v_2 \\ \dot{v}_2 &= -\omega v_1 , \end{aligned}$$

whose associated Hamiltonian corresponds to that of a harmonic oscillator, namely $\mathcal{H}_3(v_1, v_2) = \frac{\omega}{2}(v_1^2 + v_2^2) + \dots$. Using action–angle variables (I, φ) for the harmonic oscillator, we can write the Hamiltonian function corresponding to (9.20) in the form

$$\mathcal{H}_4(I, \varphi, t) = \omega I + \gamma F(I, \varphi, t) ,$$

for a suitable function $F = F(I, \varphi, t)$. A Birkhoff normal form can now be implemented in the style of Section 6.5 to reduce the perturbation and to get a better approximation in the neighborhood of the periodic orbit.

9.2 The Lindstedt–Poincaré technique

Convergent series approximations of periodic solutions can be found through the *Lindstedt–Poincaré technique*, also known as the *continuation method.* Consider a dynamical system described by the second–order differential equation

$$\ddot{x} + \omega_0^2 x = \varepsilon f(x, \dot{x}) \ , \qquad x \in \mathbf{R} \ , \tag{9.21}$$

where $\varepsilon \geq 0$ is a small real parameter and $f : \mathbf{R}^2 \to \mathbf{R}$ is a regular function. For $\varepsilon = 0$ the system reduces to a harmonic oscillator, which has periodic solutions with period $T_0 = \frac{2\pi}{\omega_0}$. The Lindstedt–Poincaré technique allows us to find periodic solutions for ε different from zero by taking into account that the frequency of the motion can change due to the non–linear terms. In fact, when ε is different from zero the period T is equal to T_0 only up to terms of order ε. Basically one expands the solution $x(t)$ and the (unknown) frequency ω of the periodic orbit as a function of ε:

$$\begin{aligned} x(t) &= x_0(t) + \varepsilon x_1(t) + \varepsilon^2 x_2(t) + \ldots \\ \omega &= \omega_0 + \varepsilon \omega_1 + \varepsilon^2 \omega_2 + \ldots \ , \end{aligned} \tag{9.22}$$

where we impose that $x_j(T) = x_j(0)$, being the quantities $x_j(t)$, ω_j, $j \geq 0$, unknown. Under the change of variables $s = \omega t$, the equation (9.21) becomes

$$\omega^2 x'' + \omega_0^2 x = \varepsilon f(x, \omega x') \ , \tag{9.23}$$

where x' and x'' denote the first and second derivatives with respect to s. Let us expand the perturbation in powers of ε as

$$\begin{aligned} f(x, \omega x') &= f(x_0, \omega_0 x_0') + \varepsilon \left[x_1 \frac{\partial f(x_0, \omega_0 x_0')}{\partial x} + x_1' \frac{\partial f(x_0, \omega_0 x_0')}{\partial x'} \right. \\ &+ \left. \omega_1 \frac{\partial f(x_0, \omega_0 x_0')}{\partial \omega} \right] + O(\varepsilon^2) \ . \end{aligned}$$

Inserting the series expansion (9.22) in (9.23) and equating terms of the same order in ε, one obtains

$$\begin{aligned} \omega_0^2 x_0'' + \omega_0^2 x_0 &= 0 \\ \omega_0^2 x_1'' + \omega_0^2 x_1 &= f(x_0, \omega_0 x_0') - 2\omega_0 \omega_1 x_0'' \\ &\ldots \end{aligned}$$

These equations can be solved recursively and the quantities ω_j can be found by imposing the periodicity conditions $x_j(s + 2\pi) = x_j(s)$, $j = 0, 1, 2, \ldots$

As a concrete example we consider the Duffing equation [146]

$$\ddot{x} + \omega_0^2 x = -\varepsilon \omega_0^2 x^3 \ .$$

Changing time as $s = \omega t$, one gets

$$\omega^2 x''(s) + \omega_0^2 x(s) = -\varepsilon \omega_0^2 x(s)^3 \ .$$

Let us expand the solution $x(s)$ and the unknown frequency ω as in (9.22) and assume that $x_j'(0) = 0$ for any $j \geq 0$. To the zeroth order in ε one obtains the equation

$$x_0'' + x_0 = 0$$

and, taking into account the initial conditions, one finds $x_0(s) = A\cos s$ for some real constant A. To the first order in ε one obtains the equation

$$x_1'' + x_1 = A\left(2\frac{\omega_1}{\omega_0} - \frac{3}{4}A^2\right)\cos s - \frac{1}{4}A^3\cos 3s\ ;$$

secular terms are avoided provided $\frac{\omega_1}{\omega_0} = \frac{3}{8}A^2$, thus yielding the first–order solution $x_1(s) = \frac{1}{32}A^3\cos 3s$. The solution at all subsequent orders can be obtained implementing iteratively the above procedure.

9.3 The KBM method

The Krylov–Bogoliubov–Mitropolsky (KBM) method allows us to find periodic solutions for systems of the form (9.21); for $\varepsilon = 0$ such systems admit the solution

$$x(t) = A\cos\xi(t) \qquad \text{with}\quad \xi(t) \equiv \omega_0 t + \varphi\ ,$$

for some constants A, φ depending on the initial conditions. For ε different from zero, one can write the solution as

$$x(t) = A\cos\xi + \varepsilon x_1(A,\xi) + \varepsilon^2 x_2(A,\xi) + \dots\ , \tag{9.24}$$

where $x_j(A,\xi)$ are 2π–periodic functions. The quantities A, ξ satisfy the equations

$$\begin{aligned} \dot A &= \varepsilon\alpha_1(A) + \varepsilon^2\alpha_2(A) + \dots \\ \dot\xi &= \omega_0 + \varepsilon\beta_1(A) + \varepsilon^2\beta_2(A) + \dots \end{aligned} \tag{9.25}$$

for some unknown functions $\alpha_j(A)$, $\beta_j(A)$. Inserting (9.24) and (9.25) in the left hand side of (9.21), one obtains

$$\ddot x + \omega_0^2 x = \varepsilon\left[-2\omega_0\alpha_1\sin\xi - 2\omega_0 A\beta_1\cos\xi + \omega_0^2\left(\frac{\partial^2 x_1}{\partial\xi^2} + x_1\right)\right] + O(\varepsilon^2)\ .$$

Concerning the right–hand side of (9.21) one has

$$\varepsilon f(x,\dot x) = \varepsilon f(x_0,\dot x_0) + O(\varepsilon^2)\ ,$$

where $x_0 = A\cos\xi$, $\dot x_0 = -A\omega_0\sin\xi$, being x_0 the lowest–order approximation in which A and $\dot\xi$ are constant. Equating same powers of ε, the first order is given by

$$\omega_0^2\left(\frac{\partial^2 x_1}{\partial\xi^2} + x_1\right) = f(x_0,\dot x_0) + 2\omega_0\alpha_1\sin\xi + 2\omega_0 A\beta_1\cos\xi \tag{9.26}$$

and similarly for higher orders which can be solved recursively. For the first order, let us expand x_1 and f in Fourier series as

$$\begin{aligned} f(A,\xi) &= f_0(A) + \sum_{j=1}^{\infty} \left[f_j^{(c)}(A) \cos j\xi + f_j^{(s)}(A) \sin j\xi \right] \\ x_1(A,\xi) &= x_0^{(1)}(A) + \sum_{j=2}^{\infty} \left[x_j^{(1c)}(A) \cos j\xi + x_j^{(1s)}(A) \sin j\xi \right] , \end{aligned} \tag{9.27}$$

for suitable functions $f_0(A)$, $f_j^{(c)}(A)$, $f_j^{(s)}(A)$, $x_0^{(1)}(A)$, $x_j^{(1c)}(A)$, $x_j^{(1s)}(A)$. To avoid secular terms we impose that

$$\int_0^{2\pi} x_1(A,\xi) \cos\xi d\xi = 0 , \qquad \int_0^{2\pi} x_1(A,\xi) \sin\xi d\xi = 0 ,$$

which yield $x_1^{(1c)}(A) = x_1^{(1s)}(A) = 0$. Inserting (9.27) in (9.26) and equating Fourier coefficients of the same order, one obtains

$$f_1^{(c)}(A) + 2\omega_0 A\beta_1(A) = 0 , \qquad f_1^{(s)}(A) + 2\omega_0\alpha_1(A) = 0 ,$$

which provide explicit expressions for $\alpha_1(A)$ and $\beta_1(A)$. Moreover, one has

$$x_0^{(1)}(A) = \frac{f_0(A)}{\omega_0^2} , \qquad x_j^{(1c)}(A) = \frac{f_j^{(c)}(A)}{\omega_0^2(1-j^2)} , \qquad x_j^{(1s)}(A) = \frac{f_j^{(s)}(A)}{\omega_0^2(1-j^2)} , \qquad j \geq 2 ,$$

which provide the Fourier coefficients appearing in (9.27). Similar computations can be performed to determine iteratively the solution to higher orders.

9.4 Lyapunov's theorem

A remarkable result due to Lyapunov allows us to determine a family of periodic solutions around an equilibrium position. We sketch the proof of the theorem, referring to [162] for complete details. As an illustrative example, we consider the J_2–problem introduced in Section 5.6.

9.4.1 Families of periodic orbits

We consider an n–dimensional Hamiltonian system described by the Hamiltonian function $\mathcal{H} = \mathcal{H}(\underline{w})$, $\underline{w} \in \mathbf{R}^{2n}$, which is assumed to be regular in a suitable neighborhood of the origin. We assume that the origin is an equilibrium position; let $\pm\lambda_1, \ldots, \pm\lambda_n$ be distinct eigenvalues of the linearized matrix L associated to the Hamiltonian $\mathcal{H}$ around the origin.

Lyapunov's Theorem. *Let λ_1 be purely imaginary, not identically zero, and assume that the ratios $\frac{\lambda_2}{\lambda_1}, \ldots, \frac{\lambda_n}{\lambda_1}$ are not integers; then, there exists a family of periodic solutions around the equilibrium position, depending analytically on a real* *parameter ρ, such that $\rho = 0$ corresponds to the equilibrium solution and that the period $T(\rho)$ is analytic in ρ with $T(0) = \frac{2\pi}{|\lambda_1|}$.*

Proof. We look for a solution $\underline{w} = \underline{w}(\xi, \eta)$ as a power series of some unknown functions $\xi = \xi(t)$, $\eta = \eta(t)$. Then, Hamilton's equations $\dot{\underline{w}} = J\mathcal{H}_{\underline{w}}(\underline{w})$ become

$$\underline{w}_\xi \dot{\xi} + \underline{w}_\eta \dot{\eta} = J\mathcal{H}_{\underline{w}}(\underline{w}) . \tag{9.28}$$

We assume that ξ, η satisfy the relations

$$\dot{\xi} = \alpha\xi , \qquad \dot{\eta} = \beta\eta \tag{9.29}$$

with α, β being suitable power series in ξ, η. We next perform a linear canonical transformation, say $\underline{w} = C\underline{z}$, for some constant matrix C, such that the linearized matrix is transformed into $C^T JLC = \Lambda$, where Λ is the diagonal matrix with non–zero elements λ_1, ..., λ_n, $-\lambda_1$, ..., $-\lambda_n$. Moreover, the matrix C is chosen to be symplectic and such that its components are suitably normalized according to [162]. With this transformation, equation (9.28) takes the form

$$\underline{z}_\xi \xi\alpha + \underline{z}_\eta \eta\beta - \Lambda\underline{z} = \underline{g}(\underline{z}) , \tag{9.30}$$

where

$$\underline{g}(\underline{z}) \equiv C^{-1}J(C^{-1})^T \, \mathcal{H}_{\underline{z}}(C\underline{z}) - \Lambda\underline{z} .$$

In order to determine uniquely the power series $z_k(\xi, \eta)$ $(k = 1, \dots, 2n)$, $\alpha(\xi, \eta)$, $\beta(\xi, \eta)$, by comparison of the coefficients in (9.30), one needs to impose the following compatibility conditions:

$(C1)$ $z_1 - \xi$, $z_2 - \eta$, z_3, ..., z_{2n} start with quadratic terms;
$(C2)$ there are no terms of the form $\xi(\xi\eta)^\ell$ in $z_1 - \xi$ and no terms of the form $\eta(\xi\eta)^\ell$ in $z_2 - \eta$;
$(C3)$ the series for α and β depend only on the quantity $\omega \equiv \xi\eta$.

By induction, one easily proves that equation (9.29) can be effectively solved and that the coefficients are uniquely determined. The constant terms of α and β are, respectively, λ_1 and $-\lambda_1$. Moreover, it can be shown (see [162]) that α and β satisfy the relation

$$\alpha + \beta = 0 \tag{9.31}$$

and that the Hamiltonian $\mathcal{H}$ becomes a series of $\omega = \xi\eta$. Referring to [162] for the proof of the convergence of the series $z_k(\xi, \eta)$ $(k = 1, \dots, 2n)$, α, β for sufficiently small values of $|\xi|$, $|\eta|$, by (9.31) one finds that

$$\frac{d\omega}{dt} = \dot{\xi}\eta + \xi\dot{\eta} = (\alpha + \beta)\xi\eta = (\alpha + \beta)\omega = 0 ;$$

therefore ω, α, β do not depend on the time and consequently from (9.29) one obtains

$$\xi = \xi_0 e^{\alpha t} , \qquad \eta = \eta_0 e^{\beta t} , \tag{9.32}$$

where ξ_0, η_0 are the initial conditions. The value $|\xi_0|$ should be taken sufficiently small, say $|\xi_0| \leq \rho$ for some positive real parameter ρ, to ensure the convergence. By (9.32) one obtains a family of periodic orbits with periods $T(\rho) = \frac{2\pi}{|\alpha|}$; since to the lowest order α coincides with λ_1, the period of the equilibrium position is $T(0) = \frac{2\pi}{|\lambda_1|}$. □

9.4.2 An example: the J_2–problem

As an application of Lyapunov's theorem, we consider the motion of a homogeneous rigid body $\mathcal{S}$ moving around an oblate planet $\mathcal{P}$. Assuming that the central planet is axially symmetric, using spherical coordinates the potential function governing the motion of the satellite is provided in Section 5.6. The J_2–problem consists in retaining only the lowest–order term in the series expansion of the potential as a series of the Legendre's polynomials (see equation (5.27)):

$$U(r, \varphi) = \frac{\mu}{r} + \frac{\mu J_2 R_e^2}{r^3} \left(\frac{1}{2} - \frac{3}{2} \sin^2 \theta \right) ,$$

where J_2 is constant, $\mu = \mathcal{G}M$, M being the mass of $\mathcal{P}$, R_e is the equatorial radius of $\mathcal{P}$, while r and θ are, respectively, the radius and the latitude of the satellite $\mathcal{S}$ with respect to the central body $\mathcal{P}$. The Hamiltonian function describing the J_2–problem is derived as follows. In a reference frame with the origin coinciding with the barycenter $\mathcal{O}$ of $\mathcal{P}$, the spherical coordinates of $\mathcal{S}$ are:

$$\begin{aligned} x_S &= r \cos\phi \cos\theta \\ y_S &= r \sin\phi \cos\theta \\ z_S &= r \sin\theta \ , \end{aligned}$$

where $r \geq 0$, $0 \leq \phi \leq 2\pi$, $0 \leq \theta \leq \pi$. Assuming that the mass of $\mathcal{S}$ is normalized to one, the Lagrangian function is given by:

$$\mathcal{L}(\dot{r}, \dot{\phi}, \dot{\theta}, r, \phi, \theta,) = \frac{1}{2}(\dot{r}^2 + r^2\dot{\phi}^2\cos^2\theta + r^2\dot{\theta}^2) + U(r, \theta) \ . \tag{9.33}$$

Due to the cylindrical symmetry of the problem, the variable ϕ is cyclic; therefore the vertical component of the angular momentum (coinciding with the momentum p_ϕ conjugated to ϕ) is constant, say equal to $\overline{g}$, providing

$$p_\phi = r^2\dot{\phi} \cos^2\theta \equiv \overline{g} \ .$$

Since the Lagrangian (9.33) does not depend explicitly on the time, another constant of the motion is given by the total energy. Using the first integral $\overline{g}$, the energy E becomes:

$$E = \frac{1}{2}\left(\dot{r}^2 + r^2\dot{\theta}^2\right) + \frac{\overline{g}^2}{2r^2}(1 + \tan^2\theta) - \frac{\mu}{r} + \frac{\mu J_2 R_e^2}{r^3}\left(\frac{3}{2}\sin^2\theta - \frac{1}{2}\right) .$$

We introduce a new pair of coordinates (ρ, z) defined as

$$\begin{aligned}\rho &= r\cos\theta\\ z &= r\sin\theta\ .\end{aligned}$$

Adopting the units of measure so that $\mu = 1$ and $R_e = 1$, the Lagrangian becomes

$$\mathcal{L}(\dot\rho, \dot z, \rho, z) = \frac{1}{2}(\dot\rho^2 + \dot z^2) - \frac{\overline{g}^2}{2\rho^2} + U(\rho, z) \tag{9.34}$$

with

$$U(\rho, z) = (\rho^2 + z^2)^{-\frac{1}{2}} - \frac{J_2}{2}(\rho^2 + z^2)^{-\frac{5}{2}}(2z^2 - \rho^2)\ .$$

Let $p_\rho = \dot\rho$, $p_z = \dot z$; the Hamiltonian associated to the Lagrangian (9.34) is given by:

$$\mathcal{H}(p_\rho, p_z, \rho, z) = \frac{1}{2}({p_\rho}^2 + {p_z}^2) + \frac{\overline{g}^2}{2\rho^2} - U(\rho, z)\ .$$

The corresponding equations of motion are:

$$\begin{aligned}\dot\rho &= p_\rho\\ \dot z &= p_z\\ \dot{p_\rho} &= \tfrac{\overline{g}^2}{\rho^3} + U_\rho(\rho, z)\\ \dot{p_z} &= U_z(\rho, z)\ ,\end{aligned} \tag{9.35}$$

where $U_\rho(\rho, z)$ and $U_z(\rho, z)$ denote the derivatives of U with respect to ρ and z:

$$U_\rho(\rho, z) \;=\; -\rho(\rho^2 + z^2)^{-\frac{3}{2}} + \frac{5}{2}J_2\rho(\rho^2 + z^2)^{-\frac{7}{2}}(2z^2 - \rho^2) + J_2\rho(\rho^2 + z^2)^{-\frac{5}{2}}\ ,$$

$$U_z(\rho, z) \;=\; -z(\rho^2 + z^2)^{-\frac{3}{2}} + \frac{5}{2}J_2 z(\rho^2 + z^2)^{-\frac{7}{2}}(2z^2 - \rho^2) - 2J_2 z(\rho^2 + z^2)^{-\frac{5}{2}}\ .$$

In order to compute the equilibrium points, we set equal to zero the right–hand side of (9.35). Selecting the solution with $z = 0$, one easily obtains that ρ must be a root of the equation $2\rho^2 - 2\overline{g}^2\rho + 3J_2 = 0$. Therefore, two equilibrium points of (9.35) are given by $P_0 \equiv (\rho_0, z_0, p_{\rho 0}, p_{z0}) = (\frac{\overline{g}^2 + \sqrt{\overline{g}^4 - 6J_2}}{2}, 0, 0, 0)$ and $P_1 = (\frac{\overline{g}^2 - \sqrt{\overline{g}^4 - 6J_2}}{2}, 0, 0, 0)$ provided $\overline{g}^4 > 6J_2$ so as to have real positive values of ρ. In the following sections we focus our attention on the equilibrium position P_0.

9.4.3 Linearization of the Hamiltonian around the equilibrium point

We proceed to linearize the equations of motion in a neighborhood of the equilibrium point P_0. First of all, through the transformation $\tilde z = z$, $\tilde\rho = \rho - \rho_0$ (which shifts P_0 to the origin of the reference frame), we get the Hamiltonian function

$$\begin{aligned}\tilde{\mathcal{H}}(p_{\tilde\rho}, p_{\tilde z}, \tilde\rho, \tilde z) &= \frac{1}{2}({p_{\tilde\rho}}^2 + {p_{\tilde z}}^2) + \frac{\overline{g}^2}{2(\tilde\rho + \rho_0)^2} - [(\tilde\rho + \rho_0)^2 + \tilde z^2]^{-\frac{1}{2}} \\ &+ \frac{J_2}{2}[(\tilde\rho + \rho_0)^2 + \tilde z^2]^{-\frac{5}{2}}(2\tilde z^2 - (\tilde\rho + \rho_0)^2)\ ,\end{aligned}$$

where $p_{\tilde\rho}$, $p_{\tilde z}$ are the momenta conjugated to $\tilde\rho$, $\tilde z$; the corresponding equations of motion are:

$$\begin{aligned}&\dot{\tilde\rho} = p_{\tilde\rho} \\ &\dot{\tilde z} = p_{\tilde z} \\ &\dot p_{\tilde\rho} = \tfrac{\overline{g}^2}{(\tilde\rho + \rho_0)^3} + U_{\tilde\rho}(\tilde\rho + \rho_0, \tilde z) \\ &\dot p_{\tilde z} = U_{\tilde z}(\tilde\rho + \rho_0, \tilde z)\ .\end{aligned}$$

Next, we expand the equations of motion by means of a Taylor power series around the equilibrium point up to the second order. The linearized system becomes

$$\begin{pmatrix} \dot{\tilde\rho} \\ \dot{\tilde z} \\ \dot p_{\tilde\rho} \\ \dot p_{\tilde z} \end{pmatrix} = L \begin{pmatrix} \tilde\rho \\ \tilde z \\ p_{\tilde\rho} \\ p_{\tilde z} \end{pmatrix}\ ,$$

where

$$L = \begin{pmatrix} 0 & 0 & 1 & 0 \\ 0 & 0 & 0 & 1 \\ \gamma & 0 & 0 & 0 \\ 0 & \delta & 0 & 0 \end{pmatrix}$$

is the matrix corresponding to the linearization and

$$\begin{aligned}\gamma &= \frac{1}{{\rho_0}^3}\Big(-\frac{3\overline{g}^2}{\rho_0} + 2 + \frac{6J_2}{{\rho_0}^2}\Big) \\ \delta &= \frac{1}{2{\rho_0}^5}(-2{\rho_0}^2 - 9J_2)\ .\end{aligned}$$

Notice that δ is negative for any initial condition. Up to constant terms, the linearized Hamiltonian in a neighborhood of $\tilde P_0 = (0,0,0,0)$ is given by

$$\mathcal{H}_L(p_{\tilde\rho}, p_{\tilde z}, \tilde\rho, \tilde z) = \frac{1}{2}({p_{\tilde\rho}}^2 + {p_{\tilde z}}^2 - \gamma\tilde\rho^2 - \delta\tilde z^2) + \mathcal{H}_3(p_{\tilde\rho}, p_{\tilde z}, \tilde\rho, \tilde z)\ , \tag{9.36}$$

where $\mathcal{H}_3(p_{\tilde\rho}, p_{\tilde z}, \tilde\rho, \tilde z)$ denotes terms of order higher than three.

9.4.4 Application of Lyapunov's theorem

In this section we apply Lyapunov's theorem to the existence of families of periodic orbits starting from the Hamiltonian (9.36). To this end the following conditions must be satisfied by the linearized system associated to (9.36):

(*i*) the eigenvalues $\lambda_1, \lambda_2, -\lambda_1, -\lambda_2$ of the matrix L must be distinct;
(*ii*) let λ_1 be purely imaginary; then the ratio $\frac{\lambda_2}{\lambda_1}$ must not be an integer.

The eigenvalues associated to (9.36) are obtained as follows. Let

$$\underline{w} \equiv \begin{pmatrix} \tilde{\rho} \\ \tilde{z} \\ p_{\tilde{\rho}} \\ p_{\tilde{z}} \end{pmatrix}, \qquad J \equiv \begin{pmatrix} 0 & 0 & 1 & 0 \\ 0 & 0 & 0 & 1 \\ -1 & 0 & 0 & 0 \\ 0 & -1 & 0 & 0 \end{pmatrix},$$

and let $\underline{w}^T$ be the transposed of $\underline{w}$. Then, (9.36) can be written as $\mathcal{H}_L(\underline{w}) = -\frac{1}{2}\underline{w}^T \cdot JL\underline{w} + \mathcal{H}_3(\underline{w})$ and the eigenvalues of the linearization L are $\lambda_1 = \sqrt{\delta}$, $\lambda_2 = \sqrt{\gamma}$, $\lambda_3 = -\sqrt{\delta}$, $\lambda_4 = -\sqrt{\gamma}$. Excluding degenerate cases (see the remarks below), conditions (i)–(ii) above are satisfied, so that Lyapunov's theorem applies.

Remarks.

(1) Since $\delta < 0$ for each initial condition, λ_1, λ_3 are always purely imaginary. Moreover if $3J_2 < \overline{g}^2 \rho_0$, then λ_2, λ_4 are also purely imaginary. Therefore, if we assume that $3J_2 < \overline{g}^2 \rho_0$, then the four eigenvalues are equal to $\lambda_{1,3} = \pm i\sqrt{|\delta|}$, $\lambda_{2,4} = \pm i\sqrt{|\gamma|}$; using the relation $2{\rho_0}^2 = 2\overline{g}^2 \rho_0 - 3J_2$, we obtain

$$\begin{aligned} \gamma &= \frac{1}{2{\rho_0}^5}(-2\overline{g}^2 \rho_0 + 6J_2) , \\ \delta &= \frac{1}{2{\rho_0}^5}(-2\overline{g}^2 \rho_0 - 6J_2) . \end{aligned}$$

Their ratio is given by

$$\frac{\gamma}{\delta} = \frac{-2\overline{g}^2 \rho_0 + 6J_2}{-2\overline{g}^2 \rho_0 - 6J_2}.$$

(2) If $J_2 \neq 0$, $\overline{g} \neq 0$, $\rho_0 \neq 0$ and $3J_2 < \overline{g}^2 \rho_0$, then λ_j $(j = 1, \ldots, 4)$ are purely imaginary and $\frac{\gamma}{\delta}$ is not an integer. Therefore, by Lyapunov's theorem there exist two families of periodic orbits with periods $\frac{2\pi}{\sqrt{|\gamma|}}$ and $\frac{2\pi}{\sqrt{|\delta|}}$.

(3) If $\overline{g} = 0$, then $\frac{\gamma}{\delta} = -1$ and Lyapunov's theorem cannot be applied.

(4) If $J_2 = 0$, then $\gamma = \delta$ and Lyapunov's theorem cannot be applied. Notice that in this case the system is integrable.

(5) The main condition for the applicability of Lyapunov's theorem is $6J_2 < \overline{g}^4$, which guarantees that ρ_0 is real. One can easily see that this condition implies the inequality $3J_2 < \overline{g}^2 \rho_0$, which ensures that γ is negative.

10 Regularization theory

The theory of regularization aims to reduce singular differential equations to regular differential equations. Regularizing transformations are often used in Celestial Mechanics, when two or more bodies approach a collision [171]. Using Hamiltonian formalism we review the most elementary regularizing methods, known as Levi–Civita, Kustaanheimo–Stiefel and Birkhoff transformations [166,167].

The Levi–Civita regularization (Section 10.1) is first introduced in the framework of the two–body problem and later extended to the circular, planar, restricted three–body problem. It is based on a suitable change of coordinates (the Levi–Civita transformation), the introduction of a fictitious time and of the extended phase space. By means of this technique one obtains a theory which allows us to regularize a single collision in the plane. The extension to the spatial problem is the content of the Kustaanheimo–Steifel (KS) regularization, which is proven to be canonical (Section 10.2).

To deal with a simultaneous regularization at both attracting centers, the Birkhoff transformation (in the plane and in the space) is introduced. Like in the Levi–Civita transformation, a fictitious time is considered and a regularizing function is defined such that a global regularization is accomplished, so that collisions with both primaries can be investigated (Section 10.3).

10.1 The Levi–Civita transformation

10.1.1 The two–body problem

We consider a two–body problem whose interacting bodies $\mathcal{P}_1$, $\mathcal{P}_2$ have masses, respectively, m_1, m_2. With reference to (3.3), we set $\underline{r} \equiv (q_1, q_2)$ and we denote by p_1, p_2 the momenta conjugated to q_1, q_2; adopting the units of measure so that $\mu = 1$ in (3.3), then the Hamiltonian describing the motion of $\mathcal{P}_1$ and $\mathcal{P}_2$ can be written in the form

$$\mathcal{H}(p_1, p_2, q_1, q_2) = \frac{1}{2}(p_1^2 + p_2^2) - \frac{1}{(q_1^2 + q_2^2)^{\frac{1}{2}}} \ .$$

We select a canonical transformation from (p_1, p_2, q_1, q_2) to (P_1, P_2, Q_1, Q_2) with a generating function linear in the momenta:

$$W(p_1, p_2, Q_1, Q_2) = p_1 f(Q_1, Q_2) + p_2 g(Q_1, Q_2) \ ;$$

the functions f and g corresponding to the *Levi–Civita transformation* are defined through the relation

$$f(Q_1,Q_2)+ig(Q_1,Q_2)\equiv(Q_1+iQ_2)^2=Q_1^2-Q_2^2+i\cdot 2Q_1Q_2\ ,$$

namely

$$f(Q_1,Q_2)\equiv Q_1^2-Q_2^2\ ,\qquad\qquad g(Q_1,Q_2)\equiv 2Q_1Q_2\ .$$

The corresponding characteristic equations are (notice that in (A.7) of Appendix A an equivalent form with opposite sign for the generating function has been introduced):

$$\begin{aligned}
q_1 &= \frac{\partial W}{\partial p_1}=f(Q_1,Q_2)=Q_1^2-Q_2^2\\
q_2 &= \frac{\partial W}{\partial p_2}=g(Q_1,Q_2)=2Q_1Q_2\\
P_1 &= \frac{\partial W}{\partial Q_1}=p_1\frac{\partial f}{\partial Q_1}+p_2\frac{\partial g}{\partial Q_1}=2p_1Q_1+2p_2Q_2\\
P_2 &= \frac{\partial W}{\partial Q_2}=p_1\frac{\partial f}{\partial Q_2}+p_2\frac{\partial g}{\partial Q_2}=-2p_1Q_2+2p_2Q_1\ .
\end{aligned}\qquad(10.1)$$

We refer to (q_1,q_2) as the *physical plane,* while (Q_1,Q_2) is the *parametric plane;* the relation between the two planes is given by

$$q_1+iq_2=f+ig=(Q_1+iQ_2)^2\ .$$

We remark that the Levi–Civita transformation has the effect that the angles at the origin are doubled, namely if $q_1=\sqrt{\rho}\cos\vartheta$, $q_2=\sqrt{\rho}\sin\vartheta$ and $Q_1=\sqrt{\sigma}\cos\varphi$, $Q_2=\sqrt{\sigma}\sin\varphi$, then one finds

$$q_1+iq_2=\sqrt{\rho}(\cos\vartheta+i\sin\vartheta)=\sqrt{\rho}e^{i\vartheta}=(Q_1+iQ_2)^2=(\sqrt{\sigma}e^{i\varphi})^2=\sigma e^{2i\varphi}\ .$$

From the last equation we obtain $\vartheta=2\varphi$ or equivalently $\varphi=\frac{1}{2}\vartheta$: a point making a revolution around the center of mass is transformed to a point of the parametric plane which has made half a revolution.

The last two equations of the transformation (10.1) can be written as

$$\underline{P}=2A_0^T\,\underline{p}\qquad\text{with}\qquad A_0=\begin{pmatrix}Q_1 & -Q_2\\ Q_2 & Q_1\end{pmatrix}\ ,$$

where $\underline{P}=(P_1,P_2)$, $\underline{p}=(p_1,p_2)$; the determinant of A_0 amounts to $\det A_0=Q_1^2+Q_2^2>0$ and the inverse is given by $A_0^{-1}=\frac{1}{\det A_0}A_0^T$. Let $D\equiv 4(Q_1^2+Q_2^2)$; the above relations imply that

$$P_1^2+P_2^2=4(Q_1^2+Q_2^2)(p_1^2+p_2^2)=D(p_1^2+p_2^2)\ .$$

Therefore the transformed Hamiltonian takes the form

$$\tilde{\mathcal{H}}(P_1,P_2,Q_1,Q_2)=\frac{1}{2D}(P_1^2+P_2^2)-\frac{1}{(f(Q_1,Q_2)^2+g(Q_1,Q_2)^2)^{\frac{1}{4}}}\ .$$

The associated Hamilton's equations are

$$\begin{aligned}
\dot{Q}_1 &= \frac{P_1}{D} \\
\dot{Q}_2 &= \frac{P_2}{D} \\
\dot{P}_1 &= \frac{1}{2D^2}(P_1^2+P_2^2)\frac{\partial D}{\partial Q_1} - \frac{1}{2}\frac{1}{(f^2+g^2)^{\frac{3}{2}}}\frac{\partial(f^2+g^2)}{\partial Q_1} \\
\dot{P}_2 &= \frac{1}{2D^2}(P_1^2+P_2^2)\frac{\partial D}{\partial Q_2} - \frac{1}{2}\frac{1}{(f^2+g^2)^{\frac{3}{2}}}\frac{\partial(f^2+g^2)}{\partial Q_2} .
\end{aligned}$$

In the *extended phase space* we introduce a new pair of variables (T, t) and we write the extended Hamiltonian as

$$\mathcal{H}_{ext}(P_1, P_2, T, Q_1, Q_2, t) = \frac{1}{2D}(P_1^2+P_2^2) + T - \frac{1}{(f(Q_1,Q_2)^2 + g(Q_1,Q_2)^2)^{\frac{1}{2}}} ,$$

where $\dot{t} = \frac{\partial \mathcal{H}_{ext}}{\partial T} = 1$ and $\dot{T} = -\frac{\partial \mathcal{H}_{ext}}{\partial t} = 0$. Therefore T is constant and in particular, along a solution, one obtains $T(t) = -\tilde{\mathcal{H}}$ (compare with the following remark).

Remark. The extended phase space is introduced in order to obtain a transformation which involves also the time. In general, if $\tilde{\mathcal{H}} = \tilde{\mathcal{H}}(\underline{P}, \underline{Q}, t)$ ($\underline{P} \equiv (P_1, P_2)$, $\underline{Q} \equiv (Q_1, Q_2)$) depends explicitly on the time, the time–independent Hamiltonian $\mathcal{H}_{ext} = \mathcal{H}_{ext}(\underline{P}, T, \underline{Q}, t) = \tilde{\mathcal{H}}(\underline{P}, \underline{Q}, t) + T$, with T conjugated to t, is identically zero. To prove this statement, let $T(0) = -\tilde{\mathcal{H}}(\underline{P}(0), \underline{Q}(0), 0)$. Then, one finds $T(t) = -\tilde{\mathcal{H}}(t)$ along a solution for any time. In fact, we first observe that $\frac{d\tilde{\mathcal{H}}}{dt} = \frac{\partial \tilde{\mathcal{H}}}{\partial t}$, since using Hamilton's equations one has

$$\frac{d\tilde{\mathcal{H}}}{dt} = \frac{\partial \tilde{\mathcal{H}}}{\partial t} + \sum_{j=1}^{2} \frac{\partial \tilde{\mathcal{H}}}{\partial Q_j}\frac{dQ_j}{dt} + \sum_{j=1}^{2} \frac{\partial \tilde{\mathcal{H}}}{\partial P_j}\frac{dP_j}{dt} = \frac{\partial \tilde{\mathcal{H}}}{\partial t} + \sum_{j=1}^{2} \frac{\partial \tilde{\mathcal{H}}}{\partial Q_j}\frac{\partial \tilde{\mathcal{H}}}{\partial P_j} - \sum_{j=1}^{2} \frac{\partial \tilde{\mathcal{H}}}{\partial P_j}\frac{\partial \tilde{\mathcal{H}}}{\partial Q_j} = \frac{\partial \tilde{\mathcal{H}}}{\partial t} .$$

Therefore one has

$$\frac{dT}{dt} = -\frac{\partial \tilde{\mathcal{H}}}{\partial t} = -\frac{d\tilde{\mathcal{H}}}{dt} ,$$

which gives

$$T(t) = T(0) + \int_0^t \frac{dT(\tau)}{d\tau} d\tau = T(0) - \int_0^t \frac{d\tilde{\mathcal{H}}(\tau)}{d\tau} d\tau = -\tilde{\mathcal{H}}(0) - \int_0^t \frac{d\tilde{\mathcal{H}}(\tau)}{d\tau} d\tau = -\tilde{\mathcal{H}}(t) ,$$

as we claimed.

After introducing the Levi–Civita transformation and the extended Hamiltonian, we define a *fictitious* or *regularized* time s through the relation

$$dt = D(Q_1, Q_2)ds \qquad \text{or, equivalently,} \qquad \frac{d}{dt} = \frac{1}{D}\frac{d}{ds} .$$

From $\dot{\underline{Q}} = \frac{\partial \mathcal{H}_{ext}}{\partial \underline{P}}$ and

$$\dot{\underline{Q}} = \frac{d\underline{Q}}{dt} = \frac{d\underline{Q}}{ds}\frac{ds}{dt} = \frac{1}{D}\frac{d\underline{Q}}{ds} \ ,$$

it follows that $\frac{d\underline{Q}}{ds} = \frac{\partial \mathcal{H}^*_{ext}}{\partial \underline{P}}$ with $\mathcal{H}^*_{ext} \equiv D\mathcal{H}_{ext}$. As for $\underline{P}$, one has $\dot{\underline{P}} = -\frac{\partial \mathcal{H}_{ext}}{\partial \underline{Q}}$ and

$$\dot{\underline{P}} = \frac{d\underline{P}}{dt} = \frac{d\underline{P}}{ds}\frac{ds}{dt} = \frac{1}{D}\frac{d\underline{P}}{ds} \ ,$$

so that one obtains

$$\frac{d\underline{P}}{ds} = -\frac{\partial \mathcal{H}^*_{ext}}{\partial \underline{Q}} \ ,$$

since

$$\frac{\partial \mathcal{H}^*_{ext}}{\partial \underline{Q}} = \frac{\partial D}{\partial \underline{Q}}\mathcal{H}_{ext} + D\frac{\partial \mathcal{H}_{ext}}{\partial \underline{Q}} = D\frac{\partial \mathcal{H}_{ext}}{\partial \underline{Q}} \ ,$$

being $\mathcal{H}_{ext} = 0$ along a solution. The new Hamiltonian $\mathcal{H}^*_{ext}$ becomes

$$\mathcal{H}^*_{ext} \equiv D\mathcal{H}_{ext} = DT + \frac{1}{2}(P_1^2 + P_2^2) - \frac{D}{(f^2+g^2)^{\frac{1}{2}}}$$

with associated Hamilton's equations ($j = 1, 2$):

$$\begin{aligned}
\frac{dQ_j}{ds} &= P_j \\
\frac{dP_j}{ds} &= -\frac{\partial}{\partial Q_j}\left[DT - \frac{D}{(f^2+g^2)^{\frac{1}{2}}}\right] \\
\frac{dt}{ds} &= D \\
\frac{dT}{ds} &= 0 \ .
\end{aligned}$$

Notice that the singularity of the problem corresponds to the term

$$\frac{D}{(f^2+g^2)^{\frac{1}{2}}} = \frac{4(Q_1^2+Q_2^2)}{(Q_1^4+Q_2^4-2Q_1^2Q_2^2+4Q_1^2Q_2^2)^{\frac{1}{2}}} = 4 \ .$$

Denoting by a prime the derivative with respect to s, the equations of motion are given by ($j = 1, 2$):

$$\begin{aligned}
Q_j' &= P_j \\
P_j' &= -T\frac{\partial D}{\partial Q_j} \ ,
\end{aligned}$$

namely

$$\begin{aligned}
Q_1' &= P_1 \\
Q_2' &= P_2 \\
P_1' &= -T \cdot 8Q_1 = 8\tilde{\mathcal{H}}Q_1 \\
P_2' &= -T \cdot 8Q_2 = 8\tilde{\mathcal{H}}Q_2 \ ,
\end{aligned}$$

being $T = -\tilde{\mathcal{H}}$. From the above expressions one gets the second–order differential equations

$$Q_j'' = 8\tilde{\mathcal{H}} Q_j \ , \qquad j = 1,2 \ .$$

If $\tilde{\mathcal{H}} < 0$ (corresponding to an elliptic orbit), one obtains the equation of a harmonic oscillator. Having found the solution $Q_j = Q_j(s)$, one can determine the relation between t and s through $dt = 4(Q_1^2+Q_2^2)ds$, from which one computes the quantities $Q_j = Q_j(t)$. Finally one obtains the solution through the expressions

$$q_1 = q_1(t) = Q_1(t)^2 - Q_2(t)^2 \ , \qquad q_2 = q_2(t) = 2Q_1(t)Q_2(t) \ .$$

10.1.2 The planar, circular, restricted three–body problem

We consider the planar, circular, restricted three–body problem introduced in Chapter 4, modeling the motion of a body $\mathcal{P}_2$ in the gravitational field of two primaries $\mathcal{P}_1$ and $\mathcal{P}_3$. The Hamiltonian describing the dynamics can be written, in a synodic reference frame, as (compare with equations (4.12))

$$\mathcal{H}(p_1, p_2, q_1, q_2) = \frac{1}{2}(p_1^2 + p_2^2) + q_2 p_1 - q_1 p_2 - V(q_1, q_2) \ , \tag{10.2}$$

where $V(q_1, q_2) = \frac{\mu_1}{r_1} + \frac{\mu_3}{r_3}$ with $\mu_1 \equiv \mathcal{G} m_1$, $\mu_3 \equiv \mathcal{G} m_3$ and

$$r_1 = [(q_1 + \mu_3)^2 + q_2^2]^{\frac{1}{2}} \ , \qquad r_3 = [(q_1 - \mu_1)^2 + q_2^2]^{\frac{1}{2}} \ .$$

Let us introduce a canonical transformation from (p_1, p_2, q_1, q_2) to the new set of variables (P_1, P_2, Q_1, Q_2) by defining a generating function of the form

$$W(p_1, p_2, Q_1, Q_2) = p_1 f(Q_1, Q_2) + p_2 g(Q_1, Q_2) \ ,$$

for some functions $f = f(Q_1, Q_2)$, $g = g(Q_1, Q_2)$ to be defined as follows. In order to regularize collisions with the primary $\mathcal{P}_1$ one defines

$$f(Q_1, Q_2) = Q_1^2 - Q_2^2 - \mu_3 \ , \qquad g(Q_1, Q_2) = 2Q_1 Q_2 \ ;$$

otherwise collisions with $\mathcal{P}_3$ are regularized taking $f(Q_1, Q_2) = Q_1^2 - Q_2^2 + \mu_1$ and $g(Q_1, Q_2) = 2Q_1 Q_2$. The associated characteristic equations are:

$$\begin{aligned}
q_1 &= \frac{\partial W}{\partial p_1} = f(Q_1, Q_2) \\
q_2 &= \frac{\partial W}{\partial p_2} = g(Q_1, Q_2) \\
P_1 &= \frac{\partial W}{\partial Q_1} = p_1 \frac{\partial f}{\partial Q_1} + p_2 \frac{\partial g}{\partial Q_1} \\
P_2 &= \frac{\partial W}{\partial Q_2} = p_1 \frac{\partial f}{\partial Q_2} + p_2 \frac{\partial g}{\partial Q_2} \ .
\end{aligned}$$

As for the two–body problem, the term $p_1^2 + p_2^2$ becomes $\frac{1}{D}(P_1^2 + P_2^2)$, while the term $q_2 p_1 - p_2 q_1$ is transformed into

$$q_2p_1 - p_2q_1 = \frac{1}{2D}\left[P_1\frac{\partial}{\partial Q_2}(f^2+g^2) - P_2\frac{\partial}{\partial Q_1}(f^2+g^2)\right] .$$

In fact, from $D \equiv \left(\frac{\partial f}{\partial Q_1}\right)^2 + \left(\frac{\partial g}{\partial Q_1}\right)^2 = 4(Q_1^2+Q_2^2)$, it is

$$\begin{aligned}
&\frac{1}{8(Q_1^2+Q_2^2)}\left[2P_1f\frac{\partial f}{\partial Q_2} + 2P_1g\frac{\partial g}{\partial Q_2} - 2P_2f\frac{\partial f}{\partial Q_1} - 2P_2g\frac{\partial g}{\partial Q_1}\right] \\
&= \frac{1}{4(Q_1^2+Q_2^2)}\left[f\frac{\partial f}{\partial Q_2}\frac{\partial g}{\partial Q_1}p_2 + g\frac{\partial g}{\partial Q_2}\frac{\partial f}{\partial Q_1}p_1 - f\frac{\partial f}{\partial Q_1}\frac{\partial g}{\partial Q_2}p_2 - g\frac{\partial g}{\partial Q_1}\frac{\partial f}{\partial Q_2}p_1\right] \\
&= \frac{1}{4(Q_1^2+Q_2^2)}\left[fp_2(\frac{\partial f}{\partial Q_2}\frac{\partial g}{\partial Q_1} - \frac{\partial f}{\partial Q_1}\frac{\partial g}{\partial Q_2}) + gp_1(\frac{\partial f}{\partial Q_1}\frac{\partial g}{\partial Q_2} - \frac{\partial f}{\partial Q_2}\frac{\partial g}{\partial Q_1})\right] \\
&= -fp_2 + gp_1 = p_1q_2 - p_2q_1 .
\end{aligned}$$

The transformed Hamiltonian becomes

$$\tilde{\mathcal{H}}(P_1,P_2,Q_1,Q_2) = \frac{1}{2D}\left[P_1^2+P_2^2+P_1\frac{\partial}{\partial Q_2}(f^2+g^2)-P_2\frac{\partial}{\partial Q_1}(f^2+g^2)\right]-\tilde{V}(Q_1,Q_2) ,$$

where $\tilde{V}(Q_1,Q_2) = V(f(Q_1,Q_2),g(Q_1,Q_2))$. The equations of motion take the form

$$\begin{aligned}
\dot{Q}_1 &= \frac{1}{2D}\left[2P_1 + \frac{\partial}{\partial Q_2}(f^2+g^2)\right] \\
\dot{Q}_2 &= \frac{1}{2D}\left[2P_2 - \frac{\partial}{\partial Q_1}(f^2+g^2)\right] \\
\dot{P}_1 &= -\frac{\partial\tilde{\mathcal{H}}}{\partial Q_1} \\
\dot{P}_2 &= -\frac{\partial\tilde{\mathcal{H}}}{\partial Q_2} .
\end{aligned}$$

In the extended phase space the Hamiltonian is given by

$$\mathcal{H}_{ext} = T + \frac{1}{2D}\left[P_1^2+P_2^2+P_1\frac{\partial}{\partial Q_2}(f^2+g^2) - P_2\frac{\partial}{\partial Q_1}(f^2+g^2)\right] - \tilde{V}(Q_1,Q_2) .$$

Next we introduce the fictitious time s as related to the ordinary time t by

$$dt = D\,ds ;$$

this transformation yields the Hamiltonian

$$\begin{aligned}
\mathcal{H}_{ext}^* = D\mathcal{H}_{ext} &= DT + \frac{1}{2}\left[P_1^2+P_2^2+P_1\frac{\partial}{\partial Q_2}(f^2+g^2) - P_2\frac{\partial}{\partial Q_1}(f^2+g^2)\right] \\
&- D\tilde{V}(Q_1,Q_2) .
\end{aligned}$$

Define the function $\Phi(Q_1,Q_2) \equiv f(Q_1,Q_2) + ig(Q_1,Q_2)$ and let $|\Phi|^2 = f^2+g^2$; Hamilton's equations with respect to the fictitious time are given by

$$
\begin{aligned}
Q_1' &= P_1 + \frac{1}{2}\frac{\partial}{\partial Q_2}|\Phi|^2 \\
Q_2' &= P_2 - \frac{1}{2}\frac{\partial}{\partial Q_1}|\Phi|^2 \\
t' &= D \\
P_1' &= -T\frac{\partial D}{\partial Q_1} - \frac{1}{2}\left[P_1\frac{\partial^2|\Phi|^2}{\partial Q_1 \partial Q_2} - P_2\frac{\partial^2|\Phi|^2}{\partial Q_1^2}\right] + \frac{\partial}{\partial Q_1}(D\tilde{V}) \\
P_2' &= -T\frac{\partial D}{\partial Q_2} - \frac{1}{2}\left[P_1\frac{\partial^2|\Phi|^2}{\partial Q_2^2} - P_2\frac{\partial^2|\Phi|^2}{\partial Q_2 \partial Q_1}\right] + \frac{\partial}{\partial Q_2}(D\tilde{V}) \\
T' &= 0 \,.
\end{aligned}
$$

From the last equation it follows that T is constant with $T = -\tilde{\mathcal{H}}$. The singularities appear in the terms $\frac{\partial}{\partial Q_j}(D\tilde{V})$ $(j = 1, 2)$, that we are going to rewrite as follows. Let $U = \frac{1}{2}(q_1^2 + q_2^2) + V = \frac{1}{2}(Q_1^2 + Q_2^2) + \tilde{V} = \frac{1}{2}|\Phi|^2 + \tilde{V}$. From the definition (4.16) of the Jacobi integral and the expression of the Hamiltonian (10.2) in terms of the coordinates and of their derivatives, one finds

$$\tilde{\mathcal{H}} = -T = -\frac{C_J}{2}$$

or

$$\frac{1}{2}|\Phi|^2 - T + \tilde{V} = U - \frac{C_J}{2} \,.$$

Finally $D\tilde{V} = D(U - \frac{C_J}{2}) - \frac{1}{2}D|\Phi|^2 + DT$, showing that the critical term is $D(U - \frac{C_J}{2})$.

Denote by $z = q_1 + iq_2$, $w = Q_1 + iQ_2$ the complex coordinates in the physical and parametric plane, respectively. In the physical plane the primaries are located at $z_1 = -\mu_3$ and $z_3 = \mu_1$; the transformation $z = -\mu_3 + w^2$ regularizes the singularity at $\mathcal{P}_1$, while the transformation $z = \mu_1 + w^2$ regularizes the singularity at $\mathcal{P}_3$. Indeed, the above transformations are said to be *local,* since only one of the two singularities is eliminated.

In order to see if the term $D(U - \frac{C_J}{2})$ still contains singularities, we proceed as follows. Consider the function

$$z \equiv \tilde{f}(w) = w^2 - \mu_3 \,,$$

which transforms the point $\mathcal{P}_1(-\mu_3, 0)$ belonging to the physical plane into the origin of the w–plane, while $\mathcal{P}_3$ takes the coordinates $w_{1,2} = \pm 1$, since $w^2 = \mu_1 + \mu_3 = 1$. In order to transform $U - \frac{C_J}{2}$ in terms of the new complex variable w, one needs the expressions of r_1 and r_3 in terms of w. Since $r_1 = |z + \mu_3|$ and $r_3 = |z - \mu_1|$, one has $r_1 = |w|^2$, $r_3 = |w^2 - 1|$; from

$$\mu_1 r_1^2 + \mu_3 r_3^2 = \mu_1(z + \mu_3)^2 + \mu_3(z - \mu_1)^2 = z^2 + \mu_1\mu_3 \,,$$

it follows that

$$
\begin{aligned}
U-\frac{C_J}{2} &= \frac{1}{2}(q_1^2+q_2^2)+V-\frac{C_J}{2} \\
&= \frac{1}{2}(\mu_1 r_1^2+\mu_3 r_3^2)-\frac{1}{2}\mu_1\mu_3+\frac{\mu_1}{r_1}+\frac{\mu_3}{r_3}-\frac{C_J}{2} \\
&= \frac{1}{2}\Big[\mu_1|w|^4+\mu_3|w^2-1|^2\Big]-\frac{1}{2}\mu_1\mu_3+\frac{\mu_1}{|w|^2}+\frac{\mu_3}{|w^2-1|}-\frac{C_J}{2}\,.
\end{aligned}
$$

Taking into account that $D=4(Q_1^2+Q_2^2)=4|w|^2$, we conclude that the term $D(U-\frac{C_J}{2})$ does not contain singularities at $\mathcal{P}_1$.

We add a remark on the values of the velocities at the location of the primaries. From the Jacobi integral in the physical space we get $|\dot z|^2=2(U-\frac{C_J}{2})$; since $z=w^2-\mu_3$, then $\dot z=2w\dot w=\frac{2}{D}ww'$ or $|w'|^2=\frac{D^2}{4|w^2|}|\dot z|^2$, so that in the parametric space we obtain

$$
|w'|^2=8|w|^2\left(U-\frac{C_J}{2}\right)\,.
$$

Therefore, we find

$$
|w'|^2=8\mu_1+|w|^2\left[\frac{8\mu_3}{|w^2-1|}+4\mu_1|w|^4+4\mu_3|w^2-1|^2-4C_J-4\mu_1\mu_3\right]\,.
$$

We remark that in $\mathcal{P}_1$ one has $r_1=0$, namely $w=0$, so that $|w'|^2=8\mu_1$ and the velocity is *finite*. In $\mathcal{P}_3$ one has $r_3=0$, namely $w=\pm 1$, so that $|w'|^2=\infty$ and the velocity is *infinite*.

10.2 The Kustaanheimo–Stiefel regularization

10.2.1 The restricted, spatial three–body problem

We consider the three–body gravitational interaction of $\mathcal{P}_1$, $\mathcal{P}_2$, $\mathcal{P}_3$ in the spatial case, in which $\mathcal{P}_2$ is allowed to move in any direction. The masses of $\mathcal{P}_1$ and $\mathcal{P}_3$ are, respectively, μ_1 and μ_3, while $\mathcal{P}_2$ has infinitesimal mass. We assume that the primaries move in the xy–plane around their common barycenter; their coordinates in the synodic reference frame are given by $\mathcal{P}_1(-\mu_3,0,0)$, $\mathcal{P}_3(\mu_1,0,0)$. The Hamiltonian function of this model is given by

$$
\mathcal{H}(p_1,p_2,p_3,q_1,q_2,q_3)=\frac{1}{2}(p_1^2+p_2^2+p_3^2)+q_2p_1-q_1p_2-V(q_1,q_2,q_3)\;, \qquad (10.3)
$$

where $V(q_1,q_2,q_3)\equiv\frac{\mu_1}{r_1}+\frac{\mu_3}{r_3}$ and $r_1^2\equiv(q_1+\mu_3)^2+q_2^2+q_3^2$, $r_3^2\equiv(q_1-\mu_1)^2+q_2^2+q_3^2$. The equations of motion of $\mathcal{P}_2$ are (see (4.12))

$$
\begin{aligned}
\ddot q_1-2\dot q_2 &= U_{q_1} \\
\ddot q_2+2\dot q_1 &= U_{q_2} \\
\ddot q_3 &= U_{q_3}\;,
\end{aligned} \qquad (10.4)
$$

where $U=\frac{1}{2}(q_1^2+q_2^2)+\frac{\mu_1}{r_1}+\frac{\mu_3}{r_3}$ and $U_{q_i}\equiv\frac{\partial U}{\partial q_i}$ for $i=1,2,3$.

10.2.2 The KS–transformation

Starting from (10.4), which can be written as

$$\begin{aligned}\ddot{q}_1 - 2\dot{q}_2 &= q_1 - \frac{\mu_1}{r_1^3}(q_1+\mu_3) - \frac{\mu_3}{r_3^3}(q_1-\mu_1)\\ \ddot{q}_2 + 2\dot{q}_1 &= q_2 - \frac{\mu_1}{r_1^3}q_2 - \frac{\mu_3}{r_3^3}q_2\\ \ddot{q}_3 &= -\frac{\mu_1}{r_1^3}q_3 - \frac{\mu_3}{r_3^3}q_3\ ,\end{aligned}$$

we proceed to perform a *time* transformation and then a *coordinate* transformation. We define the fictitious time s through the relation

$$dt = D\ ds\ ,$$

namely $\frac{d}{dt} = \frac{1}{D}\frac{d}{ds}$, where D will be defined later. The second derivatives with respect to t and s are related by

$$\frac{d^2}{dt^2} = \frac{d}{dt}\left(\frac{1}{D}\frac{d}{ds}\right) = \frac{1}{D}\frac{d}{ds}\left(\frac{1}{D}\frac{d}{ds}\right) = \frac{1}{D^2}\frac{d^2}{ds^2} - \frac{1}{D^3}\frac{dD}{ds}\frac{d}{ds}\ .$$

The equations of motion with respect to the new time s are

$$\begin{aligned}\frac{1}{D^2}q_1'' - \frac{1}{D^3}D'q_1' - \frac{2}{D}q_2' &= U_{q_1}\\ \frac{1}{D^2}q_2'' - \frac{1}{D^3}D'q_2' + \frac{2}{D}q_1' &= U_{q_2}\\ \frac{1}{D^2}q_3'' - \frac{1}{D^3}D'q_3' &= U_{q_3}\end{aligned}$$

and multiplying by D^3 one obtains

$$\begin{aligned}Dq_1'' - D'q_1' - 2D^2q_2' &= D^3U_{q_1}\\ Dq_2'' - D'q_2' + 2D^2q_1' &= D^3U_{q_2}\\ Dq_3'' - D'q_3' &= D^3U_{q_3}\ .\end{aligned} \tag{10.5}$$

The singular terms are the right–hand sides of (10.5).

In the planar case we denoted by (q_1, q_2) the physical plane and by (Q_1, Q_2) the parametric plane. The Levi–Civita transformation of the coordinates was written in the form (compare with (10.1))

$$\begin{pmatrix} q_1\\ q_2\end{pmatrix} = A_0(\underline{Q})\begin{pmatrix} Q_1\\ Q_2\end{pmatrix} = \begin{pmatrix} Q_1^2 - Q_2^2\\ 2Q_1Q_2\end{pmatrix}\ ,$$

where, setting $\underline{Q} \equiv (Q_1, Q_2)$ we have $A_0(\underline{Q}) \equiv \begin{pmatrix} Q_1 & -Q_2\\ Q_2 & Q_1\end{pmatrix}$. The matrix $A_0(\underline{Q})$ has the property that every element is linear and that it is orthogonal. In the n–dimensional case we investigate whether there exists a generalization $A(\underline{u})$ with $\underline{u} \in \mathbf{R}^n$ of the regularizing matrix with the following properties:

(i) the elements of $A(\underline{u})$ are linear homogeneous functions of $u_1, \ldots, u_n$;
(ii) the matrix is orthogonal, namely
  (a) the scalar product of different rows vanishes,
  (b) each row has norm $u_1^2 + \cdots + u_n^2$.

A result by A. Hurwitz [167] states that such a matrix can only be produced whenever $n = 1, 2, 4$ or 8, but not in the spatial case $n = 3$. As a consequence it is necessary to map the three–dimensional physical space into a four–dimensional parametric space by defining the transformation matrix

$$A(\underline{u}) = \begin{pmatrix} u_1 & -u_2 & -u_3 & u_4 \\ u_2 & u_1 & -u_4 & -u_3 \\ u_3 & u_4 & u_1 & u_2 \\ u_4 & -u_3 & u_2 & -u_1 \end{pmatrix} .$$

As for the coordinates, we extend the three–dimensional physical space to a four–dimensional space by imposing that the fourth component is equal to zero, namely $(q_1, q_2, q_3, 0)$.

The Kustaanheimo–Stiefel (KS) transformation [167] for the case of a collision with $\mathcal{P}_1$ is introduced by defining the following change of coordinates:

$$\begin{pmatrix} q_1 \\ q_2 \\ q_3 \\ 0 \end{pmatrix} = A(\underline{u}) \begin{pmatrix} u_1 \\ u_2 \\ u_3 \\ u_4 \end{pmatrix} - \begin{pmatrix} \mu_3 \\ 0 \\ 0 \\ 0 \end{pmatrix} = \begin{pmatrix} u_1 & -u_2 & -u_3 & u_4 \\ u_2 & u_1 & -u_4 & -u_3 \\ u_3 & u_4 & u_1 & u_2 \\ u_4 & -u_3 & u_2 & -u_1 \end{pmatrix} \begin{pmatrix} u_1 \\ u_2 \\ u_3 \\ u_4 \end{pmatrix} - \begin{pmatrix} \mu_3 \\ 0 \\ 0 \\ 0 \end{pmatrix} , \tag{10.6}$$

namely

$$\begin{aligned} q_1 &= u_1^2 - u_2^2 - u_3^2 + u_4^2 - \mu_3 \\ q_2 &= 2u_1u_2 - 2u_3u_4 \\ q_3 &= 2u_1u_3 + 2u_2u_4 \ . \end{aligned}$$

Notice that the fourth component of the right–hand side of (10.6) is trivially zero.

Remarks.
(1) Setting $u_3 = u_4 = 0$ the KS–transformation (10.6) reduces to the Levi–Civita transformation, which cannot be trivially extended to the spatial case introducing only three parameters (say u_1, u_2, u_3).
(2) The norms of each row of the matrix A are equal to the square of the norm of the vector $\underline{u}$: $u_1^2 + u_2^2 + u_3^2 + u_4^2$.
(3) To regularize collisions with $\mathcal{P}_3$, one needs to replace the constant vector $(-\mu_3, 0, 0, 0)$ with $(\mu_1, 0, 0, 0)$.
(4) The matrix A is orthogonal: $A^T(\underline{u})A(\underline{u}) = (\underline{u}, \underline{u}) \cdot I_4$ (where I_4 is the 4×4–dimensional identity matrix). Denoting by $\underline{X} \equiv (q_1 + \mu_3, q_2, q_3, 0)$ it follows that under the transformation $\underline{X} = A(\underline{u})\underline{u}$, it is

$$r_1^2 = (\underline{X}, \underline{X}) = \underline{X}^T \underline{X} = \underline{u}^T A^T(\underline{u}) A(\underline{u}) \underline{u} = \underline{u}^T \underline{u} (\underline{u}, \underline{u}) = (\underline{u}, \underline{u})^2 \ ,$$

namely $r_1 = (\underline{u}, \underline{u}) = |\underline{u}|^2 = u_1^2 + u_2^2 + u_3^2 + u_4^2$.

(5) A direct computation shows that $A(\underline{u})' = A(\underline{u}')$. From $\underline{X} = A(\underline{u})\underline{u}$, it follows that $\underline{X}' = 2A(\underline{u})\underline{u}'$. In fact,

$$\begin{aligned} q_1' &= 2u_1u_1' - 2u_2u_2' - 2u_3u_3' + 2u_4u_4' \\ q_2' &= 2u_2u_1' + 2u_1u_2' - 2u_4u_3' - 2u_3u_4' \\ q_3' &= 2u_3u_1' + 2u_1u_3' + 2u_4u_2' + 2u_2u_4' \ , \end{aligned} \tag{10.7}$$

which can be written as

$$\begin{pmatrix} q_1' \\ q_2' \\ q_3' \\ 0 \end{pmatrix} = 2A(\underline{u})\underline{u}' = 2\begin{pmatrix} u_1u_1' - u_2u_2' - u_3u_3' + u_4u_4' \\ u_2u_1' + u_1u_2' - u_4u_3' - u_3u_4' \\ u_3u_1' + u_4u_2' + u_1u_3' + u_2u_4' \\ u_4u_1' - u_3u_2' + u_2u_3' - u_1u_4' \end{pmatrix} ,$$

provided the last row, which is named the *bilinear relation,* is identically zero:

$$u_4u_1' - u_3u_2' + u_2u_3' - u_1u_4' = 0 \ .$$

(6) The second derivative is given by

$$\underline{X}'' = 2A(\underline{u})\underline{u}'' + 2A(\underline{u}')\underline{u}' \ ,$$

namely

$$\begin{aligned} q_1'' &= 2(u_1u_1'' - u_2u_2'' - u_3u_3'' + u_4u_4'') + 2(u_1'^2 - u_2'^2 - u_3'^3 + u_4'^3) \\ q_2'' &= 2(u_2u_1'' + u_1u_2'' - u_4u_3'' - u_3u_4'') + 4(u_1'u_2' - u_3'u_4') \\ q_3'' &= 2(u_3u_1'' + u_2''u_4 + u_1u_3'' + u_2u_4'') + 4(u_1'u_3' + u_2'u_4') \\ 0 &= 2(u_4u_1'' - u_3u_2'' + u_2u_3'' - u_1u_4'') \end{aligned} \tag{10.8}$$

(last equation follows from the bilinear relation).

Let us now conclude by showing that the KS–transformation (10.6) provides the desired regularization. We select the scale factor D as

$$D \equiv r_1 = (\underline{u}, \underline{u}) = u_1^2 + u_2^2 + u_3^2 + u_4^2 \ ;$$

one finds that $D' = r_1' = 2(u_1u_1' + u_2u_2' + u_3u_3' + u_4u_4')$. As a consequence the equations of motion are given by (10.5), where q_1, q_2, q_3, as well as their first and second derivatives, are expressed in terms of $\underline{u}$, $\underline{u}'$, $\underline{u}''$ through (10.6), (10.7), (10.8). The singular parts of equations (10.5) are given by $D^3U_{q_1}$ (or equivalently by $D^3U_{q_2}$, $D^3U_{q_3}$). Since U_{q_1} is proportional to $\frac{1}{r_1^3}$ and D is proportional to r_1, it follows that $D^3\Omega_{q_1}$ does not contain singularities; a similar result holds for the terms $D^3\Omega_{q_2}$, $D^3\Omega_{q_3}$, thus yielding the required regularization of the equations of motion at $\mathcal{P}_1$.

10.2.3 Canonicity of the KS–transformation

Following [167], we proceed to prove that the KS–transformation is canonical. In the planar case the KS–transformation reduces to the Levi–Civita transformation (here we use (u_1, u_2) instead of (Q_1, Q_2)):

$$\begin{pmatrix} q_1 \\ q_2 \end{pmatrix} = \begin{pmatrix} u_1 & -u_2 \\ u_2 & u_1 \end{pmatrix} \begin{pmatrix} u_1 \\ u_2 \end{pmatrix} = \begin{pmatrix} u_1^2 - u_2^2 \\ 2u_1 u_2 \end{pmatrix} \equiv A_0(\underline{u})\underline{u}$$

with $A_0(\underline{u}) = \begin{pmatrix} u_1 & -u_2 \\ u_2 & u_1 \end{pmatrix}$, such that if $\underline{U} = (U_1, U_2)$ is conjugated to $\underline{u} = (u_1, u_2)$, then

$$\begin{pmatrix} p_1 \\ p_2 \end{pmatrix} = \frac{1}{2}(A_0^T(\underline{u}))^{-1}\underline{U} = \frac{1}{2(u_1^2 + u_2^2)} A_0(\underline{u}) \begin{pmatrix} U_1 \\ U_2 \end{pmatrix} . \tag{10.9}$$

In the spatial case let us define

$$\Lambda(\underline{u}) \equiv \begin{pmatrix} u_1 & -u_2 & -u_3 & u_4 \\ u_2 & u_1 & -u_4 & -u_3 \\ u_3 & u_4 & u_1 & u_2 \end{pmatrix} ,$$

obtained from $A(\underline{u})$ eliminating the last row. Let

$$\begin{pmatrix} q_1 \\ q_2 \\ q_3 \end{pmatrix} = \Lambda(\underline{u}) \begin{pmatrix} u_1 \\ u_2 \\ u_3 \\ u_4 \end{pmatrix} - \begin{pmatrix} \mu_3 \\ 0 \\ 0 \end{pmatrix} . \tag{10.10}$$

By analogy with (10.9) we define

$$\begin{pmatrix} p_1 \\ p_2 \\ p_3 \end{pmatrix} \equiv \frac{1}{2(u_1^2 + u_2^2 + u_3^2 + u_4^2)} \Lambda(\underline{u}) \begin{pmatrix} U_1 \\ U_2 \\ U_3 \\ U_4 \end{pmatrix} . \tag{10.11}$$

Setting $u_0 \equiv t$, $U_0 \equiv T$, where T is the variable conjugated to t, we proceed to verify that (10.10), (10.11) define a canonical transformation. Let

$$r_1 \equiv \sqrt{(q_1 + \mu_3)^2 + q_2^2 + q_3^2} = u_1^2 + u_2^2 + u_3^2 + u_4^2 = |\underline{u}|^2 ;$$

introducing the scalar product

$$\sigma(\underline{U}, \underline{u}) \equiv (\underline{U}, \underline{u}^*)$$

with $\underline{u}^* \equiv (u_4, -u_3, u_2, -u_1)$ (coinciding with the last row of $A(\underline{u})$), one obtains

$$\begin{pmatrix} p_1 \\ p_2 \\ p_3 \\ \rho \end{pmatrix} = \frac{1}{2r_1} A(\underline{u}) \begin{pmatrix} U_1 \\ U_2 \\ U_3 \\ U_4 \end{pmatrix} ,$$

where the auxiliary variable ρ is defined as

$$\rho \equiv \frac{1}{2r_1}(U_1u_4 - U_2u_3 + U_3u_2 - U_4u_1) = \frac{1}{2r_1}\,\sigma(\underline{U},\underline{u}) \ .$$

From

$$\begin{aligned}(p_1,p_2,p_3,\rho)\begin{pmatrix}p_1\\ p_2\\ p_3\\ \rho\end{pmatrix} &= (U_1,U_2,U_3,U_4)\frac{1}{2r_1}A(\underline{u})^T\cdot\frac{1}{2r_1}A(\underline{u})\begin{pmatrix}U_1\\ U_2\\ U_3\\ U_4\end{pmatrix}\\ &= \frac{1}{4r_1^2}r_1|\underline{U}|^2 = \frac{1}{4r_1}|\underline{U}|^2 \ ,\end{aligned}$$

it follows that $(p_1^2+p_2^2+p_3^2+\rho^2)r_1 = \frac{1}{4}|\underline{U}|^2$, namely

$$(p_1^2+p_2^2+p_3^2)r_1 \ = \ \frac{1}{4}|\underline{U}|^2 - \frac{1}{4r_1}\sigma(\underline{U},\underline{u})^2 \ . \tag{10.12}$$

Putting into evidence the term $\frac{1}{r_1}$ corresponding to the interaction with $\mathcal{P}_1$, let us write the Hamiltonian describing the spatial case in a fixed reference frame within the framework of the extended phase space in the form

$$\mathcal{H}_0 = \frac{1}{2}(p_1^2+p_2^2+p_3^2) + T - \frac{1}{r_1} - V_1(q_1,q_2,q_3)$$

for a suitable function $V_1(q_1,q_2,q_3,t)$. The introduction of the fictitious time defined through $dt = r_1 ds$ provides the Hamiltonian

$$\mathcal{H}_1 = \frac{1}{2}(p_1^2+p_2^2+p_3^2)r_1 + Tr_1 - 1 - r_1V_1(q_1,q_2,q_3) \ .$$

Using $r_1 = |\underline{u}|^2$ and (10.12) one gets the Hamiltonian

$$\mathcal{H}_2 = \frac{1}{8}|\underline{U}|^2 + T|\underline{u}|^2 - 1 - |\underline{u}|^2V_1(\underline{u}) - \frac{1}{8|\underline{u}|^2}\sigma(\underline{U},\underline{u})^2 \ . \tag{10.13}$$

We now show that $\sigma = \sigma(\underline{U},\underline{u})$ is constant along the solutions of the equations of motion.

Lemma. *The bilinear quantity*

$$\sigma(\underline{U},\underline{u}) = U_1u_4 - U_2u_3 + U_3u_2 - U_4u_1$$

is a first integral *of the canonical equations*

$$\frac{du_k}{ds} = \frac{\partial\mathcal{H}_2}{\partial U_k} \ , \qquad \frac{dU_k}{ds} = -\frac{\partial\mathcal{H}_2}{\partial u_k} \ , \qquad k = 0,1,2,3,4 \ ,$$

where $u_0 \equiv t$, $U_0 \equiv T$.

Proof. One finds

$$\frac{d\sigma}{ds} = \sum_{j=1}^{4} \frac{\partial\sigma}{\partial U_j}\frac{dU_j}{ds} + \frac{\partial\sigma}{\partial u_j}\frac{du_j}{ds} = \left(\underline{u}^*, \frac{d\underline{U}}{ds}\right) - \left(\underline{U}^*, \frac{d\underline{u}}{ds}\right) .$$

The new canonical equations are

$$\begin{aligned} \frac{d\underline{u}}{ds} &= \frac{\partial\mathcal{H}_2}{\partial\underline{U}} = \frac{1}{4}\underline{U} - \frac{1}{4|\underline{u}|^2}\underline{u}^*\sigma(\underline{U},\underline{u}) \\ \frac{d\underline{U}}{ds} &= -\frac{\partial\mathcal{H}_2}{\partial\underline{u}} = -2U_0\underline{u} + \frac{\partial}{\partial\underline{u}}(|\underline{u}|^2V_1) - \frac{1}{4|\underline{u}|^4}\underline{u}\,\sigma(\underline{U},\underline{u})^2 \\ &- \frac{1}{4|\underline{u}|^2}\underline{U}^*\sigma(\underline{U},\underline{u}) . \end{aligned} \tag{10.14}$$

Since $(\underline{u}^*,\underline{u}) = (\underline{U}^*,\underline{U}) = 0$, one has

$$\begin{aligned} \frac{d\sigma}{ds} &= -2U_0(\underline{u}^*,\underline{u}) + \left(\underline{u}^*, \frac{\partial}{\partial\underline{u}}(|\underline{u}|^2V_1)\right) - \frac{1}{4|\underline{u}|^4}(\underline{u}^*,\underline{u})\sigma(\underline{U},\underline{u})^2 \\ &- \frac{1}{4|\underline{u}|^2}(\underline{u}^*,\underline{U}^*)\sigma(\underline{U},\underline{u}) - \frac{1}{4}(\underline{U}^*,\underline{U}) + \frac{1}{4|\underline{u}|^2}(\underline{U}^*,\underline{u}^*)\sigma(\underline{U},\underline{u}) \\ &= \left(\underline{u}^*, \frac{\partial}{\partial\underline{u}}[(\underline{u},\underline{u})V_1]\right) = 2(\underline{u}^*,\underline{u})V_1 + |\underline{u}|^2\left(\underline{u}^*, \frac{\partial V_1}{\partial\underline{u}}\right) = |\underline{u}|^2\left(\underline{u}^*, \frac{\partial V_1}{\partial\underline{u}}\right) . \end{aligned}$$

Direct computations show that $\frac{\partial V_1}{\partial\underline{u}} = 2A^T(\underline{u})\frac{\partial V_1}{\partial\underline{q}}$ (since $\frac{\partial V_1}{\partial\underline{u}} = \frac{\partial V_1}{\partial\underline{q}}\frac{\partial\underline{q}}{\partial\underline{u}}$). Moreover one has that

$$\left(\underline{u}^*, 2|\underline{u}|^2A^T(\underline{u})\frac{\partial V_1}{\partial\underline{q}}\right) = 0 . \tag{10.15}$$

In fact, if $\underline{y} \equiv \frac{\partial V_1}{\partial\underline{q}}$ and $\underline{v} \equiv 2|\underline{u}|^2A^T(\underline{u})\underline{y}$, then $(\underline{u}^*,\underline{v}) = (\underline{v},\underline{u}^*) = 0$, since $(\underline{v},\underline{u}^*) = \sigma(\underline{v},\underline{u}) = u_4v_1 - u_3v_2 + u_2v_3 - u_1v_4$, which is zero. In fact, the latter expression is equal to the fourth component of $A(\underline{u})\underline{v} = 2A(\underline{u})A^T(\underline{u})\underline{y}\ |\underline{u}|^2 = 2(\underline{u},\underline{u})\underline{y}\ |\underline{u}|^2$, where the fourth component of $\underline{y} \equiv \frac{\partial V_1}{\partial\underline{q}}$ (which represents the derivative with respect to the time in the physical space) is zero by definition. Finally from (10.14) and $(\underline{u}^*, \frac{\partial V_1}{\partial\underline{u}}) = 0$, we obtain that $\frac{d\sigma}{ds} = 0$ along the equations of motion. □

Lemma. *Assume that the initial values* $\underline{u}(0)$, $\underline{U}(0)$ *of* $\underline{u}(s)$, $\underline{U}(s)$ *satisfy at* $s = 0$ *the bilinear relation*

$$\sigma(\underline{U}(0),\underline{u}(0)) = U_1(0)u_4(0) - U_2(0)u_3(0) + U_3(0)u_2(0) - U_4(0)u_1(0) = 0 .$$

Then one has:
(a) the value of the first integral σ *is zero:* $\sigma(\underline{U},\underline{u}) = U_1u_4 - U_2u_3 + U_3u_2 - U_4u_1 = 0$;
(b) the Hamiltonian (10.13) *is equivalent to the reduced Hamiltonian*

$$\mathcal{H}_3 = \frac{1}{8}|\underline{U}|^2 + U_0|\underline{u}|^2 - 1 - |\underline{u}|^2V_1(\underline{u},u_0) .$$

Proof.
(a) It is a consequence of the fact that $\sigma(\underline{U},\underline{u})$ is a first integral.
(b) The canonical equations associated to the Hamiltonian $\mathcal{H}_2$ and $\mathcal{H}_3$ have the same solutions as a consequence of (a), due to the fact that the term $\sigma^2(\underline{U},\underline{u})$ can be factored out in $\mathcal{H}_2-\mathcal{H}_3$ and that taking the derivatives there is always a factor σ which is zero along the given solutions. □

Next we need to associate to each set of initial conditions in the physical space, say $q_1(0)$, $q_2(0)$, $q_3(0)$, $p_1(0)$, $p_2(0)$, $p_3(0)$, a new set of initial conditions in the parametric space, say $\underline{u}(0)$, $u_0(0)$, $\underline{U}(0)$, $U_0(0)$, which are obtained by means of (10.10), (10.11), so that

(i) $u_0(0)=0$, $U_0(0)=-\mathcal{H}(q_1(0),q_2(0),q_3(0),p_1(0),p_2(0),p_3(0))$;
(ii) $\sigma(\underline{U}(0),\underline{u}(0))=U_1(0)u_4(0)-U_2(0)u_3(0)+U_3(0)u_2(0)-U_4(0)u_1(0)=0$.

We remark that to obtain (ii) one can proceed as in (10.15) by setting $\underline{v}=A^T(\underline{u})\underline{y}$ with $\underline{y}=\underline{q}'(0)$ and defining $\underline{U}(0)\equiv\frac{1}{2|\underline{u}(0)|^2}A^T(\underline{u}(0))\underline{q}'(0)$.

Theorem. *Assume that* $\underline{u}(0)$, $u_0(0)$, $\underline{U}(0)$, $U_0(0)$ *satisfy* (i) *and* (ii); *then, the solutions of*

$$\frac{du_k}{ds}=\frac{\partial\mathcal{H}_3}{\partial U_k}\ ,\qquad \frac{dU_k}{ds}=-\frac{\partial\mathcal{H}_3}{\partial u_k}\ ,\qquad k=0,1,2,3,4\ ,$$

give by (10.10) *and* (10.11) *the solution of Hamilton's equations associated to* $\mathcal{H}$ *defined in* (10.3) *with the corresponding initial data, namely if* $\underline{q}=(q_1,q_2,q_3)$, $\underline{p}=(p_1,p_2,p_3)$, *one has*

$$\frac{d\underline{q}}{dt}=\frac{\partial\mathcal{H}}{\partial\underline{p}}\ ,\qquad \frac{d\underline{p}}{dt}=-\frac{\partial\mathcal{H}}{\partial\underline{q}}\ .$$

Remark. The above theorem implies that the transformation from $(\underline{q},\underline{p})$ to $(\underline{u},\underline{U})$ is *canonical,* since it preserves the Hamiltonian structure.

Proof. The canonicity is proved using the criterion based on Poisson brackets, namely that

$$(a)\qquad \{q_k,q_l\}=\sum_{j=0}^{4}\left(\frac{\partial q_k}{\partial U_j}\frac{\partial q_l}{\partial u_j}-\frac{\partial q_k}{\partial u_j}\frac{\partial q_l}{\partial U_j}\right)=0$$

$$(b)\qquad \{p_k,q_l\}=\sum_{j=0}^{4}\left(\frac{\partial p_k}{\partial U_j}\frac{\partial q_l}{\partial u_j}-\frac{\partial p_k}{\partial u_j}\frac{\partial q_l}{\partial U_j}\right)=\delta_{kl}$$

$$(c)\qquad \{p_k,p_l\}=\sum_{j=0}^{4}\left(\frac{\partial p_k}{\partial U_j}\frac{\partial p_l}{\partial u_j}-\frac{\partial p_k}{\partial u_j}\frac{\partial p_l}{\partial U_j}\right)=0\ ,$$

for $k=0,1,2,3$, $l=0,1,2,3$. Cases $k=0$ or $l=0$ are trivial; therefore the sum can be restricted to $j=1,2,3,4$. Since the variables $\underline{U}$ do not enter in the KS–transformation (10.10), one has $\frac{\partial q_k}{\partial U_j}=0$ providing (a). Moreover, (b) becomes

$$\{p_k, q_l\} = \sum_{j=1}^{4} \frac{\partial p_k}{\partial U_j} \frac{\partial q_l}{\partial u_j} = \delta_{kl} \ , \qquad k, l = 1, 2, 3 \ . \tag{10.16}$$

Using matrix notation one obtains

$$\begin{aligned} \left(\frac{\partial q_l}{\partial u_j}\right) &= 2\Lambda(\underline{u}) \ , \qquad \left(\frac{\partial p_k}{\partial U_j}\right) = \frac{1}{2|\underline{u}|^2}\Lambda(\underline{u}) \ , \\ \left(\frac{\partial p_k}{\partial u_j}\right) &= -\frac{1}{|\underline{u}|^4}\Lambda(\underline{u}) \begin{pmatrix} U_1 \\ U_2 \\ U_3 \\ U_4 \end{pmatrix} (u_1, u_2, u_3, u_4) + \frac{1}{2|\underline{u}|^2}\Lambda(\underline{U}) \end{aligned}$$

for $l, k = 1, 2, 3$, $j = 1, 2, 3, 4$. Therefore, due to the orthogonality of Λ, the relation (10.16) is equivalent to

$$\{p_k, q_l\} = \frac{1}{|\underline{u}|^2}[\Lambda(\underline{u})\Lambda^T(\underline{u})]_{kl} = \frac{1}{|\underline{u}|^2}(\underline{u}, \underline{u}) \ [I_4]_{kl} = \delta_{kl} \ ,$$

where $[A]_{kl}$ denotes the element at the crossing of the kth row and of the lth column of the matrix A and I_4 is the 4×4–dimensional identity matrix. To prove (c) we use the bilinear relation, which is valid along the solution due to (a) of the previous lemma. Let A, B be the matrices defined as

$$\begin{aligned} A &= \frac{1}{2}|\underline{u}|^2\Lambda(\underline{U})\Lambda^T(\underline{U}) \\ B &= \Lambda(\underline{U}) \begin{pmatrix} U_1 \\ U_2 \\ U_3 \\ U_4 \end{pmatrix} \left[\Lambda(\underline{U}) \begin{pmatrix} u_1 \\ u_2 \\ u_3 \\ u_4 \end{pmatrix}\right]^T \ , \end{aligned}$$

so that

$$A - B = \left(\sum_{j=1}^{4} \frac{\partial p_k}{\partial u_j} \frac{\partial p_l}{\partial U_j}\right) 2|\underline{u}|^6 \ .$$

The proof of the symmetry of the matrix $A - B$ is not trivial and we refer the reader to [167] for complete details. This latter result proves (c). □

10.3 The Birkhoff regularization

The problem of regularizing both singularities simultaneously, namely to find a transformation to regularize at once both collisions with the primaries, is the content of the *Birkhoff transformation.* Such *global* regularization is obtained as follows.

In the framework of the restricted, planar, circular, three–body problem, we consider a synodic reference frame with origin at the barycenter of the primaries. Let the primaries be $\mathcal{P}_1$, $\mathcal{P}_3$, while the small body is denoted by $\mathcal{P}_2$. We normalize to unity the sum of the masses of the primaries, which we denote as $\mu_1 = 1 - \mu$ and $\mu_3 = \mu$. Moreover we set to unity the distance between the primaries as well as the gravitational constant. Let (x_1, x_2) be the synodic coordinates of $\mathcal{P}_2$ and let (p_1, p_2) be the corresponding momenta. The Hamiltonian function is given by

$$\mathcal{H}(p_1, p_2, q_1, q_2) = \frac{1}{2}(p_1^2 + p_2^2) + q_2 p_1 - q_1 p_2 - \frac{\mu_1}{r_1} - \frac{\mu_3}{r_3} , \qquad (10.17)$$

where

$$r_1 = \sqrt{(q_1 + \mu_3)^2 + {q_2}^2} , \qquad r_3 = \sqrt{(q_1 - \mu_1)^2 + {q_2}^2} .$$

Birkhoff regularization transforms the physical plane into the parametric plane with the aim of removing the singularities in (10.17). To this end we shift the origin of the reference frame by choosing the midpoint between the primaries through a transformation (q_1, q_2) to (q_x, q_y) such that $q_x + iq_y = q_1 + iq_2 - \frac{1}{2} + \mu$. As a consequence $\mathcal{P}_1$ is located at $(-\frac{1}{2}, 0)$, while $\mathcal{P}_3$ is at $(\frac{1}{2}, 0)$. Denoting by (p_x, p_y) the corresponding momenta, the transformed Hamiltonian takes the form

$$\mathcal{H}_1(p_x, p_y, q_x, q_y) = \frac{1}{2}({p_x}^2 + {p_y}^2) + q_y p_x - \left(q_x + \frac{1}{2} - \mu\right) p_y - \frac{\mu_1}{r_1} - \frac{\mu_3}{r_3} ,$$

where

$$r_1 = \sqrt{\left(q_x + \frac{1}{2}\right)^2 + {q_y}^2} , \qquad r_3 = \sqrt{\left(q_x - \frac{1}{2}\right)^2 + {q_y}^2} .$$

Denoting by $q = q_x + iq_y$ and by ∇_q the gradient with respect to q, we can write the equations of motion in complex form as

$$\ddot{q} + 2i\dot{q} = \nabla_q U(q) , \qquad (10.18)$$

where $U(q) \equiv \frac{1}{2}[(q_x + \frac{1}{2} - \mu)^2 + q_y^2] + U_c(q)$, being $U_c(q) = \frac{\mu_1}{r_1} + \frac{\mu_3}{r_3}$ the *critical part* of U. Setting $w = w_1 + iw_2$, we define the regularizing transformation through the complex change of variables $q = h(w) = \alpha w + \frac{\beta}{w}$ for suitable constants α and β. We introduce also a time transformation through the expression

$$\frac{dt}{d\tau} = g(w) \equiv |k(w)|^2$$

for a suitable complex function $k = k(w)$. In order to write the transformed equations of motion, let us compute the first and second derivatives as

$$\begin{aligned} \dot{q} &= \frac{dh}{dw}\frac{dw}{d\tau}\frac{d\tau}{dt} = h'w'\dot{\tau} \\ \ddot{q} &= h'w'\ddot{\tau} + (h''w'^2 + h'w'')\dot{\tau}^2 , \end{aligned}$$

where the prime denotes the derivative with respect to the argument. The gradient operator is transformed as

$$\overline{h}' \nabla_q U = \nabla_w U ,$$

where $\overline{h}$ is the complex conjugate of h. Finally, the equation of motion (10.18) takes the form

$$w'' + 2i\frac{w'}{\dot\tau} + w'\frac{\ddot\tau}{\dot\tau^2} + w'^2\frac{h''}{h'} = \nabla_w U \frac{1}{|h'|^2\dot\tau^2} \,. \tag{10.19}$$

From the relations $\dot\tau = \frac{1}{g} = \frac{1}{k\bar k}$ and $\ddot\tau = -\frac{\dot g}{g^2}$, it follows that $\frac{\ddot\tau}{\dot\tau^2} = -\dot g$. Taking into account the expression

$$-\dot g = -\left(k\frac{\overline{dk}}{dw}\frac{\overline{dw}}{d\tau} + \bar k\frac{dk}{dw}\frac{dw}{d\tau}\right)\dot\tau = -\left(\frac{\overline{k'}\,\overline{w'}}{\bar k} + \frac{k'w'}{k}\right),$$

it follows that (10.19) takes the form

$$w'' + 2ik\bar k w' - \frac{|w'|^2}{\bar k}\bar k' + w'^2\left(\frac{h''}{h'} - \frac{k'}{k}\right) = \frac{|k|^4}{|h'|^2}\nabla_w U \,. \tag{10.20}$$

Setting $\tilde U(q) = U(q) - \frac{C_J}{2}$ where C_J denotes the Jacobi integral, by (4.15) one has

$$|\dot q|^2 = 2\tilde U(q) = |h'|^2|w'|^2\frac{1}{|k|^4} \,,$$

namely

$$|w'|^2 = 2\frac{|k|^4}{|h'|^2}\tilde U \,.$$

Using (10.20) and the relation $(\frac{h''}{h'} - \frac{k'}{k}) = \frac{d}{dw}(\log\frac{h'}{k})$, one obtains

$$w'' + 2ik\bar k w' = \frac{|k|^4}{|h'|^2}\left(2\tilde U\frac{d\log\bar k}{dw} + \nabla_w U\right) - |w'^2|\frac{d}{dw}\left(\log\frac{h'}{k}\right) .$$

Choosing $k = h'$, one obtains the equation

$$w'' + 2i|h'|^2 w' = \nabla_w\left(|h'|^2\tilde U\right) .$$

Next we determine α and β by requiring that h regularizes both singularities and by imposing that the transformation leaves $\mathcal{P}_1, \mathcal{P}_3$ invariant. Due to the expressions

$$\begin{aligned} U_c(w) &= \frac{\mu_1}{r_1} + \frac{\mu_3}{r_3} = \frac{\mu_1}{|\alpha w + \frac{\beta}{w} + \frac{1}{2}|} + \frac{\mu_3}{|\alpha w + \frac{\beta}{w} - \frac{1}{2}|} \\ |h'(w)|^2 &= \frac{|\alpha w^2 - \beta|^2}{|w|^4} \,, \end{aligned}$$

it follows that

$$U_c(w)|h'(w)|^2 = \frac{1}{|w|^3}\left(\frac{\mu_1|\alpha w^2 - \beta|^2}{|\alpha w^2 + \beta + \frac{w}{2}|} + \frac{\mu_3|\alpha w^2 - \beta|^2}{|\alpha w^2 + \beta - \frac{w}{2}|}\right) .$$

The singularity at $q = \frac{1}{2}$ corresponds to the roots of $\alpha w^2 + \beta - \frac{1}{2}w = 0$, which give the positions $w_{1,2} = \frac{1}{4\alpha}(1 \pm (1 - 16\alpha\beta)^{\frac{1}{2}})$. Since the root of the numerator is $\pm(\frac{\beta}{\alpha})^{1/2}$, we require that

$$\frac{1}{4\alpha}(1 \pm (1 - 16\alpha\beta)^{\frac{1}{2}}) = \pm\left(\frac{\beta}{\alpha}\right)^{\frac{1}{2}} . \tag{10.21}$$

Equation (10.21) is satisfied provided that $16\alpha\beta = 1$, which gives $w_{1,2} = \frac{1}{4\alpha}$. If we require that $\mathcal{P}_3$ is located at $(\frac{1}{2}, 0)$, then we find $\alpha = \frac{1}{2}$ and $\beta = \frac{1}{8}$. The singularity at $\mathcal{P}_1$ of the first term of $U_c(w)$ can be removed by applying the same procedure, which provides the same values for α and β. The term $\frac{1}{|w|^3}$ is singular for $w = 0$, which corresponds to the non–realistic solution of q tending to infinity. In summary, the Birkhoff transformation is given by $q = \frac{1}{2}(w + \frac{1}{4w})$, which is equivalent to

$$
\begin{aligned}
q_1 &= \frac{1}{2}\left(w_1 + \frac{w_1}{4(w_1^2 + w_2^2)}\right) \\
q_2 &= \frac{1}{2}\left(w_2 - \frac{w_2}{4(w_1^2 + w_2^2)}\right) .
\end{aligned}
$$

10.3.1 The B_3 regularization

Consider the restricted, circular, spatial three–body problem, where it is assumed that the primaries $\mathcal{P}_1$, $\mathcal{P}_3$ move on circular orbits on the same plane, while the small body $\mathcal{P}_2$ is allowed to move in the space. The B_3 *regularization* allows us to remove in the spatial case both singularities with the primaries. We introduce a synodic reference frame rotating with the angular velocity of the primaries. Let (q_1, q_2, q_3) be the coordinates of $\mathcal{P}_2$. The equations of motion are given by

$$
\begin{aligned}
\ddot{q}_1 - 2\dot{q}_2 &= \tilde{\Omega}_{q_1} \\
\ddot{q}_2 + 2\dot{q}_1 &= \tilde{\Omega}_{q_2} \\
\ddot{q}_3 + q_3 &= \tilde{\Omega}_{q_3}
\end{aligned}
$$

with

$$
\tilde{\Omega}(q_1, q_2, q_3) = \frac{1}{2}(q_1^2 + q_2^2 + q_3^2) + \frac{\mu_1}{r_1} + \frac{\mu_3}{r_3}
$$

and $r_1 = \sqrt{(q_1 + \mu_3)^2 + q_2^2 + q_3^2}$, $r_3 = \sqrt{(q_1 - \mu_1)^2 + q_2^2 + q_3^2}$. In the spatial case, the Hamiltonian in the synodic reference frame is given by

$$
\mathcal{H}(p_1, p_2, p_3, q_1, q_2, q_3) = \frac{1}{2}(p_1^2 + p_2^2 + p_3^2) + q_2 p_1 - q_1 p_2 - \frac{\mu_1}{r_1} - \frac{\mu_3}{r_3} .
$$

The Birkhoff regularizing transformation in the spatial case can be obtained as follows. Let us define the coordinate's transformation $q_1 = \frac{1}{2}\tilde{q}_1 + (\frac{1}{2} - \mu)$, $q_2 = \tilde{q}_2$, $q_3 = \tilde{q}_3$, so that the primaries are located at $(-1, 0, 0)$ and $(1, 0, 0)$. Then, we implement a transformation by reciprocal radii with center at $(1, 0, 0)$ and radius equal to $\sqrt{2}$:

$$
\begin{aligned}
\hat{q}_1 &= 1 + \frac{2(\tilde{q}_1 - 1)}{(\tilde{q}_1 - 1)^2 + \tilde{q}_2^2 + \tilde{q}_3^2} \\
\hat{q}_2 &= \frac{2\tilde{q}_2}{(\tilde{q}_1 - 1)^2 + \tilde{q}_2^2 + \tilde{q}_3^2} \\
\hat{q}_3 &= \frac{2\tilde{q}_3}{(\tilde{q}_1 - 1)^2 + \tilde{q}_2^2 + \tilde{q}_3^2} .
\end{aligned}
$$

This transformation places $\mathcal{P}_1$ at the origin, while $\mathcal{P}_3$ is ejected at infinity. A KS–transformation is implemented to regularize both singularities, at the origin and at infinity. To this end, a four–dimensional parametric space is introduced with coordinates (u_1, u_2, u_3, u_4) such that

$$\begin{aligned}\hat{q}_1 &= u_1^2 - u_2^2 - u_3^2 + u_4^2 \\ \hat{q}_2 &= 2(u_1 u_2 - u_3 u_4) \\ \hat{q}_3 &= 2(u_1 u_3 + u_2 u_4) \ .\end{aligned}$$

A new transformation by reciprocal radii is implemented, which brings $\mathcal{P}_1$ at $(-1, 0, 0)$ and $\mathcal{P}_3$ at $(1, 0, 0)$:

$$\begin{aligned}u_1 &= 1 + \frac{2(v_1 - 1)}{(v_1 - 1)^2 + v_2^2 + v_3^2 + v_4^2} \\ u_k &= \frac{2v_k}{(v_1 - 1)^2 + v_2^2 + v_3^2 + v_4^2} \ , \qquad\qquad k = 2, 3, 4 \ .\end{aligned}$$

Finally we place $\mathcal{P}_1$ and $\mathcal{P}_3$ at $(-\frac{1}{2}, 0, 0)$ and $(\frac{1}{2}, 0, 0)$, respectively. The relation between the physical plane (q_1, q_2, q_3) and the parametric plane (v_1, v_2, v_3, v_4) is given by

$$\begin{aligned}q_1 &= \frac{1}{2} - \mu + \frac{1}{2}\left[v_1 + \frac{v_1(v_4^2 + \frac{1}{4})}{v_1^2 + v_2^2 + v_3^2}\right] \\ q_2 &= \frac{1}{2}\left[v_2 + \frac{v_2(v_4^2 - \frac{1}{4}) - v_3 v_4}{v_1^2 + v_2^2 + v_3^2}\right] \\ q_3 &= \frac{1}{2}\left[v_3 + \frac{v_3(v_4^2 - \frac{1}{4}) + v_2 v_4}{v_1^2 + v_2^2 + v_3^2}\right] \ .\end{aligned}$$

The fictitious time s is defined through $dt = Dds$, where

$$\begin{aligned}D &= \frac{r_1 r_3}{v_1^2 + v_2^2 + v_3^2} \\ r_1 &= \frac{1}{2}\frac{(v_1 + \frac{1}{2})^2 + v_2^2 + v_3^2 + v_4^2}{\sqrt{v_1^2 + v_2^2 + v_3^2}} \ , \qquad r_3 = \frac{1}{2}\frac{(v_1 - \frac{1}{2})^2 + v_2^2 + v_3^2 + v_4^2}{\sqrt{v_1^2 + v_2^2 + v_3^2}} \ .\end{aligned}$$

The equations of motion with respect to the fictitious time are given by

$$\begin{aligned}Dq_1'' - D'q_1' - 2D^2 q_2' &= D^3 \tilde{\Omega}_{q_1} \\ Dq_2'' - D'q_2' + 2D^2 q_1' &= D^3 \tilde{\Omega}_{q_2} \\ Dq_3'' - D'q_3' &= D^3 \tilde{\Omega}_{q_3} \ . \end{aligned} \tag{10.22}$$

The singularities appearing in (10.22) through the functions $\tilde{\Omega}_{q_1}$, $\tilde{\Omega}_{q_2}$, $\tilde{\Omega}_{q_3}$ are due to the terms $\frac{1}{r_1^3}$ and $\frac{1}{r_3^3}$. Since D^3 is proportional to $r_1^3 r_3^3$, the contributions of $D^3\tilde{\Omega}_{q_1}$, $D^3\tilde{\Omega}_{q_2}$, $D^3\tilde{\Omega}_{q_3}$ do not contain singularities at the attracting centers. We conclude by remarking that the origin is still a singular point, but it corresponds to place $\mathcal{P}_3$ at infinity.

A Basics of Hamiltonian dynamics

A.1 The Hamiltonian setting

Consider a mechanical system with n degrees of freedom[1]; let $T = T(\dot{\underline{q}})$ be the kinetic energy and let $V = V(\underline{q})$ be the potential energy, where $\dot{\underline{q}} \in \mathbf{R}^n$, $\underline{q} \in \mathbf{R}^n$. The corresponding Lagrangian function is defined as

$$\mathcal{L}(\dot{\underline{q}}, \underline{q}) \equiv T(\dot{\underline{q}}) - V(\underline{q}) \ .$$

Let us introduce the *momenta* conjugated to the coordinates through the expression

$$\underline{p} \equiv \frac{\partial \mathcal{L}(\dot{\underline{q}}, \underline{q})}{\partial \dot{\underline{q}}} \ . \tag{A.1}$$

From the Lagrange equations

$$\frac{d}{dt} \frac{\partial \mathcal{L}(\dot{\underline{q}}, \underline{q})}{\partial \dot{\underline{q}}} = \frac{\partial \mathcal{L}(\dot{\underline{q}}, \underline{q})}{\partial \underline{q}} \ ,$$

one obtains

$$\dot{\underline{p}} = \frac{\partial \mathcal{L}(\dot{\underline{q}}, \underline{q})}{\partial \underline{q}} \ .$$

Therefore it follows that

$$d\mathcal{L} = \frac{\partial \mathcal{L}}{\partial \dot{\underline{q}}} d\dot{\underline{q}} + \frac{\partial \mathcal{L}}{\partial \underline{q}} d\underline{q} = \underline{p} d\dot{\underline{q}} + \dot{\underline{p}} d\underline{q} = d(\underline{p}\ \dot{\underline{q}}) - \dot{\underline{q}} d\underline{p} + \dot{\underline{p}} d\underline{q} \ ,$$

namely

$$d(\underline{p}\ \dot{\underline{q}} - \mathcal{L}) = -\dot{\underline{p}} d\underline{q} + \dot{\underline{q}} d\underline{p} \ . \tag{A.2}$$

Let us introduce the *Hamiltonian function* as

$$\mathcal{H}(\underline{p}, \underline{q}) \equiv \underline{p}\ \dot{\underline{q}} - \mathcal{L}(\dot{\underline{q}}, \underline{q}) \ ,$$

where the variables $\dot{\underline{q}}$ are intended to be expressed in terms of $\underline{p}$ and $\underline{q}$, by inverting the relation (A.1). From (A.2) one obtains:

$$d\mathcal{H}(\underline{p}, \underline{q}) = -\dot{\underline{p}} d\underline{q} + \dot{\underline{q}} d\underline{p} \ ;$$

[1] The *number of degrees of freedom* is the minimum number of independent coordinates which are necessary to describe the mechanical system.

being

$$d\mathcal{H}(\underline{p},\underline{q}) = \frac{\partial \mathcal{H}}{\partial \underline{p}} d\underline{p} + \frac{\partial \mathcal{H}}{\partial \underline{q}} d\underline{q} ,$$

one finds the equations

$$\begin{aligned} \dot{\underline{q}} &= \frac{\partial \mathcal{H}(\underline{p},\underline{q})}{\partial \underline{p}} \\ \dot{\underline{p}} &= -\frac{\partial \mathcal{H}(\underline{p},\underline{q})}{\partial \underline{q}} . \end{aligned} \tag{A.3}$$

The equations (A.3) are called *Hamilton's equations.* While in the Lagrangian case one needs to solve a differential equation of the second order, in the Hamiltonian setting one is led to find the solution of two differential equations of the first order. In terms of the components of the vectors $\underline{p}$ and $\underline{q}$, Hamilton's equations are written as

$$\begin{aligned} \dot{q}_i &= \frac{\partial \mathcal{H}(\underline{p},\underline{q})}{\partial p_i} \\ \dot{p}_i &= -\frac{\partial \mathcal{H}(\underline{p},\underline{q})}{\partial q_i} , \qquad i = 1, \dots, n . \end{aligned}$$

Example. Given the Lagrangian function

$$\mathcal{L}(\dot{q}, q) = \frac{1}{2}\dot{q}^2 + q\dot{q} + 3q^2 ,$$

the corresponding Hamiltonian function and the solution of Hamilton's equations are found as follows.

The momentum conjugated to q is

$$p = \frac{\partial \mathcal{L}}{\partial \dot{q}} = \dot{q} + q ,$$

which yields

$$\dot{q} = p - q .$$

Therefore

$$\begin{aligned} \mathcal{H}(p,q) &= p\dot{q} - \mathcal{L} \\ &= \frac{1}{2}p^2 - pq - \frac{5}{2}q^2 . \end{aligned}$$

The corresponding Hamilton's equations are

$$\begin{aligned} \dot{p} &= -\frac{\partial \mathcal{H}}{\partial q} = p + 5q \\ \dot{q} &= \frac{\partial \mathcal{H}}{\partial p} = p - q . \end{aligned}$$

Differentiating the second equation with respect to time one has

$$\ddot{q} = \dot{p} - \dot{q} = 6q \ ,$$

namely

$$\ddot{q} - 6q \ = \ 0 \ ,$$

whose solution is given by

$$q(t) = A_1 e^{\sqrt{6}t} + A_2 e^{-\sqrt{6}t} \ ,$$

where A_1 and A_2 are arbitrary constants depending on the initial data. From $p = q + \dot{q}$ one finds the solution for the momentum:

$$p(t) = \left(A_1 + \sqrt{6}A_1\right) e^{\sqrt{6}t} + \left(A_2 - \sqrt{6}A_2\right) e^{-\sqrt{6}t} \ .$$

A.2 Canonical transformations

Given the Hamiltonian $\mathcal{H} = \mathcal{H}(\underline{p}, \underline{q})$ with n degrees of freedom ($\underline{p} \in \mathbf{R}^n$, $\underline{q} \in \mathbf{R}^n$), let us consider the coordinate transformation

$$\begin{aligned} \underline{P} &= \underline{P}(\underline{p}, \underline{q}) \\ \underline{Q} &= \underline{Q}(\underline{p}, \underline{q}) \ , \end{aligned} \tag{A.4}$$

where $\underline{P} \in \mathbf{R}^n$, $\underline{Q} \in \mathbf{R}^n$. The coordinate change (A.4) is said to be *canonical,* if the equations of motion in the variables $(\underline{P}, \underline{Q})$ keep the Hamiltonian structure, namely the transformed variables satisfy Hamilton's equations with respect to a new Hamiltonian, say $\mathcal{H}_1 = \mathcal{H}_1(\underline{P}, \underline{Q})$.

Let us derive the conditions under which the transformation (A.4) is canonical. We introduce the notation

$$\underline{x} = \begin{pmatrix} \underline{q} \\ \underline{p} \end{pmatrix} \ , \qquad \underline{z} = \begin{pmatrix} \underline{Q} \\ \underline{P} \end{pmatrix}$$

and let $\underline{z} = \underline{z}(\underline{x})$ be the transformation (A.4). Set

$$J \equiv \begin{pmatrix} 0 & I_n \\ -I_n & 0 \end{pmatrix} \ ,$$

where I_n is the n–dimensional identity matrix; Hamilton's equations (A.3) can be written as

$$\dot{\underline{x}} = J \frac{\partial \mathcal{H}(\underline{x})}{\partial \underline{x}} \ .$$

Let $M = \frac{\partial \underline{z}}{\partial \underline{x}}$; then, the transformed equations are

$$\dot{\underline{z}} = \frac{\partial \underline{z}}{\partial \underline{x}} \dot{\underline{x}} = M \dot{\underline{x}} = MJ \frac{\partial \mathcal{H}(\underline{x})}{\partial \underline{x}} = MJ \frac{\partial \mathcal{H}(\underline{x})}{\partial \underline{z}} \frac{\partial \underline{z}}{\partial \underline{x}} = MJM^T \ \frac{\partial \mathcal{H}(\underline{x})}{\partial \underline{z}} \ .$$

Therefore, the canonicity condition is equivalent to require that the following relation is satisfied:

$$MJM^T = J ; \tag{A.5}$$

equation (A.5) implies that the matrix M is symplectic. In that case Hamilton's equations with respect to the new variables $\underline{z}$ hold, whenever the new Hamiltonian is defined through the expression $\mathcal{H}_1(\underline{z}) = \mathcal{H}(\underline{x}(\underline{z}))$.

A canonicity criterion is obtained through the introduction of the so–called *Poisson brackets.* Let us consider the functions $f = f(\underline{p}, \underline{q})$, $g = g(\underline{p}, \underline{q})$; the *Poisson bracket* between f and g is defined as the quantity

$$\{f, g\} = \sum_{k=1}^{n} \frac{\partial f}{\partial q_k}\frac{\partial g}{\partial p_k} - \frac{\partial f}{\partial p_k}\frac{\partial g}{\partial q_k} .$$

A direct computation shows that the condition (A.5) is equivalent to the following condition [5]. The transformation (A.4) is canonical if the following relations hold:

$$\{Q_i, Q_j\} = \{P_i, P_j\} = 0 , \qquad \{Q_i, P_j\} = \delta_{ij} , \qquad i, j = 1, \ldots, n .$$

In the one–dimensional case ($n = 1$) it suffices to verify that

$$\{Q, P\} = 1 ,$$

since $\{Q, Q\}$ and $\{P, P\}$ are identically zero.

Let us introduce the *generating function* of a canonical transformation as follows. Consider the general case of a time–dependent canonical transformation

$$\begin{aligned} \underline{Q} &= \underline{Q}(\underline{q}, \underline{p}, t) \\ \underline{P} &= \underline{P}(\underline{q}, \underline{p}, t) . \end{aligned} \tag{A.6}$$

The generating function, that we extend to encompass the time–dependent transformation (A.6), is a function of the form

$$F = F(\underline{q}, \underline{Q}, t) ,$$

such that the following transformation rules hold:

$$\begin{aligned} \underline{p} &= \frac{\partial F}{\partial \underline{q}} \\ \underline{P} &= -\frac{\partial F}{\partial \underline{Q}} . \end{aligned}$$

If $\mathcal{H}_1 = \mathcal{H}_1(\underline{P}, \underline{Q}, t)$ is the Hamiltonian in the new set of variables, then

$$\mathcal{H}_1(\underline{P}, \underline{Q}, t) = \mathcal{H}(\underline{p}, \underline{q}, t) + \frac{\partial F}{\partial t} .$$

Equivalent forms of the generating functions are the following:
(i) $F = F(\underline{q}, \underline{P}, t)$ with transformation rules:

$$\begin{aligned} \underline{p} &= \frac{\partial F}{\partial \underline{q}} \\ \underline{Q} &= \frac{\partial F}{\partial \underline{P}} \ ; \end{aligned}$$

(ii) $F = F(\underline{p}, \underline{Q}, t)$ with transformation rules:

$$\begin{aligned} \underline{q} &= -\frac{\partial F}{\partial \underline{p}} \\ \underline{P} &= -\frac{\partial F}{\partial \underline{Q}} \ ; \end{aligned} \tag{A.7}$$

(iii) $F = F(\underline{p}, \underline{P}, t)$ with transformation rules:

$$\begin{aligned} \underline{q} &= -\frac{\partial F}{\partial \underline{p}} \\ \underline{Q} &= \frac{\partial F}{\partial \underline{P}} \ . \end{aligned}$$

Example. Let us compute the values of α and β for which the following transformation is canonical:

$$\begin{aligned} P &= \alpha p e^{\beta q} \\ Q &= \frac{1}{\alpha} e^{-\beta q} \ ; \end{aligned}$$

for such values we proceed to find the corresponding generating function.

Let us use the Poisson brackets to check the canonicity of the transformation; in particular, since the change of variables is one–dimensional, it suffices to verify that

$$\{Q, P\} \equiv \frac{\partial Q}{\partial q}\frac{\partial P}{\partial p} - \frac{\partial Q}{\partial p}\frac{\partial P}{\partial q} = 1 \ .$$

Therefore one has:

$$-\frac{\beta}{\alpha} e^{-\beta q} \cdot \alpha e^{\beta q} = 1 \ ,$$

which is satisfied for $\beta = -1$ and for any $\alpha \neq 0$. In this case the transformation becomes:

$$\begin{aligned} P &= \alpha p e^{-q} \\ Q &= \frac{1}{\alpha} e^{q} \ . \end{aligned} \tag{A.8}$$

Let us look for a generating function $F = F(q, P)$, whose transformation rules are given by

$$p = \frac{\partial F}{\partial q}$$
$$Q = \frac{\partial F}{\partial P} \; .$$

From (A.8) one has:

$$p = \frac{P}{\alpha} e^q$$
$$Q = \frac{1}{\alpha} e^q \; .$$

Therefore it should be

$$\frac{\partial F}{\partial q} = \frac{P}{\alpha} e^q \; , \tag{A.9}$$

namely $F(q, P) = \frac{P}{\alpha} e^q + f(P)$, where $f(P)$ is a total function of P; analogously, from the relation

$$\frac{\partial F}{\partial P} = \frac{1}{\alpha} e^q \; , \tag{A.10}$$

one finds $F(q, P) = \frac{P}{\alpha} e^q + g(q)$, where $g(q)$ depends only on the variable q. Comparing the solutions of (A.9) and (A.10) one obtains $f(P) = g(q) = 0$, thus yielding

$$F(q, P) = \frac{P}{\alpha} e^q \; .$$

A.3 Integrable systems

A Hamiltonian system with n degrees of freedom is said to be *integrable* if there exist n integrals, say $U_1, \ldots, U_n$, which satisfy the following assumptions:

(1) the integrals are in involution: $\{U_j, U_k\} = 0$ for any $j, k = 1, \ldots, n$;

(2) the integrals are independent, which implies that the matrix

$$\begin{pmatrix} \frac{\partial U_1}{\partial p_1} & \ldots & \frac{\partial U_1}{\partial p_n} & \frac{\partial U_1}{\partial q_1} & \ldots & \frac{\partial U_1}{\partial q_n} \\ \vdots & & & & & \\ \frac{\partial U_n}{\partial p_1} & \ldots & \frac{\partial U_n}{\partial p_n} & \frac{\partial U_n}{\partial q_1} & \ldots & \frac{\partial U_n}{\partial q_n} \end{pmatrix}$$

has rank n;

(3) in place of (2) one can require the non–singularity condition:

$$\det \begin{pmatrix} \frac{\partial U_1}{\partial p_1} & \ldots & \frac{\partial U_1}{\partial p_n} \\ \vdots & & \\ \frac{\partial U_n}{\partial p_1} & \ldots & \frac{\partial U_n}{\partial p_n} \end{pmatrix} \neq 0 \; ;$$

notice that this condition is stronger than the independence of item (2).

Having fixed a point $(\underline{p}_0, \underline{q}_0)$, let $\underline{\alpha}_0 = \underline{U}(\underline{p}_0, \underline{q}_0)$, where $\underline{U} \equiv (U_1, \ldots, U_n)$; for $\underline{\alpha} \in \mathbf{R}^n$ define the manifold $M_{\underline{\alpha}}$ as

$$M_{\underline{\alpha}} = \{(\underline{p}, \underline{q}) \in \mathbf{R}^{2n} \ : \ U_1(\underline{p}, \underline{q}) = \alpha_1, \ldots, U_n(\underline{p}, \underline{q}) = \alpha_n\} \ .$$

The integrability of a Hamiltonian system can be obtained through the following theorem [5].

Liouville–Arnold theorem. *Suppose that the Hamiltonian* $\mathcal{H}(\underline{p}, \underline{q})$, $\underline{p}$, $\underline{q} \in \mathbf{R}^n$, *admits* n *integrals* $U_1, \ldots, U_n$, *satisfying the above conditions of involution and non–singularity. Assume that the manifold* $M_{\underline{\alpha}}$ *is compact in a suitable neighborhood of* $\underline{\alpha}_0$. *Then, there exists a transformation of coordinates from* $(\underline{p}, \underline{q})$ *to* $(\underline{I}, \underline{\varphi})$ *with* $\underline{I} \in \mathbf{R}^n$, $\underline{\varphi} \in \mathbf{T}^n$, *such that the new Hamiltonian* $\mathcal{H}_1$ *takes the form*

$$\mathcal{H}_1(\underline{I}, \underline{\varphi}) \ \equiv \ h(\underline{I}) \ ,$$

for a suitable function $h = h(\underline{I})$.

A.4 Action–angle variables

Consider the mechanical system described by the Hamiltonian $\mathcal{H}(\underline{p}, \underline{q})$, where $\underline{p} \in \mathbf{R}^n$, $\underline{q} \in \mathbf{R}^n$. When dealing with integrable systems one can introduce a canonical transformation $\mathcal{C} : (\underline{p}, \underline{q}) \in \mathbf{R}^{2n} \to (\underline{I}, \underline{\varphi}) \in \mathbf{R}^n \times \mathbf{T}^n$, such that the transformed Hamiltonian depends only on the action variables $\underline{I}$:

$$\mathcal{H} \circ \mathcal{C}(\underline{I}, \underline{\varphi}) = h(\underline{I}) \ ,$$

for some function $h = h(\underline{I})$. The coordinates $(\underline{I}, \underline{\varphi})$ are known as *action–angle variables* [5].

The Liouville–Arnold theorem provides an explicit algorithm to construct the action–angle variables. In particular, let us introduce as transformed momenta the actions $(I_1, \ldots, I_n)$ defined through the relation

$$I_i \ = \ \oint p_i \, dq_i \ ,$$

where the integral is computed over a full cycle of motion. If the initial Hamiltonian is completely integrable, it will depend only on the action variables, say

$$h \ = h(I_1, \ldots, I_n) \ . \tag{A.11}$$

The canonical variables conjugated to I_i are named *angle variables;* they will be denoted as $(\varphi_1, \ldots, \varphi_n)$. One immediately recognizes that Hamilton's equations associated to (A.11) are integrable. Indeed, let us define the *frequency* or *rotation vector* as

$$\underline{\omega} = \underline{\omega}(\underline{I}) = \frac{\partial h(\underline{I})}{\partial \underline{I}} \ ;$$

then, one has:

$$\begin{aligned}\dot{\underline{I}} &= \frac{\partial h(\underline{I})}{\partial \underline{\varphi}} = \underline{0} \\ \dot{\underline{\varphi}} &= \frac{\partial h(\underline{I})}{\partial \underline{I}} = \underline{\omega}(\underline{I}) \ .\end{aligned}$$

Therefore the *action* $\underline{I}$ is constant along the motion, namely $\underline{I} = \underline{I}_0$, while the second equation provides the variation of the *angle* $\underline{\varphi}$ as a function of the time: $\underline{\varphi} = \underline{\omega}(\underline{I}_0)t + \underline{\varphi}_0$, where $(\underline{I}_0, \underline{\varphi}_0)$ denote the initial conditions.

In action–angle variables a system is *nearly–integrable* if it can be written as

$$\mathcal{H}(\underline{I}, \underline{\varphi}) = h(\underline{I}) + \varepsilon f(\underline{I}, \underline{\varphi}) \ ,$$

where $h(\underline{I})$ represents the integrable part, $\varepsilon f(\underline{I}, \underline{\varphi})$ is the perturbing function and ε is the perturbing parameter measuring a small variation from the integrable case.

In the three–body problem, the integrable part coincides with the Keplerian two–body interaction, while the perturbing function provides the gravitational attraction with the third body and the perturbing parameter is the mass ratio of the primaries.

Example. Let us determine the action–angle variables for the harmonic oscillator described by the Hamiltonian

$$\mathcal{H}(p, q) = \frac{1}{2m}(p^2 + \omega^2 q^2) \ .$$

Setting $\mathcal{H}(p, q) = E$, one has

$$p^2 = 2mE - \omega^2 q^2$$

and the corresponding action variable is:

$$I = \oint p dq = \oint \sqrt{2mE - \omega^2 q^2} \ dq \ .$$

Let $q = \sqrt{\frac{2mE}{\omega^2}} \sin \vartheta$; then, one has:

$$\begin{aligned}I &= \int_0^{2\pi} \sqrt{2mE - 2mE \sin^2 \vartheta} \ \sqrt{\frac{2mE}{\omega^2}} \cos \vartheta \ d\vartheta \\ &= \frac{2mE}{\omega} \int_0^{2\pi} \cos^2 \vartheta d\vartheta = \frac{2\pi mE}{\omega} \ .\end{aligned}$$

The Hamiltonian in action–angle variables becomes:

$$E = \mathcal{H}(I) = \frac{\omega}{2\pi m} I \ .$$

The associated Hamilton's equations are

$$\begin{aligned} \dot{I} &= 0 \\ \dot{\varphi} &= \frac{\omega}{2\pi m} , \end{aligned}$$

whose solution is found to be

$$\begin{aligned} I(t) &= I(0) \\ \varphi(t) &= \frac{\omega}{2\pi m} t + \varphi(0) . \end{aligned}$$

B The sphere of influence

Assume that a small body $\mathcal{P}_2$ of negligible mass moves under the influence of two point–mass bodies $\mathcal{P}_1$ and $\mathcal{P}_3$ with masses, respectively, m_1 and m_3. An estimate of the region where the motion is dominated by the influence of $\mathcal{P}_1$ or that of $\mathcal{P}_3$, namely the respective *sphere of influence,* is obtained as follows.

Let $\underline{d}_1$ be the distance vector from $\mathcal{P}_2$ to $\mathcal{P}_1$, $\underline{d}_3$ that joining $\mathcal{P}_2$ and $\mathcal{P}_3$, and let $\underline{d}$ be the distance vector between $\mathcal{P}_1$ and $\mathcal{P}_3$. Let d_1, d_3, d be their norms; if $\mathcal{P}_2$ is close to $\mathcal{P}_3$, we can assume that $\frac{d_3}{d}$ is small, so that $d_1 \simeq d$. The motion of $\mathcal{P}_2$ is governed by the equations:

$$\ddot{\underline{d}}_1 = -\frac{\mathcal{G}m_1}{d_1^3}\underline{d}_1 - \frac{\mathcal{G}m_3}{d_3^3}\underline{d}_3$$

and the ratio of the gravitational attraction exerted by $\mathcal{P}_3$ to that exerted by $\mathcal{P}_1$ is given by the term

$$\frac{m_3}{m_1}\left(\frac{d}{d_3}\right)^2 .$$

The motion of $\mathcal{P}_3$ is ruled by the equation

$$\ddot{\underline{d}} = -\frac{\mathcal{G}m_1}{d^3}\underline{d} .$$

Therefore, one obtains:

$$\begin{aligned}\ddot{\underline{d}}_3 &= \ddot{\underline{d}}_1 - \ddot{\underline{d}} = -\frac{\mathcal{G}m_3}{d_3^3}\underline{d}_3 - \frac{\mathcal{G}m_1}{d^3}(\underline{d}_1 - \underline{d}) \\ &= -\frac{\mathcal{G}m_3}{d_3^3}\underline{d}_3 - \frac{\mathcal{G}m_1}{d^3}\underline{d}_3 .\end{aligned}$$

As a consequence, the ratio of the gravitational attraction exerted by $\mathcal{P}_1$ to that exerted by $\mathcal{P}_3$ is given by the term

$$\frac{m_1}{m_3}\left(\frac{d_3}{d}\right)^3 .$$

Finally, we can conclude that the influence of $\mathcal{P}_3$ is dominant whenever

$$\frac{m_1}{m_3}\left(\frac{d_3}{d}\right)^3 < \frac{m_3}{m_1}\left(\frac{d}{d_3}\right)^2 ,$$

which leads to define the sphere of influence around $\mathcal{P}_3$ of radius r given by

$$r \equiv \left(\frac{m_3}{m_1}\right)^{\frac{2}{5}} d \ .$$

As a concrete example, we identify $\mathcal{P}_1$ with the Sun and $\mathcal{P}_3$ with the Earth. Then, $m_1 = 2 \cdot 10^{30}$ kg, $m_3 = 5.97 \cdot 10^{24}$ kg, $d = 1.5 \cdot 10^8$ km, so that the Earth's sphere of influence has a radius approximately equal to $r = 9.25 \cdot 10^5$ km, which amounts to about 2.4 times the Earth–Moon distance. As for Jupiter, taking $m_3 = 1.9 \cdot 10^{27}$ kg and $d = 7.8 \cdot 10^8$ km, one finds that $r = 4.82 \cdot 10^7$ km, amounting to more than 125 times the Earth–Moon distance.

C Expansion of the perturbing function

Let $\mathcal{P}_1$, $\mathcal{P}_2$, $\mathcal{P}_3$ be three point–mass bodies with masses, respectively, m_1, m_2, m_3. In the framework of the three–body problem, we investigate the motion of $\mathcal{P}_2$ under the influence of $\mathcal{P}_1$ and $\mathcal{P}_3$. Let $\underline{r}_2$, $\underline{r}_3$ be the radius vectors associated to $\mathcal{P}_1\mathcal{P}_2$ and to $\mathcal{P}_1\mathcal{P}_3$, and let r_2, r_3 be their sizes with $r_2 < r_3$. Let ψ be the angle formed by $\underline{r}_2$ and $\underline{r}_3$. Denote by f_2, f_3 the true anomalies of the osculating orbits of $\mathcal{P}_2$, $\mathcal{P}_3$ around $\mathcal{P}_1$ and let g_2, g_3 be their arguments of perihelion. Let $\vartheta_2 = f_2 + g_2$, $\vartheta_3 = f_3 + g_3$ and define $\gamma = \cos\psi - \cos(\vartheta_2 - \vartheta_3)$. If ε is the m_1/m_3 mass ratio, then the perturbing function associated to $\mathcal{P}_2$ can be written as

$$R = -\varepsilon \frac{\underline{r}_1 \underline{r}_3}{r_3^3} + \frac{\varepsilon}{|\underline{r}_1 - \underline{r}_3|} = -\varepsilon \frac{r_1}{r_3^2} \cos\psi + \frac{\varepsilon}{a_3} R' \ ,$$

where we denote by a_2, a_3 the semimajor axes of $\mathcal{P}_2$, $\mathcal{P}_3$ and we expand R' as (see [61])

$$R' = \sum_{i=0}^{\infty} \frac{(2i)!}{(i!)^2} \left(\frac{1}{2} \frac{r_2}{a_2} \frac{r_3}{a_3} \gamma \right)^i \frac{a_2^i a_3^{i+1}}{2} \sum_{j=-\infty}^{\infty} \left[\sum_{\ell=0}^{\infty} \frac{1}{\ell!} \sum_{k=0}^{\ell} \binom{\ell}{k} \left(\frac{r_2}{a_2} - 1 \right)^k \left(\frac{r_3}{a_3} - 1 \right)^{\ell-k} \Gamma_{i,j,k,\ell-k} \cos((\vartheta_2 - \vartheta_3) j) \right] ,$$

with

$$\Gamma_{i,j,m,n} \equiv a_2^m a_3^n \frac{\partial^{m+n}}{\partial a_2^m \partial a_3^n} \left(a_3^{-2i-1} b_{i+\frac{1}{2}}^{(j)} \left(\frac{a_2}{a_3} \right) \right) ,$$

where the Laplace coefficients $b_s^{(j)}$ are defined as

$$\frac{1}{2} b_s^{(j)} \left(\frac{a_2}{a_3}\right) = \frac{1}{2\pi} \int_0^{2\pi} \frac{\cos(j\varphi)}{(1 - 2\frac{a_2}{a_3} \cos\varphi + (\frac{a_2}{a_3})^2)^s} \, d\varphi \ .$$

Notice that the Laplace coefficients can be expanded in power series of $\frac{a_2}{a_3}$ as

$$\begin{aligned} \frac{1}{2} b_s^{(j)} \left(\frac{a_2}{a_3} \right) &= \frac{s(s+1)\ldots(s+j-1)}{1 \cdot 2 \cdot 3 \cdot \cdots \cdot j} \left(\frac{a_2}{a_3} \right)^j \\ &\cdot \left[1 + \frac{s(s+j)}{j+1} \left(\frac{a_2}{a_3} \right)^2 + \frac{s(s+1)(s+j)(s+j+1)}{2(j+1)(j+2)} \left(\frac{a_2}{a_3} \right)^4 + \ldots \right] . \end{aligned}$$

D Floquet theory and Lyapunov exponents

Floquet theory provides a tool for investigating the regular or chaotic behavior of a dynamical system and it allows us to introduce the *Lyapunov exponents.* Referring to the specialized literature for an exhaustive treatment of the topic, we briefly introduce as follows the elementary notions related to Floquet theory.

Consider the dynamical system described by the differential equations

$$\dot{\underline{x}} = A(t)\underline{x} \ , \qquad \underline{x} \in \mathbf{R}^n \ , \tag{D.1}$$

where $A = A(t)$ is an $n \times n$ periodic matrix with period T. Floquet theory [93] provides a transformation of coordinates, such that the analysis of (D.1) is reduced to the study of a differential system with constant real coefficients.

Let $\Phi(t)$ be the *principal monodromy* or *fundamental* matrix, whose columns are linear independent solutions of (D.1) and such that $\Phi(0)$ is the identity matrix. After a period T one has

$$\Phi(t + T) = \Phi(t)\Phi^{-1}(0)\Phi(T) \ .$$

By the Floquet theorem [93], there exists a constant matrix B and a periodic symplectic matrix $C(t)$ such that at any time t one has

$$\Phi(t) = e^{Bt}C(t) \ .$$

After the transformation $\underline{y} = C^{-1}(t)\underline{x}$, the system (D.1) is reduced to the following differential equation with real constant coefficients:

$$\dot{\underline{y}} = B\underline{y} \ .$$

The eigenvalues of $\Phi(T)$ are called the *characteristic multipliers;* they measure the rate of expansion or contraction of a solution. A *characteristic exponent* is a quantity ℓ such that $e^{\ell T}$ is a characteristic multiplier. The real parts of the characteristic exponents are the so–called Lyapunov exponents. When all Lyapunov exponents are negative, the solution is asymptotically stable; when the Lyapunov exponents are positive, the solution is unstable.

E The planetary problem

Consider three bodies $\mathcal{P}_1$, $\mathcal{P}_2$, $\mathcal{P}_3$ of masses m_1, m_2, m_3 subject to the mutual gravitational attraction. Assume that the three bodies move on the same plane and let $\underline{q}_i \in \mathbf{R}^2$, $i = 1,2,3$, be their coordinates in an inertial reference frame; denote by $\underline{p}_i \in \mathbf{R}^2$, $i = 1,2,3$, the conjugated momenta. Then, the Hamiltonian function is given by

$$\begin{aligned}\mathcal{H}_1(\underline{p}_1,\underline{p}_2,\underline{p}_3,\underline{q}_1,\underline{q}_2,\underline{q}_3) &= \frac{\|\underline{p}_1\|^2}{2m_1} + \frac{\|\underline{p}_2\|^2}{2m_2} + \frac{\|\underline{p}_3\|^2}{2m_3} \\ &- \frac{m_1 m_2}{\|\underline{q}_2 - \underline{q}_1\|} - \frac{m_1 m_3}{\|\underline{q}_3 - \underline{q}_1\|} - \frac{m_2 m_3}{\|\underline{q}_3 - \underline{q}_2\|} \ .\end{aligned}$$

We introduce heliocentric coordinates centered in $\mathcal{P}_1$ by means of the change of variables:

$$\begin{aligned}\underline{u}_1 &= \underline{q}_1 \ , & \underline{v}_1 &= \underline{p}_1 + \underline{p}_2 + \underline{p}_3 \\ \underline{u}_2 &= \underline{q}_2 - \underline{q}_1 \ , & \underline{v}_2 &= \underline{p}_2 \\ \underline{u}_3 &= \underline{q}_3 - \underline{q}_1 \ , & \underline{v}_3 &= \underline{p}_3 \ .\end{aligned}$$

The new Hamiltonian becomes:

$$\begin{aligned}\mathcal{H}_2(\underline{v}_2,\underline{v}_3,\underline{u}_2,\underline{u}_3) &= \left[\frac{m_1+m_2}{2m_1m_2}\|\underline{v}_2\|^2 - \frac{m_1m_2}{\|\underline{u}_2\|}\right] \\ &+ \left[\frac{m_1+m_3}{2m_1m_3}\|\underline{v}_3\|^2 - \frac{m_1m_3}{\|\underline{u}_3\|}\right] + \frac{\underline{v}_2\cdot\underline{v}_3}{m_1} - \frac{m_2m_3}{\|\underline{u}_2-\underline{u}_3\|} \ .\end{aligned}$$

As in a planetary model, we assume that one mass (say, the star) is much larger than the two others (say, the planets); as a consequence we rescale the masses as $m_1 = \mu_1$, $m_2 = \varepsilon\mu_2$, $m_3 = \varepsilon\mu_3$, where ε is a small quantity and μ_i, $i = 1,2,3$, are of order one. Applying the change of variables

$$\underline{\tilde{v}}_i = \frac{\underline{v}_i}{\varepsilon} \ , \qquad \underline{\tilde{u}}_i = \underline{u}_i \ , \qquad i = 2,3 \ ,$$

the new Hamiltonian is given by

$$\begin{aligned}\mathcal{H}_3(\underline{\tilde{v}}_2,\underline{\tilde{v}}_3,\underline{\tilde{u}}_2,\underline{\tilde{u}}_3) &= \left[\frac{\mu_1+\varepsilon\mu_2}{2\mu_1\mu_2}\|\underline{\tilde{v}}_2\|^2 - \frac{\mu_1\mu_2}{\|\underline{\tilde{u}}_2\|}\right] \\ &+ \left[\frac{\mu_1+\varepsilon\mu_3}{2\mu_1\mu_3}\|\underline{\tilde{v}}_3\|^2 - \frac{\mu_1\mu_3}{\|\underline{\tilde{u}}_3\|}\right] + \varepsilon\left[\frac{\underline{\tilde{v}}_2\underline{\tilde{v}}_3}{\mu_1} - \frac{\mu_2\mu_3}{\|\underline{\tilde{u}}_2-\underline{\tilde{u}}_3\|}\right] \ ,\end{aligned}$$

which can be written as

$$\mathcal{H}_4(\underline{\tilde{v}}_2, \underline{\tilde{v}}_3, \underline{\tilde{u}}_2, \underline{\tilde{u}}_3) = \left[\frac{\|\underline{\tilde{v}}_2\|^2}{2\mu_2} - \frac{\mu_1\mu_2}{\|\underline{\tilde{u}}_2\|} \right] + \left[\frac{\|\underline{\tilde{v}}_3\|^2}{2\mu_3} - \frac{\mu_1\mu_3}{\|\underline{\tilde{u}}_3\|} \right] + \varepsilon F(\underline{\tilde{v}}_2, \underline{\tilde{v}}_3, \underline{\tilde{u}}_2, \underline{\tilde{u}}_3)$$

for a suitable function $F = F(\underline{\tilde{v}}_2, \underline{\tilde{v}}_3, \underline{\tilde{u}}_2, \underline{\tilde{u}}_3)$ which acts as a perturbation of order ε. In fact for $\varepsilon = 0$ the Hamiltonian $\mathcal{H}_4$ is equal to the sum of two decoupled Keplerian motions.

F Yoshida's symplectic integrator

We give the recipe for Yoshida's symplectic integrator [175] in the case of a differential system of the form

$$\begin{aligned} \dot{x} &= y \\ \dot{y} &= V(x,t) \ , \end{aligned}$$

where $y \in \mathbf{R}$, $(x,t) \in \mathbf{T}^2$ and $V = V(x,t)$ is a regular periodic function. The initial conditions are $x(t_0) = x_0$, $y(t_0) = y_0$. Let h denote the integration step and define the following quantities

$$\begin{aligned} c_1 &\equiv \frac{1}{2-2^{1/3}} & c_0 &\equiv 1-2c_1 \\ \tau_1 &\equiv c_1 h & \tau_1' &\equiv c_0 h \\ \tau_2 &\equiv \frac{1}{2} c_1 h & \tau_2' &\equiv \frac{1}{2}(c_0+c_1)h \ . \end{aligned}$$

Introduce the vector functions

$$\underline{u}(\tau; x, y) \equiv (x + y\tau, y) \ , \qquad \underline{v}(\tau; x, y) \equiv (x, y + V(x,t)\tau) \ .$$

Then, Yoshida's integration algorithm allows us to find the solution at time $t_1 = t_0 + h$, say (x_1, y_1), implementing the following sequence of transformations of coordinates from (x_0, y_0):

$$\begin{aligned} (\xi_1, \eta_1) &= \underline{u}(\tau_2; x_0, y_0) \\ (\xi_2, \eta_2) &= \underline{v}(\tau_1; \xi_1, \eta_1) \\ (\xi_3, \eta_3) &= \underline{u}(\tau_2'; \xi_2, \eta_2) \\ (\xi_4, \eta_4) &= \underline{v}(\tau_1'; \xi_3, \eta_3) \\ (\xi_5, \eta_5) &= \underline{u}(\tau_2'; \xi_4, \eta_4) \\ (\xi_6, \eta_6) &= \underline{v}(\tau_1; \xi_5, \eta_5) \\ (x_1, y_1) &= \underline{u}(\tau_2; \xi_6, \eta_6) \ . \end{aligned}$$

G Astronomical data

We report some astronomical data of the planets of the solar system.

Notation: a = semimajor axis (AU)[1]; e = eccentricity; i = inclination (degrees); R_e = equatorial radius (km); M = mass ($\times 10^{24}$ kg).

Planet	a	e	i	R_e	M
Mercury	0.39	0.2056	7.00	2 440	0.33
Venus	0.72	0.0068	3.39	6 052	4.87
Earth	1.00	0.0167	0.00	6 378	5.97
Mars	1.52	0.0934	1.85	3 396	0.64
Jupiter	5.20	0.0484	1.30	71 492	1898.13
Saturn	9.54	0.0539	2.49	60 268	568.32
Uranus	19.19	0.0472	0.77	25 559	86.81
Neptune	30.07	0.0086	1.77	24 764	102.41

We report some astronomical data of the dwarf planets.

Notation: a = semimajor axis (AU); e = eccentricity; i = inclination (degrees); R = mean radius (km); M = mass (kg).

Dwarf planet	a	e	i	R	M
Ceres	2.77	0.079	10.59	487	$9.43 \cdot 10^{20}$
Eris	67.67	0.442	44.19	1300	$1.67 \cdot 10^{22}$
Haumea	43.13	0.195	28.22	1436	$4.00 \cdot 10^{21}$
Makemake	45.79	0.159	28.96	750	$4.00 \cdot 10^{21}$
Pluto	39.48	0.249	17.14	1151	$1.31 \cdot 10^{22}$

[1] One astronomical unit (AU) amounts to the average Earth–Sun distance, about equal to $1.5 \cdot 10^8$ km.

We report some astronomical data of the satellites.

Notation: d = distance from the planet (km); e = eccentricity; i = inclination (degrees); R = mean radius (km); M = mass (kg).

Planet	Satellite	d	e	i	R	M
Earth	Moon	384 400	0.0554	5.160	1737.5	$7.35 \cdot 10^{22}$
Mars	Phobos	9 376	0.0151	1.075	11.1	$1.07 \cdot 10^{16}$
	Deimos	23 458	0.0002	1.788	6.2	$1.48 \cdot 10^{15}$
Jupiter	Io	421 800	0.0041	0.036	1821.6	$8.93 \cdot 10^{22}$
	Europa	671 100	0.0094	0.466	1560.8	$4.80 \cdot 10^{22}$
	Ganymede	1 070 400	0.0013	0.177	2631.2	$1.48 \cdot 10^{23}$
	Callisto	1 882 700	0.0074	0.192	2410.3	$1.08 \cdot 10^{23}$
	Amalthea	181 400	0.0032	0.380	83.45	$2.07 \cdot 10^{18}$
	Thebe	221 900	0.0176	1.080	49.3	$1.50 \cdot 10^{18}$
	Adrastea	129 000	0.0018	0.054	8.2	$7.49 \cdot 10^{15}$
	Metis	128 000	0.0012	0.019	21.5	$1.20 \cdot 10^{17}$
	Himalia	11 461 000	0.1623	27.496	85	$6.74 \cdot 10^{18}$
	Elara	11 741 000	0.2174	26.627	43	$8.69 \cdot 10^{17}$
	Pasiphae	23 624 000	0.4090	151.431	30	$3.00 \cdot 10^{17}$
	Sinope	23 939 000	0.2495	158.109	19	$7.49 \cdot 10^{16}$
	Lysithea	11 717 000	0.1124	28.302	18	$6.29 \cdot 10^{16}$
	Carme	23 404 000	0.2533	164.907	23	$1.32 \cdot 10^{17}$
	Ananke	21 276 000	0.2435	148.889	14	$3.00 \cdot 10^{16}$
	Leda	11 165 000	0.1636	27.457	10	$1.09 \cdot 10^{16}$

Planet	Satellite	d	e	i	R	M
Saturn	Mimas	185 540	0.0196	1.572	198.2	$3.75 \cdot 10^{19}$
	Enceladus	238 040	0.0047	0.009	252.1	$1.08 \cdot 10^{20}$
	Tethys	294 670	0.0001	1.091	533.0	$6.18 \cdot 10^{20}$
	Dione	377 420	0.0022	0.028	561.7	$1.10 \cdot 10^{21}$
	Rhea	527 070	0.0010	0.331	764.3	$2.31 \cdot 10^{21}$
	Titan	1 221 870	0.0288	0.280	2575.5	$1.35 \cdot 10^{23}$
	Hyperion	1 500 880	0.0274	0.630	135.0	$5.59 \cdot 10^{18}$
	Iapetus	3 560 840	0.0283	7.489	735.6	$1.81 \cdot 10^{21}$
	Phoebe	12 947 780	0.1635	175.986	106.6	$8.29 \cdot 10^{18}$
	Janus	151 460	0.0068	0.163	89.4	$1.90 \cdot 10^{18}$
	Epimetheus	151 410	0.0098	0.351	56.7	$5.26 \cdot 10^{17}$
	Helene	377 420	0.0071	0.213	16.0	$2.55 \cdot 10^{16}$
	Telesto	294 710	0.0002	1.180	11.8	$7.19 \cdot 10^{15}$
	Calypso	294 710	0.0005	1.499	10.7	$3.60 \cdot 10^{15}$
	Atlas	137 670	0.0012	0.003	15.3	$6.59 \cdot 10^{15}$
	Prometheus	139 380	0.0022	0.008	43.1	$1.59 \cdot 10^{17}$
	Pandora	141 720	0.0042	0.050	40.3	$1.37 \cdot 10^{17}$
	Pan	133 580	0.0000	0.001	14.8	$4.95 \cdot 10^{15}$

Planet	Satellite	d	e	i	R	M
Uranus	Ariel	190 900	0.0012	0.041	578.9	$1.29 \cdot 10^{21}$
	Umbriel	266 000	0.0039	0.128	584.7	$1.22 \cdot 10^{21}$
	Titania	436 300	0.0011	0.079	788.9	$3.42 \cdot 10^{21}$
	Oberon	583 500	0.0014	0.068	761.4	$2.88 \cdot 10^{21}$
	Miranda	12 9900	0.0013	4.338	235.8	$6.59 \cdot 10^{19}$
	Cordelia	49 800	0.0003	0.085	20.1	$4.50 \cdot 10^{16}$
	Ophelia	53 800	0.0099	0.104	21.4	$5.39 \cdot 10^{16}$
	Bianca	59 200	0.0009	0.193	25.7	$9.29 \cdot 10^{16}$
	Cressida	61 800	0.0004	0.006	39.8	$3.43 \cdot 10^{17}$
	Desdemona	62 700	0.0001	0.113	32.0	$1.78 \cdot 10^{17}$
	Juliet	64 400	0.0007	0.065	46.8	$5.57 \cdot 10^{17}$
	Portia	66 100	0.0001	0.059	67.6	$1.68 \cdot 10^{18}$
	Rosalind	69 900	0.0001	0.279	36.0	$2.55 \cdot 10^{17}$
	Belinda	75 300	0.0001	0.031	40.3	$3.57 \cdot 10^{17}$
	Puck	86 000	0.0001	0.319	81.0	$2.89 \cdot 10^{18}$
Neptune	Tritone	354 759	0.0000	156.865	1352.6	$2.14 \cdot 10^{22}$
	Nereide	5 513 818	0.7507	7.090	170.0	$3.09 \cdot 10^{19}$
	Naiade	48 227	0.0003	4.691	33	$1.95 \cdot 10^{17}$
	Thalassa	50 074	0.0002	0.135	41	$3.75 \cdot 10^{17}$
	Despina	52 526	0.0002	0.068	75	$2.10 \cdot 10^{18}$
	Galatea	61 953	0.0001	0.034	88	$3.75 \cdot 10^{18}$
	Larissa	73 548	0.0014	0.205	97	$4.95 \cdot 10^{18}$
	Proteus	117 646	0.0005	0.075	210	$5.04 \cdot 10^{19}$

References

[1] H.D.I. Abarbanel, *Analysis of Observed Chaotic Data,* Springer–Verlag, New York (1996)

[2] V.I. Arnold, Proof of a Theorem by A.N. Kolmogorov on the invariance of quasi-periodic motions under small perturbations of the Hamiltonian, *Russ. Math. Surveys* **18**, 13–40 (1963)

[3] V.I. Arnold, Small divisor problems in classical and Celestial Mechanics, *Russ. Math. Surveys* **18**, 85–191 (1963)

[4] V.I. Arnold, Instability of dynamical systems with several degrees of freedom, *Sov. Math. Doklady* **5**, 342–355 (1964)

[5] V.I. Arnold, *Mathematical Methods of Classical Mechanics,* Springer–Verlag, New York (1978) (Russian original, Moscow, 1974)

[6] V.I. Arnold (editor), *Encyclopaedia of Mathematical Sciences, Dynamical Systems III,* Springer–Verlag (1988)

[7] S. Aubry, P.Y. Le Daeron, The discrete Frenkel–Kontorova model and its extensions, I, *Phys. D* **8**, 381–422 (1983)

[8] G.L. Baker, J.P. Gollub, *Chaotic Dynamics,* Cambridge University Press (1990)

[9] C. Beaugé, S. Ferraz–Mello, Resonance trapping in the primordial solar nebula: The case of a Stokes drag dissipation, *Icarus* **103**, 301–318 (1993)

[10] V.V. Beletskii, Motion of an artificial satellite about its center of mass, Israel Program for Scientific Translations, Jerusalem (1966)

[11] G. Benettin, Physical applications of Nekhoroshev theorem and exponential estimates, Lectures given at the Cetraro school 2000 "Hamiltonian Dynamics. Theory and Applications", A. Giorgilli ed., Springer, 1–73 (2005)

[12] G. Benettin, F. Fassó, M. Guzzo, Nekhoroshev–stability of L4 and L5 in the spatial restricted three–body problem, *Reg. Chaotic Dyn.* **3**, 56–72 (1998)

[13] G. Benettin, M. Guzzo, F. Fassó, Long term stability of proper rotations of the perturbed Euler rigid body, *Commun. Math. Phys.* **250**, 133–160 (2004)

[14] G. Benettin, L. Galgani, A. Giorgilli, J.-M. Strelcyn, Lyapunov characteristic exponents for smooth dynamical systems and for Hamiltonian systems; a method for computing all of them, *Meccanica* **15**, 9–20 (1980)

[15] L. Biasco, L. Chierchia, Exponential stability for the resonant D'Alembert model of Celestial Mechanics, *DCDS A* **12**, 569–594 (2005)

[16] L. Biasco, L. Chierchia, E. Valdinoci, Elliptic two–dimensional invariant tori for the planetary three–body problem, *Arch. Rational Mech. Anal.* **170**, 91–135 (2003)

[17] L. Biasco, L. Chierchia, E. Valdinoci, N–dimensional elliptic invariant tori for the planar $(N+1)$–body problem, *SIAM Journal on Mathematical Analysis* **37**, no. 5, 1560–1588 (2006)

[18] D. Boccaletti, G. Pucacco, *Theory of Orbits,* Springer–Verlag, Berlin, Heidelberg, New York (2001)

[19] H.W. Broer, G.B. Huitema, M.B. Sevryuk, *Quasi–periodic Motions in Families of Dynamical Systems,* Lecture Notes in Mathematics, Springer–Verlag, New York (1996)

[20] H.W. Broer, C. Simó, J.C. Tatjer, Towards global models near homoclinic tangencies of dissipative diffeomorphisms, *Nonlinearity* **11**, no. 3, 667–770 (1998)

[21] D. Brouwer, G.M. Clemence, *Methods of Celestial Mechanics*, Academic Press, New York and London (1961)

[22] A.A. Burov, Non–integrability of planar oscillation equation for satellite in elliptical orbit, *Vestnik Moskov. Univ. Ser. Mat. Mekh.* **1**, 71–73 (1984) (in Russian)

[23] A. Celletti, Analysis of resonances in the spin-orbit problem in Celestial Mechanics: The synchronous resonance (Part I), *J. Applied Math. and Physics (ZAMP)* **41**, 174–204 (1990)

[24] A. Celletti, Construction of librational invariant tori in the spin–orbit problem, *J. Applied Math. and Physics (ZAMP)* **45**, 61–80 (1993)

[25] A. Celletti, L. Chierchia, Rigorous estimates for a computer–assisted KAM theory, *J. Math. Phys.* **28**, 2078–2086 (1987)

[26] A. Celletti, L. Chierchia, Construction of analytic KAM surfaces and effective stability bounds, *Commun. in Math. Physics* **118**, 119–161 (1988)

[27] A. Celletti, L. Chierchia, Invariant curves for area–preserving twist maps far from integrable, *J. Stat. Phys.* **65**, 617–643 (1991)

[28] A. Celletti, L. Chierchia, A constructive theory of Lagrangian tori and computer–assisted applications, *Dynamics Reported* (C.K.R.T. Jones, U. Kirchgraber, H.O. Walther Managing Editors), Springer–Verlag, **4** (New Series), 60–129 (1995)

[29] A. Celletti, L. Chierchia, Construction of stable periodic orbits for the spin–orbit problem of Celestial Mechanics, *Reg. Chaotic Dyn.* **3**, no. 3, 107–121 (1998)

[30] A. Celletti, L. Chierchia, KAM tori for N–body problems (a brief history), *Cel. Mech. Dyn. Astr.* **95**, no. 1, 117–139 (2006)

[31] A. Celletti, L. Chierchia, KAM stability and Celestial Mechanics, *Memoirs American Mathematical Society* **187**, no. 878 (2007)

[32] A. Celletti, L. Chierchia, Quasi–periodic attractors in Celestial Mechanics, *Arch. Rat. Mech. Anal.* **191**, no. 2, 311–345 (2009)

[33] A. Celletti, C. Falcolini, Construction of invariant tori for the spin–orbit problem in the Mercury–Sun system, *Cel. Mech. Dyn. Astr.* **53**, 113–127 (1992)

[34] A. Celletti, L. Ferrara, An application of the Nekhoroshev theorem to the restricted three–body problem, *Cel. Mech. Dyn. Astr.* **64**, 261–272 (1996)

[35] A. Celletti, C. Froeschlé, On the determination of the stochasticity threshold of invariant curves, *Int. J. Bifurcation and Chaos* **5**, no. 6, 1713–1719 (1995)

[36] A. Celletti, C. Froeschlé, E. Lega, Dissipative and weakly–dissipative regimes in nearly–integrable mappings, *Discrete and Continuous Dynamical Systems – Series A* **16**, no. 4, 757–781 (2006)

[37] A. Celletti, C. Froeschlé, E. Lega, Dynamics of the conservative and dissipative spin–orbit problem, *Planetary and Space Science* **55**, 889–899 (2007)

[38] A. Celletti, A. Giorgilli, On the stability of the Lagrangian points in the spatial restricted problem of three bodies, *Cel. Mech. Dyn. Astr.* **50**, 31–58 (1991)

[39] A. Celletti, M. Guzzo, Cantori of the dissipative sawtooth map, *Chaos* **19**, 013140, pp. 6 (2009)

[40] A. Celletti, R.S. MacKay, Regions of non–existence of invariant tori for spin–orbit models, *Chaos* **17**, 043119, pp. 12 (2007)

[41] A. Celletti, V.V. Sidorenko, Some properties of the dumbbell satellite attitude dynamics, *Cel. Mech. Dyn. Astr.* **101**, 105–126 (2008)

[42] Q. Chen, R.S. MacKay, J.D. Meiss, Cantori for symplectic maps, *J. Phys. A: Math. Gen.* **23**, L1093–L1100 (1990)

[43] A. Chenciner, *Bifurcations de points fixes elliptiques. II. Orbites periodiques et ensembles de Cantor invariants,* Inventiones Mathematicae **80**, 81–106 (1985)

[44] A. Chenciner, R. Montgomery, A remarkable periodic solution of the three body problem in the case of equal masses, *Annals of Mathematics* **152**, 881–901 (2000)

[45] C.Q. Cheng, J. Yan, Existence of diffusion orbits in a priori unstable Hamiltonian systems, *J. Diff. Geom.* **67**, no. 3, 457–518 (2004)

[46] L. Chierchia, G. Gallavotti, Drift and diffusion in phase space, *Ann. Inst. H. Poincaré* **60**, no. 1, 1–144 (1994)

[47] B.V. Chirikov, A Universal Instability of Many–Dimensional Oscillator Systems, *Phys. Rep.* **52**, 264–379 (1979)

[48] G. Colombo, I.I. Shapiro, The rotation of the planet Mercury, *Astroph. J.* **145**, 296–307 (1966)

[49] G. Contopoulos, M. Harsoula, N. Voglis, Crossing of various cantori, *Cel. Mech. Dyn. Astr.* **78**, 197–210 (2000)

[50] A.C.M. Correia, J. Laskar, Mercury's capture into the 3/2 spin–orbit resonance as a result of its chaotic dynamics, *Nature* **429**, 848–850 (2004)

[51] H. Curtis, *Orbital Mechanics,* Butterworth–Heinemann, Elsevier (2005)

[52] J.M.A. Danby, Stability of the triangular points in the elliptic restricted problem of three bodies, *Astron. J.* **69**, no. 2, 165–172 (1964)

[53] C. Delaunay, Théorie du mouvement de la lune, *Mémoires de l'Académie des Sciences* **1**, Tome XXVIII, Paris (1860)

[54] A. Delshams, R. de la Llave, KAM theory and a partial justification of Greene's criterion for nontwist maps, *SIAM J. Math. Anal.* **31**, no. 6, 1235–1269 (2000)

[55] A. Deprit, Free rotation of a rigid body studied in the phase plane, *American J. of Phys.* **35**, no. 5, 424–428 (1967)

[56] A. Deprit, The secular accelerations in Gylden's problem, *Cel. Mech.* **31**, 1–22 (1983)

[57] S. D'Hoedt, A. Lemaitre, The spin–orbit resonant rotation of Mercury: a two degree of freedom Hamiltonian model, *Cel. Mech. Dyn. Astr.* **89**, no. 3, 267–283 (2004)

[58] R. Dvorak, H. Lichtenegger, On the two–body problem with variable masses, The motion of planets and natural and artificial satellites (Embu, 1981), 11–17, Math. Dynam. Astronom. Ser., 2, Univ. São Paulo, São Paulo (1983)

[59] J.P. Eckmann, S. Oliffson Kamphorst, D. Ruelle, S. Ciliberto, Liapunov exponents from time series, *Phys. Rev. A* **34**, no. 6, 4971–4979 (1986)

[60] J.-P. Eckmann, P. Wittwer, *Computer Methods and Borel Summability Applied to Feigenbaum's Equation,* Springer Lecture Notes in Physics. no. 227 (1985)

[61] K.M. Ellis, C.D. Murray, The disturbing function in solar system dynamics, *Icarus* **147**, 129–144 (2000)

[62] D.F. Escande, Stochasticity in classical Hamiltonian systems: Universal aspects, *Physics Reports* **121**, 165–261 (1985)

[63] C. Falcolini, R. de la Llave, A rigorous partial justification of Greenes criterion, *J. Stat. Phys.* **67**, 609–643 (1992)

[64] F. Fassó, M. Guzzo and G. Benettin, Nekhoroshev–stability of elliptic equilibria of Hamiltonian systems, *Commun. Math. Phys.* **197**, 347–360 (1997)

[65] M.J. Feigenbaum, The universal metric properties of nonlinear transformations, *J. Stat. Phys.* **21**, 669–706 (1979)

[66] S. Ferraz–Mello, *Canonical Perturbation Theories,* Springer–Verlag, New York (2007)

[67] S. Ferraz–Mello, On the convergence domain of the Laplacian expansion of the disturbing function, *Cel. Mech. Dyn. Astr.* **58**, 37–52 (1994)

[68] S. Ferraz–Mello, On the convergence of the disturbing function, in *From Newton to Chaos: Modern Techniques for Understanding and Coping with Chaos in N–body Dynamical Systems*, A.E. Roy, B.A. Steves eds., Plenum Press, New York, 97–98 (1995)

[69] P. Frederickson, J.L. Kaplan, E.D. Yorke, J.A. Yorke, The Liapunov dimension of strange attractors, *J. Diff. Eq.* **49**, 185–207 (1983)

[70] C. Froeschlé, Numerical study of a four–dimensional mapping, *Astron. Astroph.* **16**, 172–189 (1972)

[71] C. Froeschlé, E. Lega, R. Gonczi, Fast Lyapunov indicators. Application to asteroidal motion, *Cel. Mech. Dyn. Astr.* **67**, 41–62 (1997)

[72] T. Fukushima, A method of solving Kepler's equation for the hyperbolic case, *Cel. Mech. Dyn. Astr.* **68**, 121–137 (1997)

[73] G. Gallavotti, *The Elements of Mechanics,* Springer–Verlag, New York (1983)

[74] E.J. Gascheau, Examen d'une classe déquations différentielles et application á un cas particulier du probléme des trois corps, *Compt. Rend.* **16**, 393–394 (1843)

[75] A. Giorgilli, Effective stability in Hamiltonian systems in the light of Nekhoroshev's theorem, *Integrable Systems and Applications*, Springer–Verlag, Berlin, New York, 142–153 (1989)

[76] A. Giorgilli, A. Delshams, E. Fontich, L. Galgani, C. Simó, Effective stability for a Hamiltonian system near an elliptic equilibrium point, with an application to the restricted three body problem, *J. Diff. Eq.* **77**, 167–198 (1989)

[77] A. Giorgilli, Ch. Skokos, On the stability of the Trojan asteroids, *Astron. Astroph.* **317**, 254–261 (1997)

[78] P. Goldreich, Final spin states of planets and satellites, *Astronom. J.* **71**, no. 1, 1–7 (1966)

[79] P. Goldreich, S. Peale, Spin–orbit coupling in the solar system, *Astronom. J.* **71**, no. 6, 425–438 (1966)

[80] G. Gomez, J. Llibre, J. Masdemont, Homoclinic and heteroclinic solutions in the restricted three–body problem, *Cel. Mech.* **44**, no. 3, 239–259 (1988/89)

[81] C. Grebogi, E. Ott, J. Yorke, Basin boundary metamorphoses: changes in accessible boundary orbits, *Phys. D* **24**, 243–262 (1987)

[82] J.M. Greene, A method for determining a stochastic transition, *J. of Math. Phys.* **20**, 1183–1201 (1979)

[83] G. Gronchi, A. Milani, Averaging on Earth-crossing orbits, *Cel. Mech. Dyn. Astr.* **71**, no. 2, 109–136 (1998/99)

[84] D.M. Grossman, Homeomorphisms of systems of differential equations, *Dokl. Akad. Nauk SSSR* **128**, 880–881 (1959)

[85] M. Guzzo, O. Bernardi, F. Cardin, The experimental localization of Aubry–Mather sets using regularization techniques inspired by viscosity theory, *Chaos* **17**, no. 3 (2007)

[86] M. Guzzo, A. Morbidelli, Construction of a Nekhoroshev–like result for the asteroid belt dynamical system, *Cel. Mech. Dyn. Astr.* **66**, 255–292 (1997)

[87] M. Guzzo, E. Lega, C. Froeschlé, First numerical evidence of global Arnold diffusion in quasi–integrable systems, *DCDS B* **5**, 687–698 (2005)

[88] H. Gylden, Die Bahnbewegungen in einem Systeme von zwei Körpern in dem Falle, dass die Massen Veränderungen unterworfen sind, *Astron. Nachr.* **109**, col. 1–6 (1884)

[89] J. Hadjidemetriou, Mapping models for Hamiltonian systems with application to resonant asteroid motion, in *Predictability, Stability and Chaos in N–body Dynamical Systems*, A.E. Roy ed., Plenum Press, New York 157–175 (1991)

[90] J. Hadjidemetriou, Analytic solutions of the two–body problem with variable mass, *Icarus* **5**, 34–46 (1966)

[91] J. Hadjidemetriou, Secular variation of mass and the evolution of binary systems, *Adv. Astron. Astrophys.* **5**, 131–188 (1967)

[92] J. Hadjidemetriou, Periodic orbits, *Cel. Mech.* **34**, no. 1–4, 379–393 (1984)

[93] P. Hartman, *Ordinary Differential Equations,* John Wiley and Sons, Inc., New York (1964)

[94] M. Hénon, Explorationes numérique du problème restreint IV: Masses egales, orbites non periodique, *Bullettin Astronomique* **3**, 1, fasc. 2, 49–66 (1966)

[95] M. Hénon, A two–dimensional mapping with a strange attractor, *Comm. Math. Phys.* **50**, 69–77 (1976)

[96] M. Hénon, *Generating Families in the Restricted Three–Body Problem,* Springer–Verlag, New York (1997)

[97] J. Henrard, Capture into resonance: an extension of the use of adiabatic invariants, *Cel. Mech. Dyn. Astr.* **27**, 3–22 (1982)

[98] J.H. Jeans, Cosmogonic problems associated with a secular decrease of mass, *MNRAS* **85**, 2–11 (1924)

[99] W.H. Jefferys, J. Moser, Quasi–periodic solutions for the three–body problem, *Astron. J.* **71**, no. 7, 568–578 (1966)

[100] A. Jorba, R. de la Llave, M. Zou, Lindstedt series for lower dimensional tori, NATO Adv. Sci. Ins. Ser. C Math. Phys. Sci., C. Simó ed., Kluwer, Dordrecht, 151–167 (1999)

[101] A. Jorba, J. Villanueva, Effective Stability Around Periodic Orbits of the Spatial RTBP, NATO Adv. Sci. Ins. Ser. C, Math. Phys. Sci., C. Simó ed., Kluwer, Dordrecht, 628–632 (1999)

[102] V. Kaloshin, M. Levi, An example of Arnold diffusion for near–integrable Hamiltonians, *Bull. Amer. Math. Soc.* **45**, 409-427 (2008)

[103] W.M. Kaula, *Theory of Satellite Geodesy,* Dover Publications Inc., New York (1966)

[104] A.Ya. Khinchin, *Continued Fractions,* Dover Publications, Mineola, New York (1964)

[105] A.N. Kolmogorov, On the conservation of conditionally periodic motions under small perturbation of the Hamiltonian, *Dokl. Akad. Nauk. SSR* **98**, 527–530 (1954)

[106] O.E. Lanford III, Computer assisted proofs in analysis, *Physics A* **124**, 465–470 (1984)

[107] J. Laskar, Frequency analysis for multi–dimensional systems. Global dynamics and diffusion, *Physica D* **67**, 257–281 (1993)

[108] J. Laskar, Large scale chaos in the solar system, *Astron. Astroph.* **287**, L9–L12 (1994)

[109] J. Laskar, C. Froeschlé, A. Celletti, The measure of chaos by the numerical analysis of the fundamental frequencies. Application to the standard mapping, *Phys. D* **56**, 253–269 (1992)

[110] J. Laskar, P. Robutel, Stability of the planetary three–body problem. I. Expansion of the planetary Hamiltonian, *Cel. Mech. Dyn. Astr.* **62**, no. 3, 193–217 (1995)

[111] E. Lega, M. Guzzo, C. Froeschlé, Detection of Arnold diffusion in Hamiltonian systems, *Phys. D* **182**, 179–187 (2002)

[112] A. Lemaitre, S. D'Hoedt, N. Rambaux, The 3:2 spin–orbit resonant motion of Mercury, *Cel. Mech. Dyn. Astr.* **95**, 213–224 (2006)

[113] B.B. Lieberman, Existence of quasi–periodic solutions to the three–body problem, *Celestial Mechanics* **3**, 408–426 (1971)

[114] C. Lhotka, C. Efthymiopoulos, R. Dvorak, Nekhoroshev stability at L_4 or L_5 in the elliptic–restricted three–body problem – application to Trojan asteroids, *MNRAS* **384**, no. 3, 1165–1177 (2008)

[115] A.J. Lichtenberg, M.A. Lieberman, *Regular and Chaotic Dynamics,* Springer, New York (1983)

[116] J.E. Littlewood, Adiabatic invariance. II. Elliptic motion about a slowly varying center of force, *Ann. Physics* **26**, 131–156 (1964)

[117] R. de la Llave, A. Gonzàlez, À. Jorba, J. Villanueva, KAM theory without action–angle variables, *Nonlinearity* **18**, no. 2, 855-895 (2005)

[118] R. de la Llave, D. Rana, Accurate strategies for small divisor problems, *Bull. Amer. Math. Soc. (N. S.)* **22**, 1, 85–90 (1990)

[119] R. de la Llave, C.E. Wayne, Whiskered and low dimensional tori in nearly integrable Hamiltonian systems, *MPEJ* **10**, n. 5 (2004)

[120] U. Locatelli, A. Giorgilli, Invariant tori in the secular motions of the three–body planetary systems, *Cel. Mech. Dyn. Astr.* **78**, 47–74 (2000)
[121] U. Locatelli, A. Giorgilli, Construction of the Kolmogorov's normal form for a planetary system, *Reg. Chaotic Dyn.* **10**, no. 2, 153–171 (2005)
[122] E.N. Lorenz, *The Essence of Chaos,* University of Washington Press (1996)
[123] G.J.F. MacDonald, Tidal friction, *Rev. Geophys.* **2**, 467–541 (1964)
[124] R.S. MacKay, A renormalization approach to invariant circles in area–preserving maps, *Phys. D* **7**, 283–300 (1983)
[125] R.S. MacKay, *Transition to Chaos for Area–preserving Maps,* Springer Lecture Notes in Physics, no. 247, 390–454 (1985)
[126] R.S. MacKay, A criterion for non–existence of invariant tori for Hamiltonian systems, *Phys. D* **36**, 64–82 (1989)
[127] R.S. MacKay, On Greene's residue criterion, *Nonlinearity* **5**, 161–187 (1992)
[128] R.S. MacKay, J.D. Meiss, J. Stark, Converse KAM theory for symplectic twist maps, *Nonlinearity* **2**, 555–570 (1989)
[129] R.S. MacKay, I.C. Percival, Converse KAM: theory and practice, *Comm. Math. Phys.* **98**, no. 4, 469–512 (1985)
[130] B.B. Mandelbrot, *The Fractal Geometry of Nature,* W.H. Freeman and Company, New York (1982)
[131] C. Marchal, *The Three–body Problem,* Elsevier Science Publishers, Amsterdam (1990)
[132] J. Mather, Existence of quasi–periodic orbits for twist homomorphisms, *Topology* **21**, 457–467 (1982)
[133] R.M. May, Simple mathematical models with very complicated dynamics, *Nature* **261**, 459–467 (1976)
[134] K.R. Meyer, D.S. Schmidt, Elliptic relative equilibria in the N–body problem, *J. Diff. Eq.* **214**, 256–298 (2005)
[135] S. Mikkola, A cubic approximation for Kepler's equation, *Cel. Mech. Dyn. Astr.* **40**, 329–334 (1987)
[136] A. Morbidelli, *Modern Celestial Mechanics,* Taylor & Francis, London (2002)
[137] A. Morbidelli, A. Giorgilli, On a connection between KAM and Nekhoroshev's theorems, *Phys. D* **86**, no. 3, 514–516 (1995)
[138] J. Moser, On invariant curves of area–preserving mappings of an annulus, *Nach. Akad. Wiss. Göttingen, Math. Phys. Kl.* II **1**, 1–20 (1962)
[139] J. Moser, A rapidly convergent iteration method and non–linear partial differential equations, *Ann. Scuola Norm. Sup. Pisa* **20**, 499–535 (1966)
[140] J. Moser, Convergent series expansions for quasi-periodic motions, *Math. Ann.* **169**, 136-176 (1967)
[141] J. Moser, *Stable and Random Motions in Dynamical Systems: with Special Emphasis on Celestial Mechanics,* Hermann Weyl Lectures, the Institute for Advanced Study, Princeton, N. J. Annals of Mathematics Studies, No. 77. Princeton University Press, Princeton, N. J.; University of Tokyo Press, Tokyo (1973)
[142] C.D. Murray, S.F. Dermott, *Solar System Dynamics*, Cambridge University Press (1999)
[143] N.N. Nekhoroshev, An exponential estimate of the stability time of near–integrable Hamiltonian systems, *Russ. Math. Surveys* **32**, no. 6, 1–65 (1977)
[144] A.I. Neishtadt, Capture into resonance and scattering on resonances in two–frequency systems, *Proc. Steklov Inst. Math.* **250**, 183–203 (2005)
[145] A. Olvera, C. Simó, An obstruction method for the destruction of invariant curves, *Phys. D* **26**, 181–192 (1987)
[146] E. Ott, *Chaos in Dynamical Systems*, Cambridge University Press (1993)
[147] S.J. Peale, The free precession and libration of Mercury, *Icarus*, **178**, 4–18 (2005)

[148] I.C. Percival, Chaotic boundary of a Hamiltonian map, *Phys. D* **6**, 67–77 (1982)
[149] H. Poincarè, *Les Methodes Nouvelles de la Mechanique Celeste,* Gauthier Villars, Paris (1892)
[150] J. Pöschel, Nekhoroshev's estimates for quasi–convex Hamiltonian systems, *Math. Z.* **213**, 187–216 (1993)
[151] J. Pöschel, On Nekhoroshev's estimate at an elliptic equilibrium, *Int. Math. Res. Not.* **4**, 203–215 (1999)
[152] J.E. Prussing, B.A. Conway, *Orbital Mechanics,* Oxford University Press (1993)
[153] L.E. Reichl, *The Transition to Chaos,* Springer–Verlag, New York (1992)
[154] G.E. Roberts, Linear stability of the elliptic Lagrangian triangle solutions in the three–body problem, *J. Diff. Eq.* **182**, 191–218 (2002)
[155] P. Robutel, Stability of the planetary three–body problem. II. KAM theory and existence of quasi–periodic motions, *Cel. Mech. Dyn. Astr.* **62**, no. 3, 219–261 (1995)
[156] E.J. Routh, On Laplace's three particles with a supplement on the stability of their motion, *Proc. London Math. Soc.* **6**, 86–97 (1875)
[157] A.E. Roy, *Orbital Motion,* Adam Hilger, London (1991)
[158] D. Ruelle, *Chaotic Evolution and Strange Attractors,* Cambridge University Press (1989)
[159] A. Schenkel, H. Koch, P. Wittwer, Computer–assisted proofs in analysis and programming in logic: A case study, *SIAM Review* **38**, no. 4, 565–604 (1996)
[160] Ch. Skokos, A. Dokoumetzidis, Effective stability of the Trojan asteroids, *Astron. Astroph.* **367**, 729–736 (2001)
[161] V.V. Sidorenko, Evolution of fast rotations of an orbital cable system, *Cosmic Research* **33**, no. 1, 33–36 (1995)
[162] C.L. Siegel, J.K. Moser, *Lectures on Celestial Mechanics,* Springer–Verlag, Berlin, Heidelberg, New York (1971)
[163] C. Simó, Dynamical properties of the figure eight solution of the three–body problem, *Contemp. Math.* **292**, 209–228 (2003)
[164] C. Simó, C. Valls, A formal approximation of the splitting of separatrices in the classical Arnold's example of diffusion with two equal parameters, *Nonlinearity* **14**, no. 6, 1707–1760 (2001)
[165] J. Stark, An exhaustive criterion for the non–existence of invariant circles for area–preserving twist maps, *Comm. Math. Phys.* **117**, 177–189 (1988)
[166] E.L. Stiefel, M. Rössler, J. Waldvogel, C.A. Burdet: NASA Contractor Report CR–769, Washington, DC (1967)
[167] E.L. Stiefel, G. Scheifele, *Linear and Regular Celestial Mechanics,* Springer–Verlag, Berlin, Heidelberg, New York (1971)
[168] G.J. Sussman, J. Wisdom, Chaotic evolution of the solar system, *Science* **241**, 56–62 (1992)
[169] V. Szebehely, *Theory of Orbits,* Academic Press, New York and London (1967)
[170] P.-F. Verhulst, Recherches mathématiques sur la loi d'accroissement de la population, *Nouv. mém. de l'Academie Royale des Sci. et Belles–Lettres de Bruxelles* **18**, 1–41 (1845)
[171] J. Waldvogel, The close triple approach, *Cel. Mech.* **11**, 429–432 (1975)
[172] J. Wisdom, Rotational dynamics of irregularly shaped natural satellites, *Astron. J.* **94**, no. 5, 1350–1360 (1987)
[173] A. Wolf, J.B. Swift, H.L. Swinney, J.A. Vastano, Determining Lyapunov exponents from a time series, *Phys. D* **16**, 285–317 (1985)
[174] J. Xia, Arnold diffusion: a variational construction, *Proc. of the ICM* **2**, 867–877 (1998)
[175] H. Yoshida, Recent progress in the theory and application of symplectic integrators, *Cel. Mech. Dyn. Astr.* **56**, 27–43 (1993)

Index

MIX
Papier aus verantwortungsvollen Quellen
Paper from responsible sources
FSC® C105338

If you have any concerns about our products,
you can contact us on
ProductSafety@springernature.com

In case Publisher is established outside the EU,
the EU authorized representative is:
Springer Nature Customer Service Center GmbH
Europaplatz 3, 69115 Heidelberg, Germany

Printed by Libri Plureos GmbH
in Hamburg, Germany